Hot Quarks 2014: Workshop for Young Scientists on the Physics of Ultrarelativistic Nucleus-Nucleus Collisions

Journal of Physics: Conference Series Volume 612

Andalucia, Spain
21-28 September 2014

ISBN: 978-1-5108-0796-9
ISSN: 1742-6588

Printed from e-media with permission by:

Curran Associates, Inc.
57 Morehouse Lane
Red Hook, NY 12571

Some format issues inherent in the e-media version may also appear in this print version.

Printed by Curran Associates, Inc. (2015)

For permission requests, please contact the Institute of Physics
at the address below.

Institute of Physics
Dirac House, Temple Back
Bristol BS1 6BE UK

Phone: 44 1 17 929 7481
Fax: 44 1 17 920 0979

techtracking@iop.org

Additional copies of this publication are available from:

Curran Associates, Inc.
57 Morehouse Lane
Red Hook, NY 12571 USA
Phone: 845-758-0400
Fax: 845-758-2634
Email: curran@proceedings.com
Web: www.proceedings.com

Table of contents

Volume 612

Hot Quarks 2014: Workshop for young scientists on the physics of ultrarelativistic nucleus-nucleus collisions
21–28 September 2014, Andalucia, Spain

Accepted papers received: 7 April 2015
Published online: 19 May 2015

Preface

011001

Hot Quarks 2014: Workshop for young scientists on the physics of ultrarelativistic nucleus–nucleus collisions

011002

Peer review statement

Papers

Jets in the vacuum and in the medium

012001

Modelling jet quenching with Quenching Weights C Andrés, N Armesto, Carlos A Salgado and Yan Zhu **1**

012002

W boson studies in pPb and PbPb collisions with CMS Émilien Chapon (on behalf of the CMS Collaboration) **5**

012003

Renormalization of the jet-quenching parameter Yacine Mehtar-Tani **9**

012004

Neutral pion production in $\sqrt{s_{NN}}$=200 GeV Cu+Au collisions at PHENIX Sarah Campbell (for the PHENIX Collaboration) **13**

012005

Study of the neutral mesons in Pb-Pb collisions at $\sqrt{s_{NN}}$=2.76 TeV in the ALICE experiment at LHC Lucia Leardini (on behalf of the ALICE Collaboration) **17**

012006

Energy loss in jet suppression - what effects matter? Bojana Blagojevic and Magdalena Djordjevic **21**

012007

Constraining of jet medium interaction in high-energy heavy-ion collisions Yan Zhu and Xin-Nian Wang **25**

012008

Jet Measurements in Heavy-ion Collisions with CMS Raghav Kunnawalkam Elayavalli (on behalf of the CMS collaboration) **29**

012009

Measurement of W-boson production in p–Pb collisions at $\sqrt{s_{NN}}$= 5.02 TeV with ALICE at the LHC Jianhui Zhu (on behalf of the ALICE collaboration) **33**

012010

Decoding the quark-gluon plasma with jet quenching observables Konrad Tywoniuk **37**

012011

Study of Z boson production in pp, pPb and PbPb collisions in CMS Anna Julia Zsigmond (for the CMS Collaboration) **41**

012012

High-p_T probes of p+Pb collisions with ATLAS Dennis V Perepelitsa (for the ATLAS Collaboration) **45**

012013

Production of strange particles in charged jets in Pb–Pb and p–Pb collisions measured with ALICE Vít Kučera (for the ALICE Collaboration) **49**

Initial state effects and p-A collisions

012014

Multiplicity Dependence of (Anti-)Deuteron Production in p-Pb Collisions at $\sqrt{s_{NN}}$ = 5.02 TeV Measured with ALICE J Anielski (for the ALICE Collaboration) **53**

012015

Chromo-Weibel instabilities in classical Yang-Mills evolution Maximilian Attems **57**

012016

Probing novel long-range correlation phenomena in pPb collisions with identified particles at CMS Zhenyu Chen (on behalf of the CMS Collaboration) **61**

012017

Experimental overview of p(d)–A physics Roberto Preghenella **65**

012018

The impact of the LHC nuclear program on nPDFs H Paukkunen and P Zurita **71**

012019

The Impact of e+A Collisions on Nuclear PDFs M R Moldes **75**

Heavy flavor, dileptons and photons

012020

From light- to heavy-quark fragmentation in jets: current measurements in pp collisions with the ALICE detector and perspectives C Bianchin (for the ALICE Collaboration) **79**

012021

Bottomonium measurements in pp, pPb and PbPb using the CMS detector Nicolas Filipovic (on behalf of the CMS collaboration) **83**

012022

Measurement of Υ production in *p+p* collisions at $\sqrt{s}$ = 500 GeV in the STAR experiment Leszek Kosarzewski (for the STAR Collaboration) **87**

012023

Five things you should remember about heavy flavor measurements Daniel Kikoła **91**

012024

Study of azimuthal correlations between D mesons and charged particles with the ALICE experiment Fabio Colamaria (on behalf of the ALICE Collaboration) **99**

012025

Low mass dimuon production in pp, p–Pb and Pb–Pb collisions with the ALICE muon spectrometer Ester Anna Rita Casula **103**

012026

Effects of quark chemical equilibration on thermal photon elliptic flow Akihiko Monnai **107**

012027

J/ψ production in the statistical hadronization model - implications and limitations P B Bahavar, J Uphoff and C Greiner **111**

012028

Low-Mass Dielectron Production in pp, p–Pb and Pb–Pb Collisions with ALICE Patrick Reichelt (for the ALICE Collaboration) **115**

012029

Study of Charmonia Production vs. Charged Track Multiplicity in $p+p$ Collisions at PHENIX Haiwang Yu (for PHENIX Collaboration) **119**

012030

Inclusive J/ψ production in pp, p-Pb and Pb-Pb collisions at forward rapidity with ALICE at the LHC Javier Martín Blanco (for the ALICE Collaboration) **123**

012031

Measurement of electrons from charm and beauty-hadron decays in p-Pb collisions at $\sqrt{s_{NN}}$ 5.02 TeV with ALICE at the LHC Jan Wagner **127**

012032

Search for muonic atoms at RHIC Kefeng Xin **131**

012033

Forward rapidity $\psi(2S)$ meson production in pp, p-Pb and Pb-Pb collisions with ALICE at the LHC Marco Leoncino (for the ALICE Collaboration) **135**

012034

Measurements of non-photonic electrons with the STAR experiment Olga Rusňáková (for the STAR Collaboration) **139**

012035

Heavy flavours in high-energy nuclear collisions: quenching, flow and correlations S A Beraudo, A De Pace, M Monteno, M Nardi and F Prino **143**

012036

Υ production in hadron collisions at forward rapidity with ALICE at the LHC **147**

S Massimiliano Marchisone (for the ALICE Collaboration)

012037

Measurement of beauty-hadron decay electrons in Pb–Pb collisions at $\sqrt{s_{NN}}$ = 2.76 TeV with ALICE Martin Völk (for the ALICE Collaboration) **151**

012038

J/ψ and $\psi(2S)$ measurement in p+p collisions at $\sqrt{s}$ = 200 and 500 GeV in the STAR experiment Barbara Trzeciak (for the STAR Collaboration) **155**

012039

Dilepton Production in Transport-based Approaches Janus Weil, Stephan Endres, Hendrik van Hees, Marcus Bleicher and Ulrich Mosel **159**

Correlations and Fluctuations

012040

Multiplicity dependence of transverse momentum correlations in p-Pb and Pb-Pb collisions in ALICE Prabhat Pujahari (for the ALICE collaboration) **163**

012041

Parametrization for chemical freeze-out conditions from net-charge fluctuations measured at RHIC M Bluhm, P Alba, W Alberico, R Bellwied, V Mantovani Sarti, M Nahrgang and C Ratti **167**

012042

Overview of fluctuations at the QCD phase transition in heavy-ion collisions Marlene Nahrgang **171**

012043

Final source eccentricity measured by HBT interferometry with the event shape selection Takafumi Niida (for the PHENIX Collaboration) **178**

012044

Charge dependence of identified two-hadron correlation relative to the reaction plane in Pb-Pb collisions measured with ALICE Jacobus Onderwaater (for the ALICE Collaboration) **182**

012045

Pion-kaon femtoscopy in Au+Au collisions at STAR Katarzyna Poniatowska (for the STAR collaboration) **186**

012046

Higher order fluctuations of conserved charges in heavy ion collisions Miki Sakaida, Masayuki Asakawa and Masakiyo Kitazawa **190**

012047

Recent results on event-by-event fluctuations in ALICE at the LHC Satyajit Jena (on behalf of the ALICE Collaboration) **194**

Relativistic hydrodynamics and collective phenomena

012048

A 'soft+hard' model for Pion, Kaon, and Proton Spectra and v_2 measured in PbPb Collisions at $\sqrt{s}$ = 2.76ATeV Gergely Gábor Barnaföldi, Károly Ürmössy and Gábor Bíró **198**

012049

Identified light-flavour particle production measured with ALICE at the LHC as a probe of soft QCD and hot hadronic matter Barbara Guerzoni (for the ALICE Collaboration) **202**

012050

Transport properties of the hot and dense sQGP H Berrehrah, E Bratkovskaya, W Cassing and R Marty **206**

012051

A new shock-capturing numerical scheme for ideal hydrodynamics Z Fecková and B Tomášik **210**

012052

3+1 dimensional viscous hydrodynamics at high baryon densities Iu Karpenko, M Bleicher, P Huovinen and H Petersen **214**

012053

Identified charged hadron production in Pb-Pb collisions with event shape engineering Marco Antonio Tangaro (for the ALICE collaboration) **218**

012054

Exact solutions of the (0+1)-dimensional kinetic equation in the relaxation time approximation Ewa Maksymiuk **222**

012055

Directed flow in relativistic heavy-ion collisions within the PHSD transport approach and 3FD hydrodynamical model V P Konchakovski, W Cassing, Y B Ivanov and V D Toneev **226**

012056

Electric conductivity of the quark-gluon plasma investigated using Boltzmann transport theory Moritz Greif, Ioannis Bouras, Zhe Xu and Carsten Greiner **230**

012057

Electric Conductivity of the QGP Armando Puglisi, Salvatore Plumari and Vincenzo Greco **234**

012058

Transport coefficients in second-order non-conformal viscous hydrodynamics Radoslaw Ryblewski **238**

012059

Initial state fluctuations and final state collectivity in high energy nuclear collisions: Status and Outlook Björn Schenke **242**

012060

Soft probes of p+Pb and Pb+Pb collisions in the ATLAS experiment at the LHC **248**

S Radhakrishnan (for the ATLAS collaboration)

012061

New formulation of leading order anisotropic hydrodynamics Leonardo Tinti **252**

Baryons and Strangeness

012062

Multi-strange baryon production in pp, p–Pb and Pb–Pb collisions measured with ALICE Domenico Colella (for the ALICE Collaboration) **256**

012063

Study of strange mesons in *p+p*, *d*+Au and Cu+Cu collisions at $\sqrt{s_{NN}}$ = 200 GeV Priyanka Sett (for the PHENIX Collaboration) **260**

012064

Resonances as Probes of Heavy-Ion Collisions at ALICE A G Knospe (for the ALICE Collaboration) **264**

012065

(Hyper)nuclei and anti-(hyper)nuclei production in Pb-Pb collisions in ALICE at LHC Manuel Colocci (for the ALICE Collaboration) **268**

Journal of Physics: Conference Series **612** (2015) 011001 doi:10.1088/1742-6596/612/1/011001

The 6th edition of the Workshop for Young Scientists on the Physics of Ultrarelativistic Nucleus-Nucleus Collisions (Hot Quarks 2014) was held in Las Negras, Spain from 21-28 September 2014. Following the traditions of the conference, this meeting gathered more than 70 participants in the first years of their scientific careers. The present issue contains the proceedings of this workshop.

As in the past, the Hot Quarks workshop offered a unique atmosphere for a lively discussion and interpretation of the current measurements from high energy nuclear collisions. Recent results and upgrades at CERN's Large Hadron Collider (LHC) and Brookhaven's Relativistic Heavy Ion Collider (RHIC) were presented. Recent theoretical developments were also extensively discussed as well as the perspectives for future facilities such as the Facility for Antiproton and Ion Research (FAIR) at Darmstadt and the Electron-Ion Collider at Brookhaven. The conference's goal to provide a platform for young researchers to learn and foster their interactions was successfully met.

We wish to thank the sponsors of the Hot Quarks 2014 Conference, who supported the authors of this volume: Brookhaven National Laboratory (USA), CPAN (Spain), Czech Science Foundation (GACR) under grant 13-20841S (Czech Republic), European Laboratory for Particle Physics CERN (Switzerland), European Research Council under grant 259612 (EU), ExtreMe Matter Institute EMMI (Germany), Helmholtz Association and GSI under grant VH-NG-822, Helmholtz International Center for FAIR (Germany), National Science Foundation under grant No.1359622 (USA), Nuclear Physics Institute ASCR (Czech Republic), Patronato de la Alhambra y Generalife (Spain) and the Universidad de Granada (Spain).

Javier López Albacete, Universidad de Granada (Spain)

Jana Bielcikova, Nuclear Physics Inst. and Academy of Sciences (Czech Republic)

Rainer J. Fries, Texas A&M University (USA)

Raphaël Granier de Cassagnac, CNRS-IN2P3 and École

Journal of Physics: Conference Series **612** (2015) 011001 doi:10.1088/1742-6596/612/1/011001

polytechnique (France)

Boris Hippolyte, CNRS-IN2P3 and Université de Strasbourg (France)

Jiangyong Jia, Stony Brook University and Brookhaven National Laboratory (USA)

André Mischke, Utrecht University and Nikhef Amsterdam (The Netherlands)

Ágnes Mócsy, Pratt Institute and Brookhaven National Laboratory (USA)

Hannah Petersen, Goethe University, FIAS and GSI (Germany)

Lijuan Ruan, Brookhaven National Laboratory (USA)

Sevil Salur, Rutgers University, (USA)

Journal of Physics: Conference Series **612** (2015) 011002 doi:10.1088/1742-6596/612/1/011002

Peer review statement

All papers published in this volume of *Journal of Physics: Conference Series* have been peer reviewed through processes administered by the proceedings Editors. Reviews were conducted by expert referees to the professional and scientific standards expected of a proceedings journal published by IOP Publishing.

Modelling jet quenching with Quenching Weights

C Andrés, N Armesto, Carlos A Salgado and Yan Zhu
Departamento de Física de Partículas and IGFAE, Universidade de Santiago de Compostela, E-15782 Galicia, Spain
E-mail: carlota.andres@usc.es

Abstract.
We present a phenomenological study of the single- and double-inclusive suppression data of high-p_T particles in central Pb-Pb collisions at LHC. The analysis is based on quenching weights for medium-induced gluon radiation computed in the multiple soft scattering approximation and embedded in a hydrodynamic description of the bulk medium.

1. Introduction

Jet quenching has been established as an important tool to study the QGP created in heavy-ion collisions. The goal of this talk is to study the suppression of high-p_T particles in central Pb-Pb collisions at the LHC. This work is based on the partonic Quenching Weights (QW) for medium-induced gluon radiation computed in the multiple soft scattering approximation then embedded in a hydrodynamic description of the bulk medium.

This analysis is an extension of a previous similar one [1] for RHIC and our aim is to compare the results obtained for RHIC with the ones obtained for LHC.

2. The analysis

In heavy-ion collisions the medium evolves dynamically. We use the results of *Hirano et al.* [2] for 3D ideal hydrodynamics to describe its evolution.

The production of a hadron h at transverse momentum p_T and rapidity y can be described by

$$\frac{d\sigma^{AA\rightarrow h+X}}{dp_T dy} = \int \frac{dx_2}{x_2}\frac{dz}{z}\sum_{i,j} x_1 f_{i/A}(x_1,Q^2) x_2 f_{j/A}(x_2,Q^2)\frac{d\hat{\sigma}^{ij\rightarrow k}}{d\hat{t}} D_{k\rightarrow h}(z,\mu_F^2), \quad (1)$$

where $f_{i/A}(x_1,Q^2)$ are the parton distribution functions (PDFs) for a colliding object. In our case, EKS98 analysis [3] for nPDFs and CTEQ6L (LO) free proton parton densities are used. x is the momentum fraction of the hadron carried by the parton and $D_{k\rightarrow h}(z,\mu_F^2)$ are the fragmentation functions for a parton k into a hadron h, where z is the momentum fraction of the parton carried by the hadron. We take the factorization and renormalization scales as $Q^2 = (p_T/z)^2$.

We model our medium modified fragmentation functions as:

$$D_{k\rightarrow h}^{(med)}(z,\mu_F^2) = \int_0^1 d\epsilon P_E(\epsilon)\frac{1}{1-\epsilon} D_{k\rightarrow h}^{(vac)}\left(\frac{z}{1-\epsilon},\mu_F^2\right), \quad (2)$$

where $P_E(\epsilon)$ are the *Quenching Weights* [4], i.e. the probability distribution of a fractional energy loss, $\epsilon = \Delta E/E$, of the fast parton in the medium. The vacuum fragmentation functions, $D^{(vac)}_{k\to h}(z,\mu_F^2)$, are taken from *Florian, Sassot and Stratmann* [5].

Eq.(2) assumes that the only effect of the medium in the fragmentation functions is the energy loss taking place before hadronization. That is, hadronization is not modified by the medium, it takes place in the vacuum. Fragmentation functions are modified by medium-induced gluon radiation through the Quenching Weights (QW) [6].

The QW are Poisson distributions obtained assuming an independent gluon emission. In the multiple soft scatterting approximation and for a static medium, they only depend on two quantities, the transport coefficient, $\hat{q}$, and the length of the medium, L, (or ω_c and R), given by: $\omega_c = \frac{1}{2}\hat{q}L^2$ and $R = \omega_c L$.

In a dynamical medium we can make use of two of the following scaling relations:

$$\omega_c^{eff}(x_0, y_0, \tau_{prod}, \phi) = \int d\xi \xi \hat{q}(\xi), \quad [\hat{q}L]^{eff}(x_0, y_0, \tau_{prod}, \phi) = \int d\xi \hat{q}(\xi)$$

$$R^{eff}(x_0, y_0, \tau_{prod}, \phi) = \frac{3}{2}\int d\xi \xi^2 \hat{q}(\xi), \tag{3}$$

corresponding to an equivalent static scenario, to write the effective quantities. In contrast to the RHIC analysis [1], we choose to use ω_c^{eff} and R^{eff}. So, we need to specify the relation between the local value of the transport coefficient, $\hat{q}(\xi)$, at a given point, ξ, on the trajectory and the medium properties given by our hydrodynamic model:

$$\hat{q}(\xi) = K\hat{q}_{QGP}(\xi) \simeq K \cdot 2\epsilon^{3/4}(\xi), \tag{4}$$

where ϵ is the local energy density.

As we have seen, the transport coefficient is defined by Eq.(4). However, there is an ambiguity on its value for times smaller than the thermalization time τ_0. Consequently, we consider here three different extrapolations for the time from the hard production to the thermalization time:

- Case i): $\hat{q}(\xi) = 0$ for $\xi < \tau_0$,
- Case ii): $\hat{q}(\xi) = \hat{q}(\tau_0)$ for $\xi < \tau_0$,
- Case iii): $\hat{q}(\xi) = \hat{q}(\tau_0)/\xi^{3/4}$ for $\xi < \tau_0$.

In the case i) we are assuming that there is no energy loss for small times, which is a really strong assumption as thermalization is not necessary in the QW approach. In the second case, the transport coefficient is frozen before thermalization; there is a continuous interaction since the production time. In the last one, called the free-streaming case, the medium is assumed to have an energy density which decreases with time as $1/\xi$.

3. Results

The experimental data used in our analysis are given in terms of the nuclear modification factor for single measurements

$$R_{AA} = \frac{dN_{AA}/d^2p_T dy}{\langle N_{coll}\rangle dN_{pp}/dp_T^2 dy}. \tag{5}$$

Experimental data are all for Pb-Pb collisions at LHC energy $\sqrt{s_{NN}} = 2.76$ TeV. We use ALICE data on R_{AA} for charged particles with $p_T > 5$ GeV in the centrality class 0-5% and for $|\eta| < 0.8$ [7] and CMS data on R_{AA} for charged particles with $p_T > 5$ GeV in the centrality class 0-5% and for $|\eta| < 1$ [8].

For the double-inclusive measurements we have:

$$I_{AA} = \frac{D_{AA}(z_T, p_T^{trig})}{D_{pp}(z_T, p_T^{trig})}, \quad D_{AA}(z_T, p_T^{trig}) \equiv p_T^{trig} \frac{d\sigma_{AA}^{h_1h_2}/dy^{trig}dp_T^{trig}dy^{assoc}dp_T^{assoc}}{d\sigma_{AA}^{h_1}/dy^{trig}dp_T^{trig}} \tag{6}$$

and $z_T = p_T^{assoc}/p_T^{trig}$ and the factorization scale is taken as the p_T of the hadrons.

We compare the predictions of our model to ALICE data on I_{AA} on the away side for central (0-5%) PbPb collisions at $\sqrt{s_{NN}} = 2.76$ TeV [9].

We present the results for both, single- and double-inclusive measurements, for the three different extrapolations before thermalization. They are presented for different values of $2K = K'/0.73$, with $K' = 0.5, 1, 2, 3, ..., 20$. In figure 1 of reference [1], it is shown that the value of K obtained for RHIC was 4.1 which, as it is shown in the following figures, is bigger than the one we obtain for LHC.

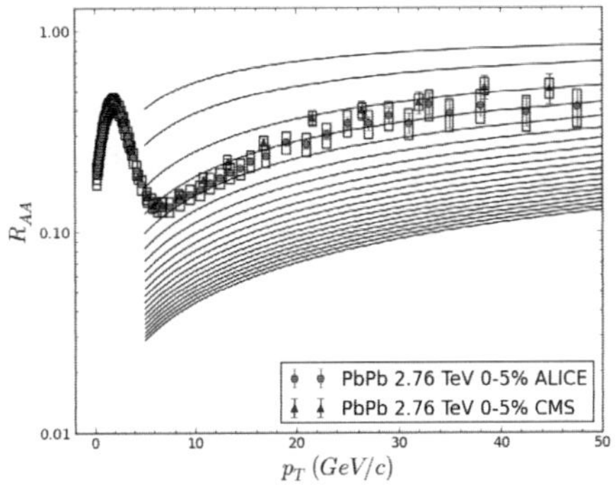

Figure 1. R_{AA} for the first extrapolation: Case i) $\hat{q}(\xi)$ =0 for $\xi < \tau_0$. Curves that best fit experimental data are the corresponding to $K = 1.37$ and $K = 2.05$.

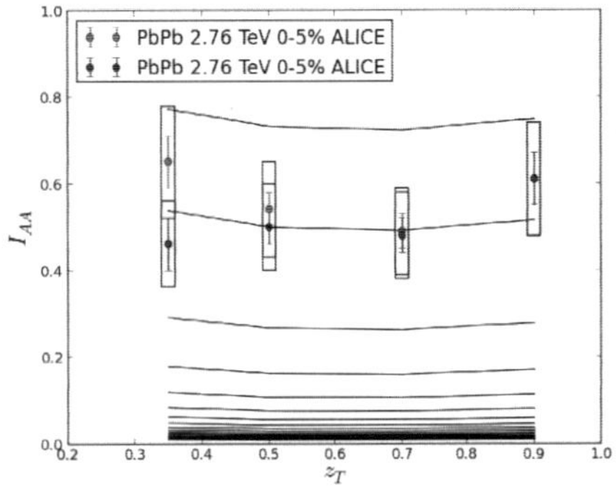

Figure 2. I_{AA} for the first extrapolation: Case i) $\hat{q}(\xi)$ =0 for $\xi < \tau_0$. The value of K obtained is $K = 1.37$.

4. Discussion and conclusions

There is a good agreement with experimental data and single- and double-inclusive results obtained are compatible. As it happened for the RHIC analysis - see Figure 2 of [1] - our results turn out to be relatively insensitive to the early time treatment except for the case i). In this case the parton travels the first 0.6 fm/c without suffering any effect of the medium and, therefore, the strong dependence of the jet quenching on the in-medium path length forces the value of K to increase.

The value of K obtained in our model is smaller than the one obtained for RHIC [1]. This means that, in our model, the medium created at LHC is more transparent than the medium created at RHIC. That is, that the behaviour of the medium is more similar to an ideal gas for LHC than for RHIC, i.e, the coupling is less strong for LHC.

Acknowledgments

The work was supported by the the European Research Council grant HotLHC ERC-2011-StG-279579, Ministerio de Ciencia e Innovación of Spain under project FPA2011-22776, Xunta de Galicia (Consellería de Educación and Consellería de Innovacil1ón e Industria - Programa

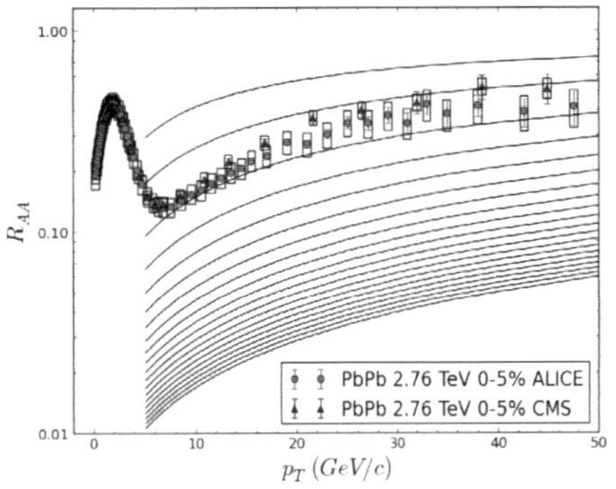

Figure 3. R_{AA} for the extrapolation $\hat{q}(\xi) = \hat{q}(\tau_0)$ for $\xi < \tau_0$. Curves that best fit experimental data are the corresponding to $K = 0.68$ and $K = 1.37$.

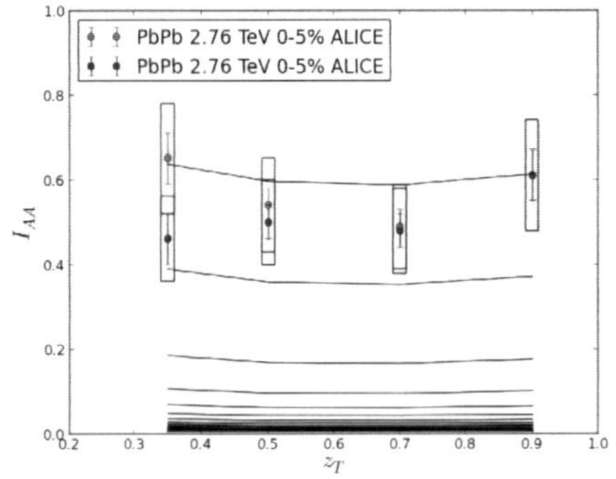

Figure 4. I_{AA} for the extrapolation $\hat{q}(\xi) = \hat{q}(\tau_0)$ for $\xi < \tau_0$. The curve that best fit experimental data is the one with $K = 0.68$.

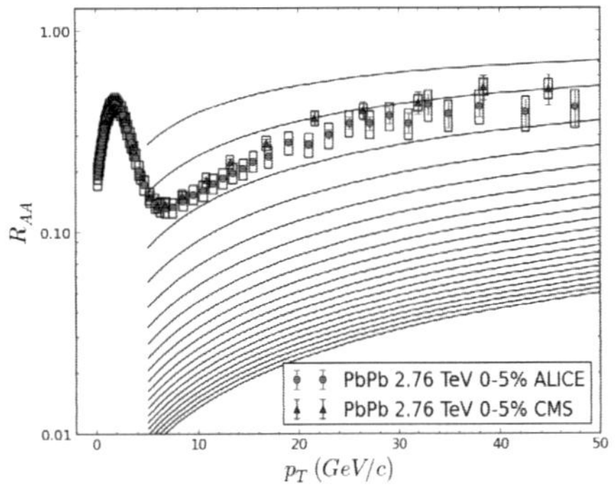

Figure 5. R_{AA} for the free streaming case $\hat{q}(\xi) = \hat{q}(\tau_0)/\xi^{3/4}$ for $\xi < \tau_0$. The value of K obtained in this case is $K = 0.68$.

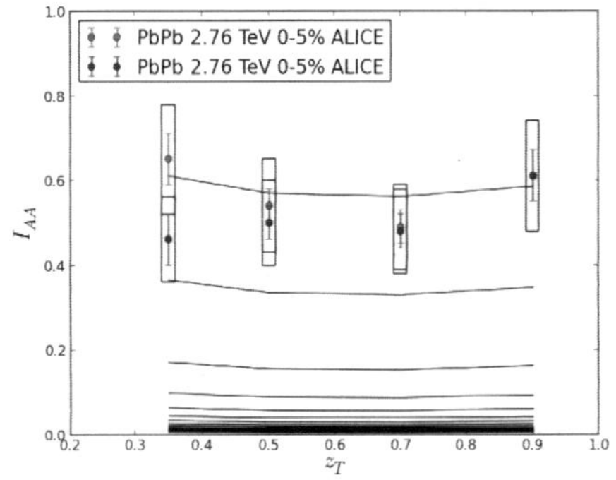

Figure 6. I_{AA} for the free streaming case $\hat{q}(\xi) = \hat{q}(\tau_0)/\xi^{3/4}$ for $\xi < \tau_0$. The value of K obtained in this case is $K = 0.68$.

Incite), the Spanish Consolider-Ingenio 2010 Programme CPAN and FEDER, and by the People Programme (Marie Curie Actions) of the European Unions Seventh Framework Programme FP7/2007-2013/ under REA grant agreement #318921 (NA). C. Andrés thanks the Spanish Ministery of Education, Culture and Sports for financial support (grant FPU2013-03558).

References

[1] N Armesto, M Cacciari, T Hirano, James L Nagle and Carlos A Salgado, J. Phys. **G37**, 025104 (2009).
[2] T Hirano, Phys. Rev. C **65**, 011901 (2002); T Hirano and K Tsuda, Phys. Rev. C **66**, 054905 (2002).
[3] K J Eskola, V J Kolhinen and C A Salgado, Eur. Phys. J. C **9**, 61 (1999).
[4] R Baier, Yuri L Dokshitzer, Alfred H Mueller and D Schiff, JHEP **0109**, 033 (2001).
[5] D de Florian, R Sassot and M Stratmann, Phys. Rev. D **75**, 114010 (2007).
[6] C A Salgado and U A Wiedemann, Phys. Rev. D **68**, 014008 (2003).
[7] ALICE Collaboration, Phys. Lett. B **720**, 52 (2013).
[8] CMS Collaboration, Eur. Phys. J. C **72**, 1945 (2012).
[9] ALICE Collaboration, Phys. Rev. Lett. **108**, 092301 (2012).

Journal of Physics: Conference Series **612** (2015) 012002 doi:10.1088/1742-6596/612/1/012002

W boson studies in pPb and PbPb collisions with CMS

Émilien Chapon, on behalf of the CMS Collaboration
Laboratoire Leprince-Ringuet, École polytechnique, 91128 PALAISEAU Cedex, France
E-mail: emilien.chapon@cern.ch

Abstract. The electroweak W bosons do not participate in the strong interaction, and thus constitute clean probes of the initial state of nuclear collisions. They provide a unique constraint on the nuclear parton distributions, in particular on the antiquarks from the sea. A first analysis of PbPb data has confirmed the medium-blind characteristic of the electroweak bosons. With the new pPb data collected at the beginning of 2013, nuclear matter without the creation of a hot medium can hence be studied. Being 10 times more prevalent than Z bosons, the yield of W bosons recorded from pPb collisions allows precise comparisons to theoretical predictions. A yield of approximately 20 000 W is observed in pPb collisions in both the muon and electron channels. In this paper the CMS measurements of W bosons in PbPb at nucleon-nucleon center-of-mass energy of $\sqrt{s_{NN}} = 2.76$ TeV and from the new pPb data at $\sqrt{s_{NN}} = 5.02$ TeV are reported. The charge asymmetry, forward/backward asymmetry and fully corrected yields will be shown.

1. Introduction

In proton-nucleus or nucleus-nucleus collisions, electroweak-boson production can be affected by initial-state conditions. Given their high mass and short lifetime, W bosons are produced and decay very early in the collision, and are not affected by the medium created during the collision. In the case of a leptonic decay of the W boson, the decay products are not affected either, thus giving direct access to initial-state effects in heavy-ion collisions. A large difference of W boson production between proton-proton and proton-nucleus or nucleus-nucleus collisions is coming from the presence of neutrons in nuclei. At leading order, the leading-order production processes are $u\bar{d} \to W^+$ and $d\bar{u} \to W^-$, mainly reflecting interactions that take place between valence quarks and sea antiquarks. Due to the mixture of protons and neutrons in the incoming nuclei, hence of u and d quarks, the W^+ and W^- rates are expected to be individually modified relative to pp collisions, but not their sum. This is often referred to as an *isospin effect*. The parton distribution functions (PDFs) can also be modified in nuclei. Such modifications include shadowing at small Bjorken x and anti-shadowing at intermediate Bjorken x, for instance. Precise measurements of W production in heavy-ion collisions can therefore constrain the nuclear PDFs (nPDFs).

2. $W \to \mu\nu$ in pp and PbPb collisions

W boson production has been studied using $(7.3 \pm 0.3)\,\mu b^{-1}$ of PbPb collisions recorded in 2010 and $(231 \pm 14)\,nb^{-1}$ of pp collisions recorded in 2011 [1], in the $W \to \mu\nu$ decay mode.

The muon and neutrino from the W boson are energetic, because of its high mass, and we use this feature to select signal events. The p_T spectrum of good quality muons in PbPb events shows a clear signal at around half the mass of the W boson, while the background is concentrated at low p_T. Furthermore, the information from the undetected neutrino can be partially accessed by reconstructing the missing transverse momentum ($\not{p}_T$) from the tracks in the silicon tracker. These characteristics motivate our selection of events, which must include one good quality, high p_T muon ($|\eta^\mu| < 2.1$, $p_T^\mu > 25\,\mathrm{GeV}/c$) and significant momentum imbalance ($\not{p}_T > 20\,\mathrm{GeV}/c$). We also cut on the transverse mass of the muon and $\not{p}_T$ system, $m_T = \sqrt{2p_T^\mu \not{p}_T(1-\cos\phi)} > 40\,\mathrm{GeV}/c^2$, to define an almost background-free sample, dominated at 98% by the W signal.

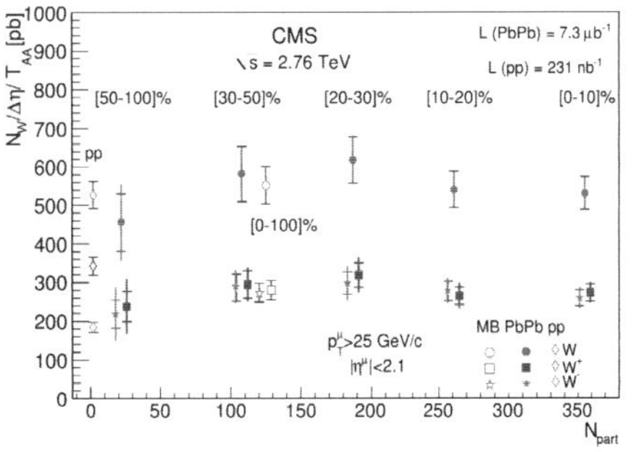

Figure 1. Collision centrality dependence of the nuclear modification factor R_{AA} for PbPb collisions at $\sqrt{s_{NN}} = 2.76$ TeV compared to pp collisions at $\sqrt{s} = 2.76$ TeV.

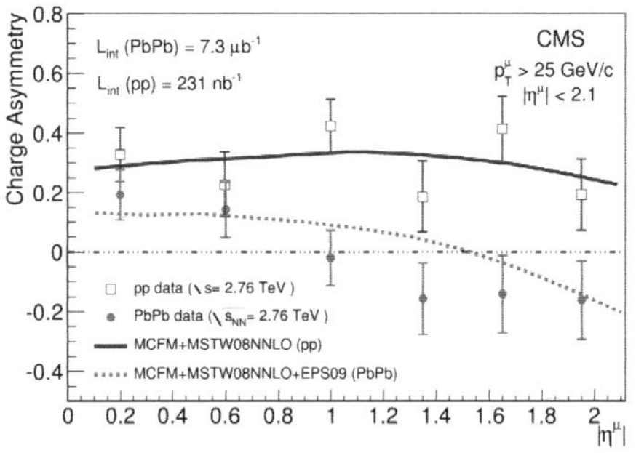

Figure 2. Muon charge asymmetry in W boson decays as a function of muon pseudo-rapidity for pp and PbPb collisions at $\sqrt{s_{NN}} = 2.76$ TeV.

The nuclear modification factor $R_{AA} = N_W/(T_{AA} \cdot \sigma_{pp})$ (where T_{AA} is the nuclear overlap function and σ_{pp} the cross section in pp collisions) has been measured for W^+ and W^- bosons separately, and for the combined charge, both inclusively and in bins of centrality, as can be seen on Fig. 1. We measure $R_{AA}(W) = 1.04 \pm 0.07(\mathrm{stat}) \pm 0.12(\mathrm{syst})$ when combining both charges, which is compatible with unity as expected. Qualitatively, such value is expected from the fact that the valence quark content of the proton is two u quarks and one d quark. Therefore we expect approximately twice as much W^+ as W^- bosons in proton-proton collisions, while the presence of neutrons in the lead nucleus, with opposite quark content, leads to the expectation of a more balanced production. The centrality dependence of the nuclear modification factor of W boson production in lead-lead collisions is consistent with being constant.

The charge asymmetry is defined as $\frac{N(W^+)-N(W^-)}{N(W^+)+N(W^-)}$. It should be noted that, due to the undetected neutrino, the rapidity of the W boson cannot be reconstructed, and hence the pseudo-rapidity of the muon is used as a proxy to this quantity. The charge asymmetry in pp and PbPb collisions is shown in Fig. 2 and compared to theoretical predictions. Again, the large difference between the two collision systems is due to the isospin effect, i.e. to the presence of neutrons in the lead nucleus. collisions, while it is closer to 0 in lead-lead collisions.

3. W $\rightarrow \ell\nu$ in pPb collisions

We have also studied W boson production in pPb collisions, using the $(34.6\pm1.2)\,\mathrm{nb}^{-1}$ integrated luminosity recorded in 2013 at $\sqrt{s_{NN}} = 5.02$ TeV [2]. Thanks to an asymmetric initial state, the study of W boson production in proton-nucleus instead of nucleus-nucleus collisions allows for the measurement of new observables, namely forward/backward asymmetries. Such asymmetries enhance the sensitivity to nPDF, due to the cancellation of uncertainties, both experimental and theoretical. The higher centre-of-mass energy available in the pPb dataset used ($\sqrt{s_{NN}} = 5.02$ TeV) also implies a higher production cross section. As a result of the energy difference of the colliding beams, the nucleon-nucleon centre-of-mass frame in the pPb collisions is not at rest with respect to the laboratory frame. Massless particles emitted at $\eta = 0$ in the nucleon-nucleon centre-of-mass frame are detected at $\eta_{\rm lab} = 0.465$ in the laboratory frame. The results presented hereafter are expressed in the laboratory frame, with the proton-going side defining the positive η direction.

The selection of the events has some differences with the one used in the analysis of pp and PbPb collisions presented above. We require a lepton with $p_T^\ell > 25\,\mathrm{GeV}/c$, but we now study both the electron and muon channels, and extend the pseudo-rapidity coverage ($|\eta_{\rm lab}| < 2.5$ for electrons, $|\eta_{\rm lab}| < 2.4$ for muons). Moreover, no explicit missing transverse energy cut is applied: instead, the signal is extracted from a template fit to the missing transverse energy distribution. However, to keep the heavy-flavour and jet backgrounds to a reasonable level, the selected lepton is required to be isolated (the lepton isolation being defined by summing the transverse energy of particle flow candidates in a cone around the lepton). The W boson yields are extracted from a fit to the missing transverse energy distribution (see Fig. 3), the signal and background contribution being estimated from simulation (except for the QCD background, parametrised using data).

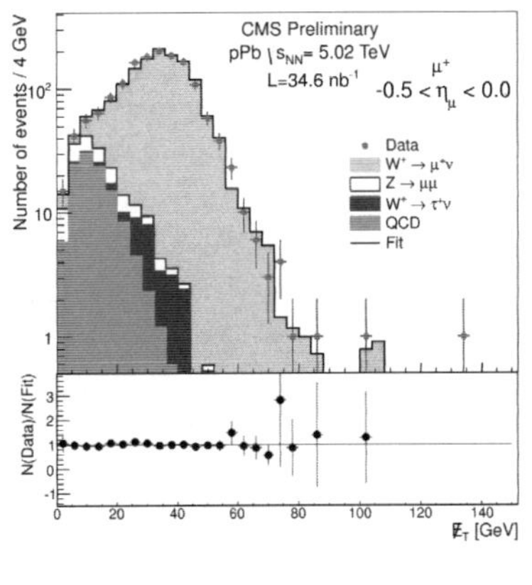

Figure 3. Missing transverse energy distribution for $W^+ \to \mu^+\nu$ in pPb collisions at $\sqrt{s_{NN}} = 5.02$ TeV within the $-0.5 < \eta_{\rm lab} < 0.0$ range.

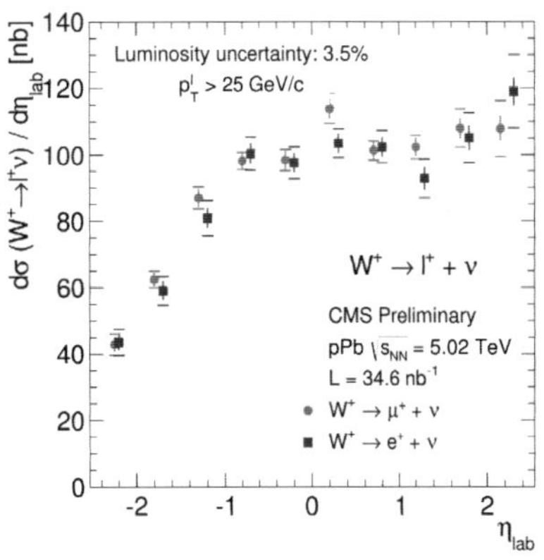

Figure 4. Production cross section for $W^+ \to \ell^+\nu$ in pPb collisions at $\sqrt{s_{NN}} = 5.02$ TeV as a function of $\eta^\ell_{\rm lab}$.

We measure the W boson production cross section in both the electron and muon decay channels (see Fig. 4) and combine the measurements, for positively and negatively charged leptons separately. The results are consistent with theoretical predictions [3], but uncertainties are too large to probe the nuclear modifications of the PDF. For this reason, we build new observables. The charge asymmetry, for instance, has been introduced before, and is shown in Fig. 5. The theoretical uncertainties are much smaller than in the case of cross sections, but the assumptions of the EPS09 nPDF [4] do not lead to a significant modification of the charge asymmetry as compared to free proton PDF such as CT10 [5]. However the data deviate from

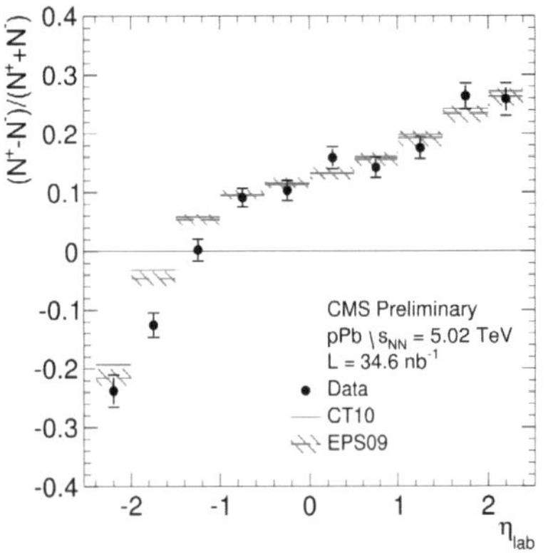

Figure 5. Charge asymmetry in W boson decays in pPb collisions at $\sqrt{s_{NN}} = 5.02$ TeV as a function of $\eta^{\ell}_{\mathrm{lab}}$.

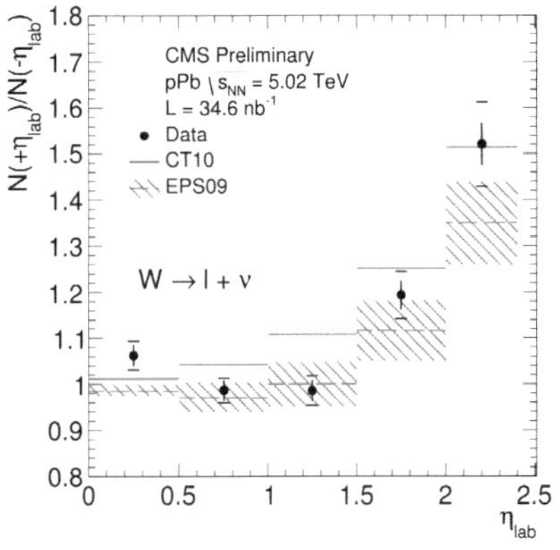

Figure 6. Forward-backward asymmetry in W boson decays in pPb collisions at $\sqrt{s_{NN}} =$ 5.02 TeV as a function of $\eta^{\ell}_{\mathrm{lab}}$.

theoretical predictions at large negative lepton pseudo-rapidity. A possible explanation for such deviation would be a different modification of u and d quarks in the nucleus, as opposed to what is assumed by EPS09.

We also measure forward-backward asymmetries: $N(+\eta_{\mathrm{lab}})/N(-\eta_{\mathrm{lab}})$. We first build them for each lepton charge separately. Again theoretical uncertainties are smaller than for cross sections, and here predictions with EPS09 differ significantly from those of CT10. The data seem to favour EPS09 for negatively charged leptons, while the conclusion is less clear for positively charged leptons. For an even better sensitivity to nPDF, we build the forward-backward asymmetry for both lepton charges combined, as shown on Fig. 6. Again the data seem to disfavour the absence of nuclear modifications of the PDFs.

4. Conclusion

We have presented the measurement of W boson production by the CMS experiment in pp, PbPb and pPb collisions at the LHC. In the case of PbPb collisions, we observe a large difference with measurements in pp collisions, which is dominated by the well-known isospin effect. The nuclear modification factor for W boson production, when combining both charges, is compatible with unity.

In the case of pPb collisions, we have built asymmetries in order to be sensitive to nuclear modifications of the PDFs. We see hints of such nuclear effects in the data. Moreover, some tension is seen between data and theory in the charge asymmetry.

Acknowledgments

The speaker received funding from the European Research Committee, grant no. 259612.

References

[1] Chatrchyan S *et al.* (CMS Collaboration) 2012 *Phys. Lett.* B **715** 66
[2] CMS Collaboration 2014 (*Preprint* CMS-PAS-HIN-13-007)
[3] Paukkunen H and Salgado P 2011 *JHEP* **1103** 071
[4] Eskola K J, Paukkunen H and Salgado C A 2009 *JHEP* **0904** 065
[5] Lai H L, Guzzi M, Huston J, Li Z, Nadolsky P M, Pumplin J and Yuan C P 2010 *Phys. Rev.* D **82**, 074024

Journal of Physics: Conference Series **612** (2015) 012003 doi:10.1088/1742-6596/612/1/012003

Renormalization of the jet-quenching parameter

Yacine Mehtar-Tani

Institut de Physique Théorique, CEA Saclay, F-91191 Gif-sur-Yvette, France

Abstract. In the context of the recently derived probabilistic picture of in-medium jet evolution, we study radiative corrections which yield potentially large double logarithms, $\alpha_s \ln^2 L$, for large enough medium length L. We show that these large corrections are universal and can be reabsorbed in a renormalization of the jet quenching parameter controlling both momentum broadening and energy loss. The renormalized jet-quenching parameter is enhanced compared to its standard perturbative estimate. As a particular consequence, the radiative energy loss scales with medium size L as $L^{2+\gamma}$, with $\gamma = 2\sqrt{\alpha_s N_c/\pi}$, as compared to the standard scaling in L^2.

1. The approximation of independent multiple scatterings and branchings

In a large QCD medium, a high energy parton might undergo multiple interactions. So long as their range, i.e., the in-medium correlation length, is much smaller than the mean free path λ of the parton in the medium they can be treated as independent from one another. The physics is then fully encoded in a local interaction which at leading order in perturbative QCD is given by the elastic cross-section $d\sigma_{\rm el}/d^2\boldsymbol{q} \simeq g^4/\boldsymbol{q}^4$, where $g \ll 1$ is the coupling constant and $\boldsymbol{q} \gg \Lambda_{\rm QCD}$ is the transverse momentum exchanged in the scattering process. In this case, the so-called quenching parameter $\hat{q}$ corresponds to a (local) diffusion coefficient in transverse momentum space with respect to the direction of its large momentum component [1], that is, $\hat{q} \equiv d\langle k_\perp^2\rangle/dt = n \int_{\boldsymbol{q}} \boldsymbol{q}^2 d\sigma_{\rm el}/d^2\boldsymbol{q}$, where n is the density of medium color charges. This yields a typical transverse momentum broadening, for a high-energy particle, $\langle k_\perp^2\rangle_{\rm typ} = \hat{q}L$, in a medium of size L. Note that, this approximation is justified in a weakly coupled quark-gluon plasma (QGP). Indeed, in that case the range of medium interactions, given by the inverse Debye mass $m_D \sim gT$, where T is the temperature of the plasma, is parametrically much smaller than the mean-free-path $\lambda \sim (g^2T)^{-1}$. In terms of microscopic scales, $\hat{q} \simeq m_D^2/\lambda$.

In addition to $p_\perp$-broadening, multiple scatterings might induce fast radiations of soft gluons off the energetic parton with constant rate along the medium, hence suppressing its energy [1, 2]. It takes a coherence time $\tau = \omega/k_\perp^2$ for a gluon of energy ω to form. When it is formed, it has accumulated in the medium a typical transverse momentum $k_\perp^2 \sim \hat{q}\tau$, via multiple scatterings. This yields the characteristic formation time $\tau_{\rm form}(\omega) \equiv \sqrt{2\,\omega/\hat{q}}$. The leading-order (LO) gluon radiation probability, the so-called BDMPS spectrum, reads (for $\tau_{\rm form} \ll L$) $\omega dN/d\omega \sim C_R\alpha_s L/\tau_{\rm form}$, and yields a mean-energy loss $\Delta E \sim \alpha_s C_R\,\hat{q}\,L^2$ where C_R is the color charge of the emitter. This mechanism provides a natural explanation for the observed suppression of jets in heavy ion collisions compared to proton-proton collisions at RHIC and the LHC [3, 4]. When $\tau_{\rm form} \lesssim \alpha_s L$, the BDMPS spectrum can no longer be identified with a probability, and higher order corrections become non-negligible. In the limit of large media, these corrections can be resummed [5]. Multiple gluon emissions can then be treated as independent when the formation time is much smaller than the radiative mean free path, i.e., the typical time between two successive emissions, $\lambda_{\rm rad} \equiv \tau_{\rm form}(\omega_{\rm BH})/\alpha_s \simeq \lambda/\alpha_s$ and $\omega_{\rm BH} \sim m_D^2 \lambda$ is the Bethe-Heitler infrared cut-off (we have used in fact that $\hat{q} \sim m_D^2/\lambda$). We observe that this is indeed realized in the small coupling limit when $\omega \ll \omega_{\rm BH}/\alpha_s^2$. Combining the two inequalities we get the condition for the validity of multiple independent interaction approximation: $\tau_{\rm form} \ll \lambda_{\rm rad} \ll L \lesssim \lambda/\alpha_s^2$. Not only gluon radiation can be treated as quasi-local compared to the size L of the system but further evolutions of the off spring gluons are independent from one another [5]. This is the basis of in-medium jet evolution. In this framework, both the local scattering and branching rates depend on medium properties solely via the quenching parameter $\hat{q}$.

Recently, radiative corrections to transverse momentum broadening have been considered [6, 7]. Potentially large corrections, enhanced by a double logarithm of the length of the medium L, were found. In light of this result, one can also expect large radiative corrections to radiative energy loss and other in-medium jet observables as well. Moreover, the large non-local radiative corrections can be expected to spoil the classical picture of independent multiple interactions and branchings. We argue that the double logarithmic corrections are universal and can be absorbed in a renormalization of the jet quenching parameter. Hence, the probabilistic approximation remains valid [8].

2. Radiative corrections and the instantaneous interaction approximation

Let us first discuss radiative corrections to $p_\perp$-broadening. The probability for a high energy parton to acquire a transverse momentum $\boldsymbol{k}$ in the medium after a time t is $\mathcal{P}(\boldsymbol{k},t) = \int d^2\boldsymbol{x}\, e^{i\boldsymbol{p}\cdot\boldsymbol{x}} S_{\boldsymbol{x}}(t,0)$, where $S_{\boldsymbol{x}}(t,0) = \langle {\rm Tr} U_{\boldsymbol{x}}^\dagger U_{\boldsymbol{0}}\rangle$ is a 2-point function involving Wilson lines that encode the eikonal propagation of the high energy parton in the background field of the medium, $A^-(t,\boldsymbol{x})$, in the amplitude and the complex conjugate amplitudes. [1] In the approximation of independent multiple interactions one can model the medium with a classical random field correlated at equal times, i.e., $\langle A^-(t)A^-(t')\rangle \propto n\,\delta(t-t')$. [2] Hence, the single instantaneous elastic interaction exponentiates as follows, $S_{\boldsymbol{x}}(t,0) = 1 - \frac{1}{2} n C_R \int_0^t dt' \sigma(\boldsymbol{x})\, S_{\boldsymbol{x}}(t',0)$, where $nC_R\,\sigma(\boldsymbol{x}) \approx \boldsymbol{x}^2\,\hat{q}\,/2$ is the dipole cross section, which is proportional to the quenching parameter. The broadening probability obeys then a Fokker-Planck equation and reads $\mathcal{P}(\boldsymbol{k},t) = 4\pi/(\hat{q}\,t)\exp[-\boldsymbol{k}^2/(\hat{q}\,t)]$.

We turn now to the first radiative correction to $S_{\boldsymbol{x}}(t,0)$. It involves the emission of a soft gluon with energy ω at time t' and absorbed at time $t'+\tau$ by the fast parton. We obtain the form

$$\Delta S_{\boldsymbol{x}}(t,0) = \alpha_s \int_0^t dt' \int_0^{t'} d\tau \int \frac{d\omega}{\omega^3} S_{\boldsymbol{x}}(t,t'+\tau)\, K(\tau,\boldsymbol{x})\, S_{\boldsymbol{x}}(t',0)\,. \tag{1}$$

where the function K is given in [8]. Let us have a close look at the structure of Eq. (1). Contrary to the instantaneous Coulomb interaction, the gluon fluctuation has a finite duration τ. However, the integration over τ diverges logarithmically at 0 as we shall see. For the time being we simply cut-off the divergence at τ_0. This divergence shows that radiative corrections are logarithmically sensitive to short distance scales, which were ignored in our instantaneous interaction model. We can then neglect the dependence on τ before and after the gluon fluctuation thanks to the logarithmic divergence, in other words the fluctuation is seen "as is if" it were instantaneous. Indeed, in this approximation one can assume $\tau \ll t, t'$ and integrate τ up to $\tau_{\rm max} \sim t'$. Therefore, the radiative correction can be "effectively" interpreted

[1] The time variable $t \equiv x^+ = (x_0 - x_3)/\sqrt{2}$ denotes the light-cone time.

[2] Recall that this correlator encodes the leading-order differential elastic cross section.

as local and thus can be absorbed in a redefinition of the dipole cross section. We finally get,

$$\Delta S_{\boldsymbol{x}}(t,0) \simeq \frac{1}{2} n C_R \int_0^t dt' S_{\boldsymbol{x}}(t,t')\, \Delta\sigma(\tau_{\max},\boldsymbol{x})\, S_{\boldsymbol{x}}(t',0) \tag{2}$$

where $n\, C_R\, \Delta\sigma(\tau_{\max},\boldsymbol{x}) = \alpha_s \int_{\tau_0}^{\tau_{\max}} d\tau \int \frac{d\omega}{\omega^3} K(\tau,\boldsymbol{x}) \approx \boldsymbol{x}^2 \Delta\hat{q}(\tau_{\max},\boldsymbol{x}^{-2})/2$ and (switching from (ω,τ) to $(\tau,\boldsymbol{q})$ variables)

$$\Delta\hat{q}(\tau_{\max},\boldsymbol{k}^2) = \bar{\alpha} \int_{\tau_0}^{\tau_{\max}} \frac{d\tau}{\tau} \int_{\hat{q}\tau}^{\boldsymbol{k}^2} \frac{d\boldsymbol{q}^2}{\boldsymbol{q}^2}\, \hat{q}(\boldsymbol{q})\,. \tag{3}$$

Where the dominant contribution is double logarithmic and corresponds to radiations that have been completed after a time τ by a single hard scattering with momentum squared $\boldsymbol{q}^2$ that is much larger than the scale $\hat{q}\,\tau$ where multiple scatterings are important. To this accuracy, we can set $\tau_{\max} \sim t$ and $\boldsymbol{k}^2 \sim \hat{q}\, t$ in Eq. (3). Finally, assuming $\hat{q}$ to be constant we perform the double logarithmic integrals and get for the renormalized $\hat{q}_{\mathrm{R}}(t) \simeq \hat{q} + \Delta\hat{q}(t)$,

$$\hat{q}_{\mathrm{R}}(t) \simeq \hat{q}\left(1 + \frac{\alpha_s N_c}{2\pi} \ln^2 \frac{t}{\tau_0}\right)\,. \tag{4}$$

Integrating Eq. (4) over t up to L, we recover to double logarithmic accuracy the result by Liou, Mueller and Wu for the typical transverse momentum squared [7].

We shall argue that this single radiative correction will exponentiate similarly to multiple scatterings. Since the integral over τ extends to $\tau_{\max} \sim t$, multiple radiative corrections associated with multiple scatterings could overlap. However, thanks to the logarithmic integrals the overlaps of the time integrals are down by at least a log compared to the independent radiation case [8]. So the result yields the exponentiation of the double logs. Multiple radiative corrections along the propagation in the medium are thus accounted for by the substitution $\sigma \to \sigma_{\mathrm{R}} \equiv \sigma + \Delta\sigma$ in $S(\boldsymbol{x},t)$.

In order to discuss the universality of the renormalization of the jet quenching parameter we shall consider now the generalization of our result to the BDMPS radiative spectrum. It can be shown that radiative corrections to the radiative spectrum exhibit the same logarithmic divergence when the duration of the additional gluon fluctuation $\tau \to 0$ which can occur anywhere in the time interval t which is typically of the order of the BDMPS formation time τ_{form}. It is accounted for to double logarithmic accuracy by shifting the dipole cross-section $\sigma(\boldsymbol{r}) \to \sigma(\boldsymbol{r}) + \Delta\sigma(\tau_{\max},\boldsymbol{r})$, with $\tau_{\max} \sim \tau_{\mathrm{form}}$. This amounts to redefining the quenching parameter in the BDMPS spectrum with its renormalized value, $\hat{q} + \Delta\hat{q}(\tau_{\mathrm{form}}, k^2_{\mathrm{form}})$, in the BDMPS spectrum. We are able now to compute the correction to the mean energy loss. Since gluons with formation times of the order of the length of the medium dominate the integral over ω, we shall set, as for $p_\perp$-broadening, $\tau_{\mathrm{form}} \sim L$ and $k^2_{\mathrm{form}} \sim \hat{q}L$. Therefore, we readily obtain to double logarithmic accuracy,

$$\Delta E \sim \hat{q}L^2\left(1 + \frac{\alpha_s N_c}{2\pi} \ln^2 \frac{L}{\tau_0}\right)\,.$$

3. Renormalization of the quenching parameter

For large media, as soon as $\alpha_s \ln^2(L/\tau_0) \sim 1$ one has to resum the DL power corrections. This resummation of the leading logarithms involves large separation of scales. The structure of the first double logarithmic corrections is based on the strong ordering of the transverse momentum of the fluctuation $\boldsymbol{q} \ll \boldsymbol{k}$ and its duration $\tau \ll L$. Hence, the fast parton cannot resolve this fluctuation from the elastic scatterings with the medium. This being set, next corrections that yield DL's will follow the same systematics, with successive gluonic fluctuations ordered in formation time, $\tau_0 \ll \tau_1 \ll ... \ll \tau_n \equiv \tau_{\rm max}$ and in transverse momentum $m_D \ll \boldsymbol{q}_1 \ll ... \ll \boldsymbol{q}_n \equiv \boldsymbol{k}$ (or in transverse size $\boldsymbol{x}_0 \gg \boldsymbol{x}_1 \gg ... \gg \boldsymbol{x}_n \equiv \boldsymbol{x}_{\rm max}$). The difference with the standard Double-Logarithmic Approximation (DLA) is the limits of the logarithmic phase-space set by the LPM effect, i.e., multiple-scatterings since in the DLA only a single scattering contributes, which imposes that the formation time of a fluctuation to be smaller than the BDMPS formation time. In terms of our variables $\boldsymbol{k} \gg \hat{q}\,\tau$. The following equation resums the DL corrections to all orders

$$\frac{\partial \hat{q}(\tau, \boldsymbol{k}^2)}{\partial \log \tau} = \int_{\hat{q}\tau}^{\boldsymbol{k}^2} \frac{d\boldsymbol{q}^2}{\boldsymbol{q}^2}\, \bar{\alpha}(\boldsymbol{q})\, \hat{q}(\tau, \boldsymbol{q}^2)\,, \tag{5}$$

given an initial condition $\hat{q}(\tau_0, \boldsymbol{k}^2)$. Here $\bar{\alpha} \equiv \alpha_s N_c/\pi$ and we have let the coupling running with the transverse scale $\boldsymbol{q}$. Note again the lower limit of the $\boldsymbol{q}$ integration which accounts for the boundary between single-scattering and multiple-scatterings. The important feature of this equation is that it predicts the evolution of the jet-quenching parameter from an initial condition $\hat{q}_0$ (which can be computed e.g. on the lattice, or to leading order in α_s). The τ_0 cut-off that was introduced to cut the logarithmic divergence in the radiative corrections, can be seen as a factorization scale. In the case where $\hat{q}(\tau_0)$ is constant and for a final $\tau = L$ and $\boldsymbol{k}^2 = \hat{q}(\tau_0)L$, merging the 2 independent variables at the end of the evolution, Eq. (5) can be solved iteratively [7]. We find then that for large L, the quenching parameter scales like L^γ, with the anomalous $\gamma = 2\sqrt{\alpha_s N_c/\pi}$ Interestingly, the resummation of large DL's modifies the L^2 scaling of the mean energy loss, $\Delta E \sim L^{2+\gamma}$, and seems to fall between the standard small coupling result, $\Delta E \sim L^2$ and the strong coupling result obtained in $\mathcal{N} = 4$ SYM theory , $\Delta E \sim L^3$ [9]. Some of these results have been recovered recently from the point of view of evolution equations [10, 11].

Acknowledgments

This research is supported by the European Research Council under the Advanced Investigator Grant ERC-AD-267258.

References

[1] R. Baier, Y. L. Dokshitzer, A. H. Mueller, S. Peigne and D. Schiff, Nucl. Phys. B **484**, 265 (1997) [hep-ph/9608322]; Nucl. Phys. B **483**, 291 (1997) [hep-ph/9607355].
[2] B. G. Zakharov, JETP Lett. **63**, 952 (1996) [hep-ph/9607440]; JETP Lett. **65**, 615 (1997) [hep-ph/9704255].
[3] D. d'Enterria, arXiv:0902.2011 [nucl-ex].
[4] Y. Mehtar-Tani, J. G. Milhano and K. Tywoniuk, Int. J. Mod. Phys. A **28** (2013) 1340013 [arXiv:1302.2579 [hep-ph]].
[5] J. -P. Blaizot, F. Dominguez, E. Iancu and Y. Mehtar-Tani, JHEP **1301**, 143 (2013) [arXiv:1209.4585 [hep-ph]]; JHEP **1406**, 075 (2014) [arXiv:1311.5823 [hep-ph]].
[6] B. Wu, JHEP **1110**, 029 (2011) [arXiv:1102.0388 [hep-ph]].
[7] T. Liou, A. H. Mueller and B. Wu, Nucl. Phys. A **916**, 102 (2013) [arXiv:1304.7677 [hep-ph]].
[8] J. P. Blaizot and Y. Mehtar-Tani, Nucl. Phys. A **929**, 202 (2014) [arXiv:1403.2323 [hep-ph]].
[9] Y. Hatta, E. Iancu and A. H. Mueller, JHEP **0801**, 063 (2008) [arXiv:0710.5297 [hep-th]].
[10] E. Iancu, JHEP **1410**, 95 (2014) [arXiv:1403.1996 [hep-ph]].
[11] B. Wu, JHEP **1412**, 081 (2014) [arXiv:1408.5459 [hep-ph]].

Neutral pion production in $\sqrt{s_{NN}}$=200 GeV Cu+Au collisions at PHENIX

Sarah Campbell for the PHENIX Collaboration
Buidling 510C, PHENIX Brookhaven National Laboratory Upton, NY 11973
E-mail: campbels@iastate.edu

Abstract. Cu+Au collisions at RHIC generate asymmetric initial geometries and densities in both azimuth and rapidity. High p_T π^0s produced in $\sqrt{s_{NN}}$ = 200 GeV Cu+Au collisions provide new environments to study parton energy loss in the Quark Gluon Plasma, including very central events where the Cu nucleus is enveloped by the Au nucleus. By measuring π^0 yields in ϕ relative to the event plane, we can probe different core-corona regions in these very central events and study the path length dependence of energy loss in various lopsided initial geometries. PHENIX has observed the suppression of π^0s as a function of the azimuthal angle with respect to the event plane in $\sqrt{s_{NN}}$ = 200 GeV Au+Au collisions and found it consistent with a larger than quadratic path length dependence suggesting a non-perturbative energy loss model applies. The unique collision geometries available in Cu+Au provide new settings to explore and possibly confirm this path length dependence. The status of the Cu+Au π^0 analysis is presented.

1. Introduction

High p_T neutral pions serve as a proxy for the jet because the leading particles of a jet are a good measure of jet energy. The suppression of high p_T pions in heavy ion collisions is understood as a consequence of jet quenching in the quark gluon plasma. By considering the reaction plane dependence of high p_T neutral pion production, we can study the path length dependence of the energy loss of the hard parton as it travels through the medium. Furthermore by comparing the neutral pion measurements to reconstructed jet measurements we can study how fragmentation is modified by the medium. In addition to highlighting the novel collision configurations available at RHIC, Cu+Au collisions create asymmetric initial geometries and densities that can test the path length dependence of parton energy loss. This proceeding first discusses the measurements of neutral pions in PHENIX and provides Au+Au results for context. Then the measurements in the Cu+Au collision system are presented.

2. Measuring neutral pions at PHENIX

Neutral pions are measured using photon showers in PHENIX's central arm electromagnetic calorimeters. The calorimeters consist of six sectors of PbSc and two sectors of PbGl calorimeters. Photon shower identification requires a minimum energy of 400 MeV with a shower shape selection in the PbSc. Photon showers in the same sector with an asymmetry, $(E1 - E2)/(E1 + E2)$, of less than 0.8 are combined to form pion candidates. Combinatorial background pairs are determined and statistically subtracted using event mixing with a side-band normalization region. Fig. 1 presents the uncorrected neutral pion yield in various fine centrality bins. While the efficiency corrections are not yet applied, this measurement extends out to p_T values of between 10 and 15 GeV/c depending on the centrality.

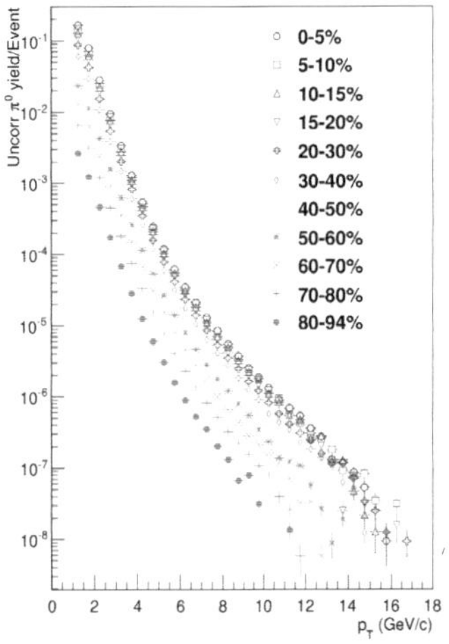

Figure 1. The uncorrected neutral pion yield per event in Cu+Au centrality bins. The centrality bins are as follows: 0-5%(black open circles), 5-10%(red squares), 10-15%(blue upward triangles), 15-20%(green downward triangles), 20-30%(purple open crosses), 30-40%(orange diamonds), 40-50%(cyan stars), 50-60%(magenta asterisk), 60-70%(grey-green x), 70-80%(brown cross), and 80-94%(grey solid circles).

One of the goals of this analysis is the reaction plane dependence of the nuclear modification factor, $R_{AA}(p_T, \Delta\phi)$ where $\Delta\phi$ is the azimuthal angle relative to the event plane. This is done by first measuring the inclusive $R_{AA}(p_T)$ and then determining the relative modulation of the yield in $\Delta\phi$, $F(p_T, \Delta\phi)$ for each p_T value, according to Equation 1.

$$R_{AA}(p_T, \Delta\phi) = R_{AA}(p_T)F(p_T, \Delta\phi) \quad (1)$$

$F(p_T, \Delta\phi)$ is the ratio of the number of counts for a given $\Delta\phi$ over the total number of counts at all $\Delta\phi$. This ratio also contains a correction for the event plane resolution determined from raw v_2^{raw} and the resolution-corrected v_2^{corr} via the factor $(1+2v_2^{corr}cos(2\Delta\phi))/(1+2v_2^{raw}cos(2\Delta\phi))$.

2.1. *Au+Au results*

PHENIX has measured the neutral pion $R_{AA}(p_T, \Delta\phi)$ in $\sqrt{s_{NN}} = 200$ GeV Au+Au collisions out to 17 GeV/c in 10% centrality bins [1]. The in-plane R_{AA}, with a $\Delta\phi$ from 0 to 15°, shows less suppression than the out-of-plane R_{AA}, with a $\Delta\phi$ from 75 to 90°. The separation between the in-plane and out-of-plane R_{AA} decreases as number of participants in the collision, N_{Part}, increases and the eccentricity of the overlap region decreases.

In Fig. 2, the PHENIX in-plane and out-of-plane π^0 R_{AA} is compared to various energy loss models, including AMY, higher twist corrections to deep inelastic scattering processes (HT), ASW calculations and ASW with AdS/CFT soft processes(ASW-AdS/CFT). AMY, HT and ASW all show a quadratic path length dependence, while ASW-AdS/CFT has a cubic path length dependence. One recent study suggests that including multiple soft processes will result in a linear path length dependence [2]. However, in this comparison the PHENIX data is best represented by the ASW-AdS/CFT curves suggesting a higher than quadratic path length dependence [1].

3. Cu+Au results

Neutral pion measurements in Cu+Au collisions provide an opportunity to further test these models in a new collision system. Asymmetric Cu+Au collisions generate unique collision

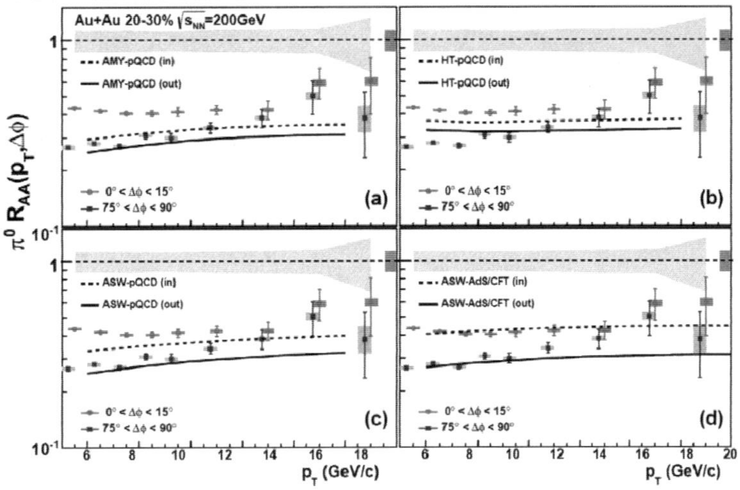

Figure 2. The π^0 R_{AA} in 20-30% Au+Au collisions. Both the in-plane and out-of-plane R_{AA} are shown and are compared to various energy loss theories, including AMY (a), higher twist corrections to deep inelastic scattering processes (b), ASW calculations (c) and ASW with AdS/CFT soft processes (d) [1].

geometries including asymmetric overlap regions, pressure gradients, density profiles and path lengths. Furthermore, measurements of the transverse energy as a function of the number of participants shows that Cu+Au collisions provide access to high energy densities, Fig. 3. The asymmetric nature of Cu+Au collisions were expected to result in angular particle distributions that contain odd flow harmonics not attributed to fluctuations in the initial state. A large positive charged pion v_1 component is measured, Fig. 4. This means that more particles are emitted from the Au-side of the overlap region. This large imbalance in particle production is not seen in symmetric collision systems.

Alternatively, the measured charged hadron v_3 in Cu+Au collisions is small. It shows no significant centrality dependence and comparisons with the charged hadron v_3 in Au+Au and Cu+Cu collisions show no clear system-size dependence. Initially this was surprising as we expected more higher-order odd harmonics, however, AMPT calculations show that the average third-order eccentricity in the Cu+Au system are similar to those in the Au+Au and Cu+Cu systems [4]. In contrast, a large v_2 is measured in Cu+Au collisions for charged hadrons and identified charged pions and protons. The pion and proton v_2 show mass ordering at low p_T and follow valence quark number scaling in all centrality bins. These characteristics, including the v_2 centrality dependence, are all similar to the behavior seen in Au+Au collisions. A clear system-size dependence in the charged hadron v_2 is seen where the v_2 in Au+Au collisions is larger than that in Cu+Au at the same number of participants, N_{Part}. The v_2 in both Au+Au and Cu+Au collisions are larger than the corresponding v_2 in Cu+Cu collisions. These charged hadron and charged pion and proton flow measurements extend out to a p_T of 3 and 2.5 GeV/c respectively. The neutral pion measurements will be able to extend our measurements out to higher p_T, specifically to a p_T of 6 GeV/c for v_1 and v_2.

The Cu+Au charged pion p_T spectrum and R_{AA} is also measured out to a p_T of 2.6 GeV/c. The suppression of charged pions in Cu+Au collisions is very similar to the suppression seen in Au+Au collisions at the same number of participants, N_{part}, and number of binary collisions,

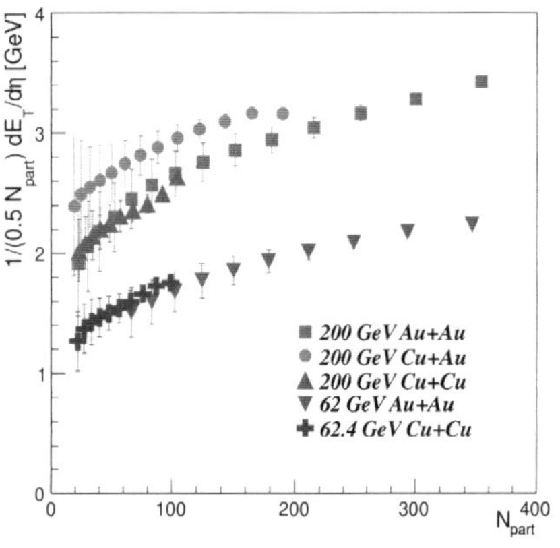

Figure 3. The $1/(0.5N_{Part})dE_T/d\eta$ as a function of the number of participants, N_{Part}, in $\sqrt{s_{NN}}$ = 200 GeV Cu+Au collisions. The Cu+Au results (circles) are compared to the measurements in Au+Au and Cu+Cu collisions at 200 and 62 GeV.

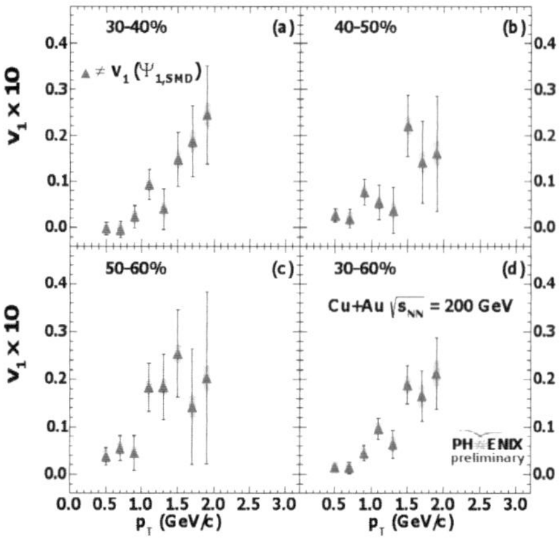

Figure 4. The charged pion v_1 as a function of p_T in the 30-40%, 40-50%, 50-60% and 0-60% centrality bins in Cu+Au collisions is shown. The v_1 is determined relative to the Ψ_1 event plane measured in the SMD, $\Psi_{1,SMD}$ [3].

N_{Coll}. The neutral pion R_{AA} will determine whether this agreement holds out at higher p_T. We expect the π^0 R_{AA} to extend out to a p_T of at least 10 GeV/c.

The unique geometries available in Cu+Au collisions consist of more than asymmetric overlap regions. It is also possible to have events where the Cu-nucleus is embedded in the Au-nucleus. The Cu-nucleus should have very few, almost zero, spectator neutrons resulting in very few hits in PHENIX's Cu-going Zero Degree Calorimeter (ZDC). Looking in the central collisions, in 1% centrality bins, we see that the concentration of events that have no hits in the Cu-going ZDC increases as events become more central. In the most central bin of 0-1%, 70% of the events have no hits in the Cu-going ZDC. Measurement of the π^0 yield in these very central bins will provide access to varying concentrations of these Cu-embedded events and shed light on the medium effects resulting from this unusual geometry.

4. Conclusions

Neutral pion production in Cu+Au collisions is capable of testing our current understanding of energy loss models and determine if the asymmetric behaviors seen at low p_T in charged hadrons extend out to high p_T. Simulations and efficiency corrections are on-going. Of particular interest is the measurement of the nuclear modification factor with respect to the event plane. The possibility of using very central centrality bins and event plane engineering to further study the novel collision geometries available in Cu+Au is very promising.

References

[1] Adare A et al. 2013 *Physical Review* C **87** 034911
[2] Christianse, P, Tywoniuk K, and Vislavicius V 2014 *Physical Review* C **89** 034912
[3] Huang S for the PHENIX Collaboration. 2012 *Nuclear Physics* A **904** 417c-420c
[4] Rihan Haque Md, Nasim Md, and Mohanty B 2011 arXiv:1111.5095

Journal of Physics: Conference Series **612** (2015) 012005 doi:10.1088/1742-6596/612/1/012005

Study of the neutral mesons in Pb-Pb collisions at $\sqrt{s_{NN}}$=2.76 TeV in the ALICE experiment at LHC

Lucia Leardini on behalf of the ALICE Collaboration
Physikalisches Institut, Heidelberg University
E-mail: lucia.leardini@cern.ch

Abstract. The π^0 and η meson production in Pb-Pb collisions at $\sqrt{s_{NN}}$=2.76 TeV is studied with the ALICE experiment at the LHC. The π^0 meson invariant yields and nuclear modification factor $R_{\rm AA}$ are presented here in six centrality classes. The results are obtained using the Photon Conversion Method (PCM) and utilizing the PHOS detector, for the transverse momentum range $0.4 < p_{\rm T} < 12$ GeV/c. The π^0 meson $R_{\rm AA}$ is studied in different centrality classes and compared with results from experiments at lower energies, as a function of transverse momentum. The η meson production is studied using the PCM and utilizing the EMCal detector. The combination of these two independent methods will make possible the measurement of the η differential invariant cross section as a function of transverse momentum from 1 to 22 GeV/c in different centrality classes.

1. Introduction

The ALICE experiment's [1][2] scientific program is focused on the analysis of heavy-ion collisions and on the study of the Quark-Gluon Plasma, a state of the matter with high temperature and large energy density, where gluon and quarks are deconfined, as predicted by QCD. As this medium has high color charge density, partons produced in the earlier stage of the collision lose energy through radiative gluon emission and multiple scattering while traversing it. The parton energy loss affects the hadron yields in the high $p_{\rm T}$ region, a modification that can be observed comparing results on particle production in Pb-Pb collisions with those from pp collisions. In order to study this effect, the nuclear modification factor $R_{\rm AA}$ is used:

$$R_{\rm AA}(p_{\rm T}) = \frac{{\rm d}^2N/dp_{\rm T}{\rm d}y|_{\rm AA}}{\langle T_{\rm AA}\rangle {\rm d}^2\sigma/{\rm d}p_{\rm T}{\rm d}y|_{\rm pp}} \quad (1)$$

where $\langle T_{\rm AA}\rangle = \langle N_{\rm coll}\rangle/\sigma^{\rm pp}_{\rm inel}$ is used to scale the pp yield to the Pb-Pb yield. Here, $\langle N_{\rm coll}\rangle$ is the average number of binary nucleon-nucleon collisions obtained from the Glauber model [3] and $\sigma^{\rm pp}_{\rm inel}$ is the pp inelastic cross section [4]. The π^0 meson $R_{\rm AA}$ has been studied through its two photon decay channel with the ALICE experiment [5][6]. The detectors involved in the measurement of photons are a lead-scintillator electromagnetic calorimeter (EMCal) and the photon spectrometer (PHOS), made of lead tungstate crystals. The EMCal covers $|\eta| < 0.7$ and $\Delta\phi = 100°$ while PHOS covers $|\eta| < 0.25$ and $\Delta\phi = 60°$. It is also possible to measure photons through their conversion into electrons and positrons in the detector material. This method is called Photon Conversion Method (PCM) and relies on the Inner Tracking System (ITS) and on the Time Projection Chamber (TPC), which have full azimuthal coverage. The material thickness, up to the middle of the TPC, is $X/X_0 = 11.4 \pm 0.5_{sys}\%$.

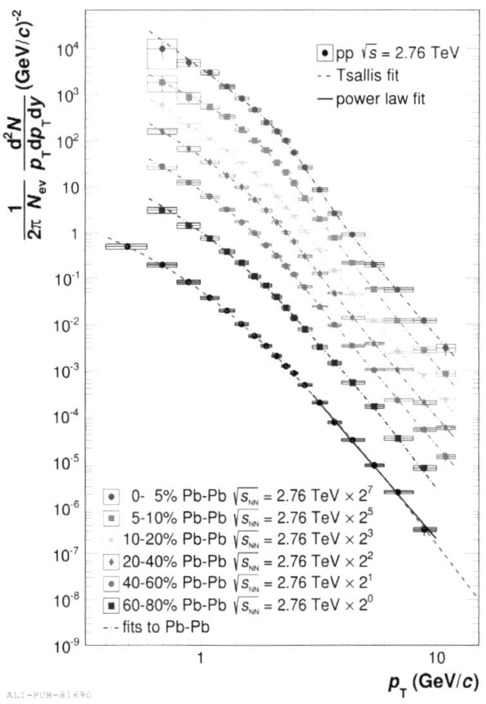

Figure 1. (Color online) Invariant differential yields of neutral pions produced in PbPb and inelastic pp collisions. The spectra are the weighted average of PHOS and PCM results. Vertical lines represent statistical uncertainties, boxes systematic uncertainties and horizontal lines indicate bin width.

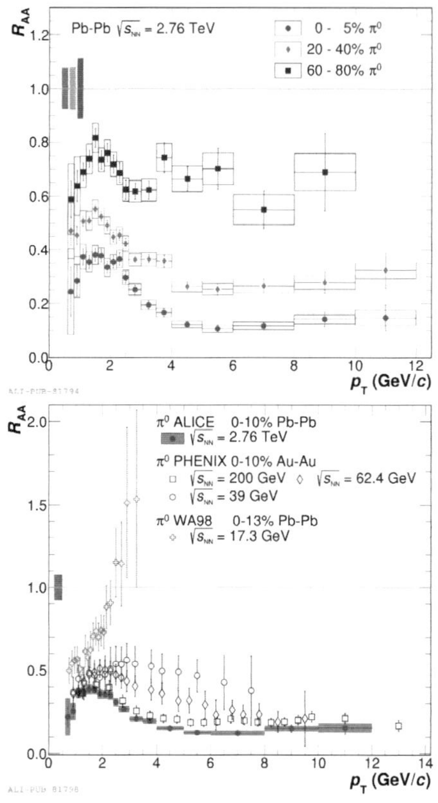

Figure 2. (Color online) Top: Neutral pion modification factor in Pb-Pb collisions. Bottom: Neutral pion modification factor in Pb-Pb for centrality class 0-10% measured by ALICE [6] is compared with the results from Au-Au collisions measured by PHENIX [10][11] and from Pb-Pb collisions at the CERN SPS [12].

2. Results

The π^0 meson invariant yields presented here have been measured independently by PHOS and PCM and were then combined into a weighted average. In Fig. 1, the combined PHOS-PCM π^0 meson yields are shown for pp collisions and Pb-Pb collisions in 6 centrality classes. For PHOS, the data sample consists of 16.1×10^6 events collected in the 2010 Pb-Pb run and of 34.7×10^6 events from the 2011 pp run at $\sqrt{s_{\rm NN}} = 2.76$ TeV, while for PCM it is 13.2×10^6 and 58×10^6 events, respectively [6]. The transverse momentum range is 0.6 to 12 GeV/c. For the pp spectrum, a Tsallis function [7] for the full $p_{\rm T}$ region and a power law $E{\rm d}^3N/{\rm d}^3p \propto 1/p_{\rm T}^n$, only for the high $p_{\rm T}$ region ($p_{\rm T} > 3$ GeV/c), are shown. The power law fit extrapolation above 8 GeV/c is used as reference in the $R_{\rm AA}$ calculation. The Pb-Pb spectra parametrization is described in the appendix of [6].

The neutral pion nuclear modification factor is calculated with Eq. 1. The nuclear overlap

function $\langle T_{AA} \rangle$ was obtained with a Monte Carlo calculation [3][8][9] and the values including their uncertainties for each centrality class can be found in [6]. In Fig. 2, top panel, the combined PCM-PHOS result for the π^0 R_{AA} is shown. The R_{AA} decreases going from peripheral to central collisions, due to different energy loss in the medium. For all the centrality classes, the R_{AA} has a maximum for $p_T < 2$ GeV/c, decreases until $p_T < 6$ GeV/c and stays flat after $p_T >$ 6 GeV/c. In this last region, particle production is expected to be dominated by fragmentation of hard-scattered partons.
The dependence of R_{AA} on the system energy has also been studied. In the bottom panel of Fig. 2 the π^0 meson R_{AA} for Pb-Pb collisions at $\sqrt{s_{NN}} = 2.76$ TeV measured with ALICE is compared to previous results at lower energies. ALICE data points are below the lower energies points and their suppression is more evident for $p_T > 2$ GeV/c. Models attribute this behavior to the higher initial energy densities created at larger $\sqrt{s_{NN}}$. This dominates over the expected increase of R_{AA} due to the harder initial p_T spectra [13]. Also, no increase at high p_T is observed. It is also noted that the position of the maximum seems to be shifting towards lower transverse momentum with increasing collision energy. The neutral pion R_{AA} has been compared with different theoretical models. These comparison plots can be found in [6]. The GLV calculation [14][15] and WHDG [16] prediction describe the interaction of hard-scattered parton with the medium using perturbative QCD. EPOS [17] and Nemchik et al. [18][19] calculations use a hydrodynamic description at low p_T and absorption of color dipoles at high p_T.

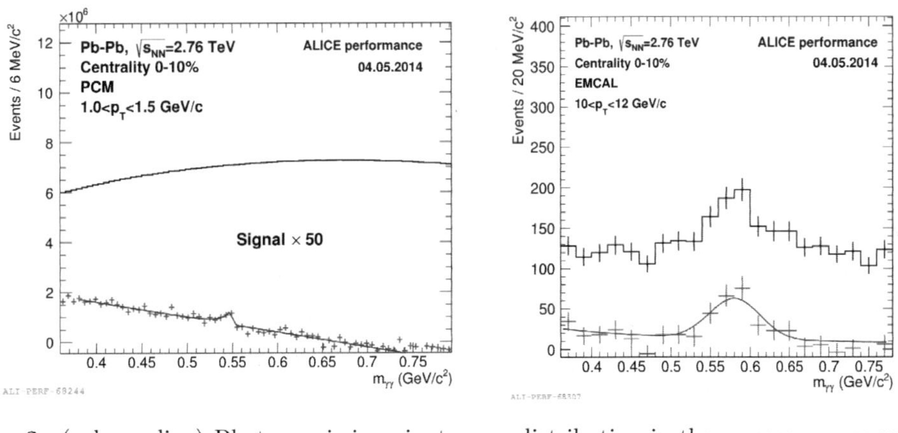

Figure 3. (color online) Photon pair invariant mass distribution in the η meson mass region in selected p_T slices for PCM (left) and EMCal (right) for 0-10% Pb-Pb collisions. The black histogram and the red points show the data before and after background subtraction, respectively. For PCM, the invariant mass spectra after background subtraction is scaled by a factor 50. The blue line shows the fit to the invariant mass spectra after the background subtraction.

The η meson can be measured in ALICE in Pb-Pb collisions at $\sqrt{s_{NN}} = 2.76$ TeV with the 2011 data sample. It has an integrated luminosity one order of magnitude larger than the corresponding one for 2010, making a reliable reconstruction of the η meson spectrum possible. We show here the first steps towards a combined PCM-EMCal η meson measurement. The data sample consists of about 17 million events in the central class 0-10% for both PCM and EMCal. The transverse momentum range for this combined measurement can go from 1 to 22 GeV/c with an overlap between the two approaches in the intermediate transverse momentum region

($4 < p_{\mathrm{T}} < 10$ GeV/c).
In Fig. 3, the photon pair invariant mass distribution is shown for two sample transverse momentum bins. The η meson peak is clearly visible. For PCM (left panel), the lowest p_{T} bin is shown. The general fit function used by PCM for all p_{T} is a Gaussian function combined with an exponential tail on left side of the peak (to account for the electron bremsstrahlung) and with a linear fit to describe the remaining combinatorial background under the peak after the combinatorial background subtraction. The fit is used only to extract the η meson peak position and width. The η meson yields are then extracted by integrating the peak area and subtracting the integral of the linear fit. In the right panel of Fig. 3, a bin in the intermediate range of the EMCal p_{T} reach is shown. The fit function used here is a Crystal-Ball function while a linear fit or a parabola, depending on the p_{T} and centrality, is adopted for the residual background hypothesis.

3. Conclusions
Measurements of the neutral pion production in Pb-Pb collisions at $\sqrt{s_{\mathrm{NN}}} = 2.76$ TeV have been shown combining two independent methods, the measurement of the photons with the PHOS detector and the reconstruction of the converted photons with the tracking system. These two independent measurements are consistent and were combined in a weighted average for the π^0 spectra. The neutral pion nuclear suppression factor was calculated from the measured neutral pion spectra separately for the two methods and then combined, in order to reduce the systematic errors. The π^0 meson R_{AA} was studied for different centrality classes and compared to low-energy results. The suppression is stronger for higher collision energy when comparing Pb-Pb collisions at $\sqrt{s_{\mathrm{NN}}} = 2.76$ TeV to results from RHIC and CERN SPS.
The η meson invariant mass distributions for two transverse momentum bins from the 2011 Pb-Pb collisions at $\sqrt{s_{\mathrm{NN}}} = 2.76$ TeV are shown. Also here two independent methods are used, the photon measurement with the EMCal detector and the photon conversion. The transverse momentum range for this combined measurement will go from 1 to 22 GeV/c with an overlap in the intermediate transverse momentum region ($4 < p_{\mathrm{T}} < 10$ GeV/c). This analysis will allow to study the η meson suppression in Pb-Pb collisions.

References
[1] B. Abelev *et al.* [ALICE Collaboration], *JINST* **3** S08002.
[2] B. Abelev *et al.* [ALICE Collaboration], *Int. J. Mod. Phys.* A **29** (2014) 1430044.
[3] B. Abelev *et al.* [ALICE Collaboration], *Phys. Rev.* C **88** (2013) 044909.
[4] B. Abelev *et al.* [ALICE Collaboration], *Eur. Phys. J.* C **73** (2013) 2456.
[5] B. Abelev *et al.* [ALICE Collaboration], *Phys. Lett.* B **717** (2012) 162.
[6] B. Abelev *et al.* [ALICE Collaboration], *Eur. Phys. J.* C **74** (2014) 3108.
[7] C. Tsallis, *J. Statist. Phys.* **52** 479-487 (1988).
[8] M. L. Miller, K. Reygers, S. J. Sanders and P. Steinberg, *Ann. Rev. Nucl. Sci.* **57** (2007) 205.
[9] B. Alver, M. Baker, C. Loizides and P.Steinberg, (2008) arXiv:0805.4411.
[10] A. Adare *et al.* [PHENIX Collaboration], *Phys. Rev. Lett.* **109** (2012) 152301.
[11] A. Adare *et al.* [PHENIX Collaboration], *Phys. Rev. Lett.* **101** (2008) 232301.
[12] M. Aggarwal *et al.* [WA89 Collaboration], *Phys. Rev. Lett.* **100** (2008) 242301.
[13] K. C. Zapp, F. Krauss and U. A. Wiedemann, JHEP **1303**, 080 (2013), arXiv:1212.1599.
[14] R. Sharma, I. Vitev and B. -W. Zhang, *Phys. Rev.* C **80** (2009) 054902.
[15] R. Neufeld, I. Vitev and B. -W. Zhang, *Phys. Lett.* B **704** (2011) 590.
[16] W. A. Horowitz, *Int. J. Mod. Phys.* E **16** (2007) 2193.
[17] K. Werner, I. Karpenko, M. Bleicher, T. Pierog and S. Porteboeuf-Houssais, *Phys. Rev.* C **85** (2012) 064907.
[18] B. Kopeliovich, J. Nemchik, I. Potashnikova adn I. Schmidt, *Phys. Rev.* C **86** (2012) 054904.
[19] J. Nemchik, I. A. Karpenko, B Kopeliovich, I. Potashnikova and Y. M. Sinyukov, (2013) arXiv:1310.3455.

Journal of Physics: Conference Series **612** (2015) 012006 doi:10.1088/1742-6596/612/1/012006

Energy loss in jet suppression - what effects matter?

Bojana Blagojevic and Magdalena Djordjevic

Institute of Physics Belgrade, Pregrevica 118, 11080 Zemun, Serbia

E-mail: bojanab@ipb.ac.rs

Abstract. Jet suppression is considered to be an excellent probe of QCD matter created in ultra-relativistic heavy ion collisions. Our theoretical predictions of jet suppression, based on our recently developed dynamical energy loss formalism, show a robust agreement with various experimental data for different probes, experiments (RHIC and LHC) and centrality regions. Our dynamical energy loss formalism includes the following key ingredients: dynamical scattering centers, collisional energy loss, finite magnetic mass and running coupling. Although all these ingredients are theoretically justified, it is currently unclear how they individually contribute to accurate suppression predictions. Natural question rises: is there one effect which is crucial for the agreement, or is the agreement a joint effect of several smaller improvements. To answer this question, we study how the above mentioned key effects affect the suppression calculations. Our results show that each energy loss effect is important and that a robust agreement between theoretical predictions and experimental data is a cumulative effect of all improvements.

1. Introduction

Suppression of high transverse momentum observables [1] is considered to be an excellent probe for mapping the properties of QCD matter created in ultra-relativistic heavy ion collisions at RHIC and LHC. Therefore comparison of available suppression experimental data with the theoretical predictions [2–4] tests different theoretical models and provides the insight into underlying QGP physics. For generating these predictions, we developed dynamical energy loss formalism which includes the following energy loss effects: *i)* dynamical scattering centers, *ii)* QCD medium of a finite size [5, 6], *iii)* both radiative [5, 6] and collisional [7] energy losses, *iv)* finite magnetic mass effects [8] and *v)* running coupling [9]. We further incorporated this energy loss formalism into a numerical procedure [9] in order to obtain suppression predictions. In the numerical procedure, accurate energy loss calculations are considered to be crucial for obtaining reliable suppression predictions.

We have shown that the suppression predictions obtained from this dynamical energy loss formalism are in a very good agreement with the available experimental data for both RHIC and LHC experiments, light and heavy flavor probes and different centrality ranges [9–11].

We here address the importance of different energy loss effects in the suppression calculations for D mesons (as a clear energy loss probe) in central 200 GeV Au+Au collisions at RHIC, because fragmentation function does not modify bare charm quark suppression [10, 12]. Our approach is to systematically include different energy loss effects. In particular, we first investigate the importance of including collisional energy loss and thus necessity of abolishing static in favor of dynamical approximation. Next we address the importance of including finite magnetic mass in the suppression calculations and finally the running coupling.

2. Theoretical and computational formalism

In this section, we give a brief description of our dynamical energy loss formalism [9] with regression on how each effect, when added, altered energy loss expression, while in Section 4 we take the reverse approach - the historical approach, starting from a static approximation and moving to systematically include all the effects.

In order to obtain quenched spectra we use generic pQCD convolution given by Eq.(1) from [9]. The initial charm quark spectrum is computed according to [13] and energy loss probability includes both radiative and collisional energy losses in a finite size dynamical QCD medium, multi-gluon [14] and path length [15, 16] fluctuations. In our calculations we do not use the fragmentation function of charm quark into D meson, as explained in Section 1.

The radiative energy loss in a finite size dynamical QCD medium is given by Eq.(2.12) from [5], while the finite magnetic mass and running coupling are introduced according to [8] and [9], respectively. For the finite magnetic mass case we use the following range of magnetic to electric mass ratio: $0.4 < \mu_M/\mu_E < 0.6$, according to non-perturbative approaches [17–21], otherwise, $\mu_M = 0$ is used. Also when the running coupling is not included, in our calculations we use $\alpha_S = \frac{g^2}{4\pi} = 0.3$ and Debye mass $\mu_E = gT$, $(g = 2)$. Collisional energy loss is calculated in accordance with Eq.(14) from [7]. Transition from the static [22] to the dynamical approximation in terms of radiative energy loss is explained in [6].

In our calculations for the charm quark mass we use $M_c = 1.2$ GeV, for 0-5% central 200 GeV Au+Au collisions we assume an average medium temperature of T=225 MeV [10] and for the number of effective light quark flavors we use $n_f = 2.5$.

3. Comparison with experimental data

As we mentioned in Section 1, our dynamical energy loss formalism [9] leads to a very good agreement with suppression experimental data for diverse probes at both RHIC [10] and LHC [9] and for different centrality regions [11]. The suppression is expressed by the nuclear modification factor R_{AA} [4], which quantifies the QCD medium effects on the yield of high-p_T particles. Fig. 1, which shows comparison of the D meson R_{AA} predictions with corresponding R_{AA} measured at the LHC and comparison of the single electron R_{AA} predictions with non-photonic single electron R_{AA} measured at RHIC, reflects the above mentioned agreement.

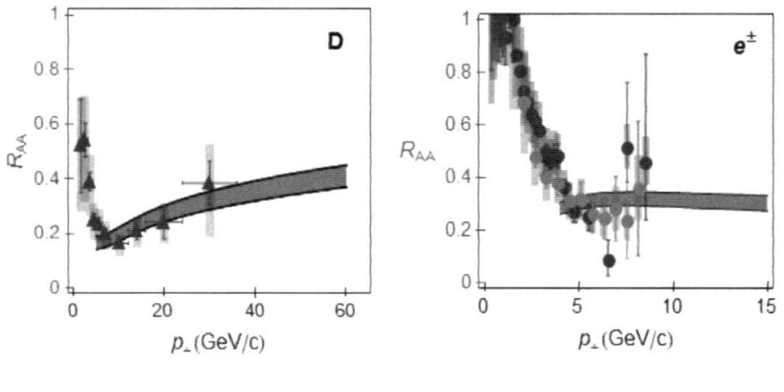

Figure 1. Theory vs. experimental data for D meson and single electron suppressions as a function of transverse momentum. Left panel shows comparison of D meson R_{AA} predictions with experimentally measured R_{AA} (triangle) in most central 2.76 TeV Pb+Pb collisions at the LHC. Right panel shows comparison of single electron R_{AA} predictions with non-photonic single electron R_{AA} (circle) measured in most central 200 GeV Au+Au collisions at RHIC. Left (right) panel is adapted from [9] ([10]).

4. Results and discussion

We start from the static approximation [22, 23] and use a constant value of the strong coupling constant and of Debye mass (as mentioned above), and no finite magnetic mass effects (μ_M=0); note that these values are used in Figs. 2 and 3, while the importance of finite magnetic mass is considered in Fig. 3. Previously, the static approximation was widely used, which assumed that collisional energy loss can be neglected compared to radiative. Left panel of Fig. 2 shows that static approximation has to be abolished, because collisional energy loss suppression is comparable or even larger than static radiative one. Therefore, central panel of Fig. 2 addresses the significance of including dynamical effects by comparing static with dynamical radiative energy loss R_{AA}. We observe a significant suppression increase in the dynamical approximation, so we conclude that dynamical effects are important. Right panel of Fig. 2 investigates whether collisional energy loss is still relevant in dynamical approximation, by comparing radiative with collisional contribution to R_{AA} in the dynamical QCD medium. We conclude that even in dynamical approximation, both radiative and collisional contributions are important, so we further include both radiative and collisional (total) energy losses in dynamical QCD medium.

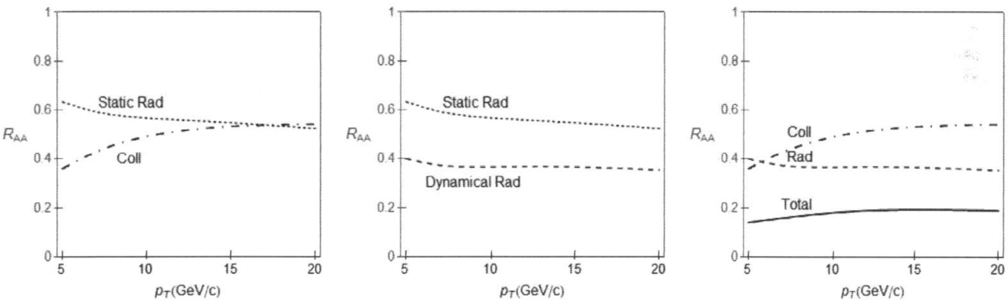

Figure 2. Static vs. dynamical approximation. D meson suppression predictions are shown as a function of transverse momentum. Left panel shows comparison of static radiative (dotted curve) with dynamical collisional (dot-dashed curve) contribution to R_{AA}. Central panel shows comparison of static radiative (dotted curve) with dynamical radiative (dashed curve) contribution to R_{AA}. Right panel shows radiative (dashed curve), collisional (dot-dashed curve) and radiative + collisional (solid curve) contribution to R_{AA} in dynamical QCD medium. Debye mass is $\mu_E = gT$, coupling constant is $\alpha_S = 0.3$ and no finite magnetic mass effects are included ($\mu_M = 0$). Adapted from [24].

Next we consider how inclusion of finite magnetic mass in radiative energy loss calculations [8] affects the R_{AA} predictions, as indicated in Section 2. By comparing R_{AA} with and without finite magnetic mass (Fig. 3), we observe significant suppression decrease due to finite magnetic mass effects. Hence, we conclude that finite magnetic mass effects are important.

Also, the importance of taking into account running coupling [9] is considered in Fig.7 from [24], where we observe suppression increase due to running coupling only at lower jet energies. Consequently running coupling is also important.

5. Conclusions

Since dynamical energy loss formalism led to a robust agreement with the suppression data for different experiments, probes and centrality ranges [9–11], we wanted to determine whether the agreement was a consequence of one dominant effect or a joint effect of several smaller

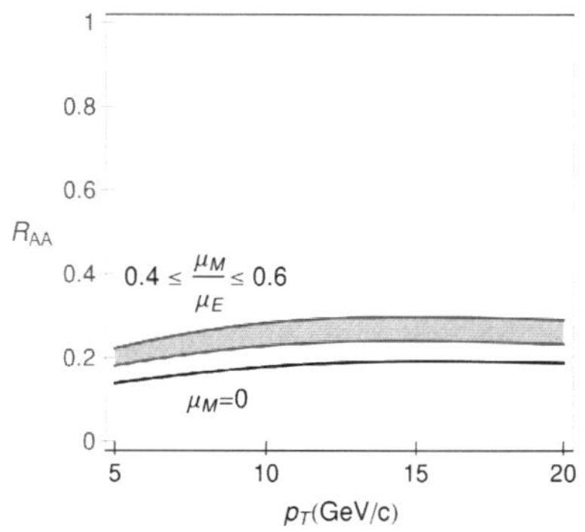

Figure 3. Magnetic mass effects on R_{AA}. D meson suppression predictions are shown, as a function of transverse momentum, for radiative and collisional energy loss in dynamical QCD medium, with (band) and without (solid curve) magnetic mass. Debye mass is $\mu_E = gT$ and coupling constant is $\alpha_S = 0.3$. The upper (lower) boundary of the band corresponds to $\mu_M/\mu_E = 0.6$ ($\mu_M/\mu_E = 0.4$). Adapted from [24].

improvements introduced to energy loss calculations. In order to examine the importance of each effect we followed first a historical approach starting from the static approximation and gradually introduced different energy loss effects in D meson suppression calculations (as a clear energy loss probe) until reaching dynamical energy loss formalism [9]. The conclusion is that each energy loss effect is important and that a robust agreement is a cumulative effect of all these improvements. Therefore, in order to obtain reliable suppression predictions we need to accurately account for all the relevant energy loss ingredients.

Acknowledgments

This work is supported by Marie Curie International Reintegration Grant within the 7th European Community Framework Programme PIRG08-GA-2010-276913 and by the Ministry of Science and Technological Development of the Republic of Serbia, under project No. ON171004.

References

[1] Bjorken J D 1982 FERMILAB-PUB-82-059-THY pp 287-92
[2] Brambilla N et al 2004 *Preprint* hep-ph/0412158
[3] Gyulassy M 2002 *Lect. Notes Phys.* **583** 37
[4] d'Enterria D and Betz B 2010 *Lect. Notes Phys.* **785** 285
[5] Djordjevic M 2009 *Phys. Rev.* C **80** 064909
[6] Djordjevic M and Heinz U 2008 *Phys. Rev. Lett.* **101** 022302
[7] Djordjevic M 2006 *Phys. Rev.* C **74** 064907
[8] Djordjevic M and Djordjevic M 2012 *Phys. Lett.* B **709** 229
[9] Djordjevic M and Djordjevic M 2014 *Phys. Lett.* B **734** 286
[10] Djordjevic M and Djordjevic M 2014 *Phys. Rev.* C **90** 034910
[11] Djordjevic M, Djordjevic M and Blagojevic B 2014 *Phys. Lett.* B **737** 298
[12] Djordjevic M 2014 *Phys. Rev. Lett.* **112** 4 042302
[13] Kang Z B, Vitev I and Xing H 2012 *Phys. Lett.* B **718** 482-7
[14] Gyulassy M, Levai P and Vitev I 2002 *Phys. Lett.* B **538** 282
[15] Wicks S, Horowitz W, Djordjevic M and Gyulassy M 2007 *Nucl. Phys.* A **784** 426
[16] Dainese A 2004 *Eur. Phys. J.* C **33** 495
[17] Maezawa Yu, Aoki S, Ejiri S, Hatsuda T, Ishii N, Kanaya K, Ukita N and Umeda T 2010 *Phys. Rev.* D **81** 091501
[18] Maezawa Yu, Aoki S, Ejiri S, Hatsuda T, Ishii N, Kanaya K, Ukita N and Umeda T 2008 *Proc. of Science Lattice 2008 (Williamsburg)* p 194 (*Preprint* hep-lat/0811.0426)
[19] Nakamura A, Saito T and Sakai S 2004 *Phys. Rev.* D **69** 014506
[20] Hart A, Laine M and Philipsen O 2000 *Nucl. Phys.* B **586** 443
[21] Bak D, Karch A and Yaffe L G 2007 *J. High Energy Phys.* JHEP0708(2007)049
[22] Djordjevic M and Gyulassy M 2004 *Nucl. Phys.* A **733** 265-98
[23] Gyulassy M and Wang X N 1994 *Nucl. Phys.* B **420** 583
Wang X N, Gyulassy M and Plumer M 1995 *Phys. Rev.* D **51** 3436
[24] Blagojevic B and Djordjevic M 2014 *Preprint* nucl-th/1411.1649

Hot Quarks 2014 IOP Publishing
Journal of Physics: Conference Series **612** (2015) 012007 doi:10.1088/1742-6596/612/1/012007

Constraining of jet medium interaction in high-energy heavy-ion collisions

Yan Zhu[1] **and Xin-Nian Wang**[2,3]

[1] Departamento de Física de Partículas and IGFAE, Universidade de Santiago de Compostela, E-15706 Santiago de Compostela, Galicia, Spain

[2] Key Laboratory of Quark and Lepton Physics (MOE) and Institute of Particle Physics, Central China Normal University, Wuhan 430079, China

[3] Nuclear Science Division Mailstop 70R0319, Lawrence Berkeley National Laboratory, Berkeley, California 94740, USA

E-mail: `yan.zhu@usc.es (presenter of the talk)`

Abstract. In this talk, we discuss jet medium interaction, in particular, γ-tagged jet energy loss and γ-jet correlation in Pb+Pb collisions at $\sqrt{s} = 2.76$ TeV within a linearized Boltzmann transport (LBT) model. It is shown that both recoiled and radiated partons have considerable contribution to the energy of a reconstructed jet. LBT results agree well with experimental data and also indicate a sizable azimuthal angle broadening of γ-jet correlation which should be observable with reduced experimental uncertainties. A γ-tagged jet fragmentation function is also shown as a better measure of jet quenching.

1. Introduction

As a result of multiple scattering and medium induced gluon radiation during the propagation of high p_T partons in the deconfined phase in high energy heavy-ion collisions, jets are strongly quenched [1] as observed in Au+Au collisions at RHIC and Pb+Pb collisions at LHC in terms of the suppression and p_T broadening of high-p_T hadrons and jets with respect to p+p collisions. Meanwhile, the medium should also be excited by the lost energy and momenta. These processes of jet-medium interaction can be simulated in a linearized Boltzmann transport (LBT) model [2, 3].

There are two building blocks in the LBT model. The first one is the simulation of the elastic scattering according to the Boltzmann transport equation,

$$p_1 \cdot \partial f_1(p_1) = -\int dp_2 dp_3 dp_4 (f_1 f_2 - f_3 f_4) |M_{12\to 34}|^2 (2\pi)^4 \delta^4(p_1 + p_2 - p_3 - p_4), \tag{1}$$

where we have neglected the Bose enhancement and Pauli blocking factor, $dp_i = d^3p_i/[2E_i(2\pi)^3]$, and use the square of elastic scattering amplitude in a small angle approximation $|M_{12\to 34}|^2 = Cg^4(\hat{s}^2+\hat{u}^2)/(-\hat{t}+\mu_D^2)^2$ with $\hat{s}$, $\hat{t}$, and $\hat{u}$ as the Mandelstam variables, and μ_D is the Debye mass. In the above, $C = 1$ $(9/4)$ is the color factor for quark-gluon (gluon-gluon) scattering, phase-space distributions $f_i = 1/(e^{p\cdot u/T} \pm 1)$ $(i = 2, 4)$ for thermal partons with local temperature T and fluid velocity $u = (1, \vec{v})/\sqrt{1-\vec{v}^2}$, and $f_i = (2\pi)^3\delta^3(\vec{p} - \vec{p_i})\delta^3(\vec{x} - \vec{x_i} - \vec{v_i}t)$ $(i = 1, 3)$ for shower partons, before and after scattering.

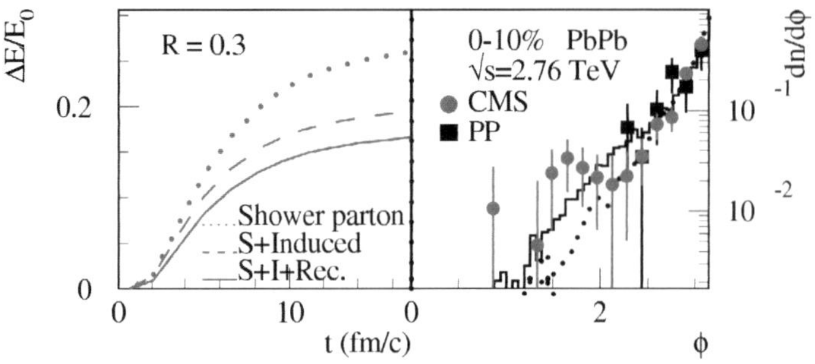

Figure 1. Averaged energy loss as a function of time (left) and azimuthal distribution relative to the γ (right) for γ-tagged jets in central (0%–10%) Pb+Pb collisions at $\sqrt{s} = 2.76$ TeV. LBT results are obtained with $\alpha_s = 0.2$ in a 3+1D hydro medium.

After each scattering, both shower (p_3) and recoiled medium partons (p_4) are propagated along classical trajectories before the next scattering. The scattering center of the next scattering is determined with the probability of scattering, $P_a = 1 - \exp[-\sum_j (\Delta x_j \cdot u) \sum_b \sigma_{ab}\rho_b(x_j)]$ for a parton a, with σ_{ab} its scattering cross section with another parton b. The local thermal parton density is ρ_b and the sum over time steps $\Delta t_j = \Delta x_j$ (in natural units) starts from the last scattering point. Moreover, the initial thermal partons (p_2) before the scattering, denoted as "negative" partons, are also propagated according to the Boltzmann equation, and are subtracted from observables as the back-reaction of medium excitation.

Another building block of LBT is the simulation of medium induced radiations. To this end, we assume that the induced radiation can accompany each elastic scattering, and the radiation probability and the spectrum are given by the higher-twist approach[4],

$$\frac{dN_g^a}{dz dk_\perp^2 dt} = \frac{6\alpha_s P(z)}{\pi k_\perp^4} \frac{p \cdot u}{E} \hat{q}_a \sin^2 \frac{t - t_i}{2\tau_f}, \tag{2}$$

with energy fraction of gluon z and transverse momentum $k_\perp$ emitted from a parent parton a with energy E. The splitting function is $P(z) = [1 + (1 - z)^2]/z$ and the formation time of the radiated gluon is $\tau_f = 2Ez(1 - z)/k_\perp^2$. $\hat{q}_a = \sum_b \rho_b \int d\hat{t} q_\perp^2 d\sigma_{ab}/d\hat{t}$ is the jet transport parameter. Multiple gluon emissions induced by a single scattering are assumed to satisfy a Poisson distribution. In the determination of energies and momenta of radiated gluons with Eq. (2), Debye mass μ_D is set as the infrared cutoff and on-shell condition is imposed to all radiated gluons. We neglect the interaction between shower, radiated and recoiled partons. The fixed value of strong coupling constant $\alpha_s = g^2/(4\pi)$ will be determined via comparisons to experimental data.

2. γ-jet Correlations

Photons created from hard scattering can be used to measure the initial energy of γ- tagged jets since they don't participate in strong interaction. In this presentation, jet quenching and γ-tagged jet modification in heavy-ion collisions at the LHC are studied within LBT model. To study jet quenching, the anti-k_t algorithm within FASTJET [5] is used to reconstruct jets.

To understand γ-jet correlation in the experimental data for Pb+Pb collisions at the LHC, we simulate propagation of γ-tagged jets within LBT in bulk medium whose evolution is given by a (3+1)D hydrodynamic model [6]. Initial γ-tagged jets at parton level are taken from HIJING[7] and embedded in LBT according to the overlap function of two nuclei with a Wood-Saxon distribution. Debye screening mass $\mu_D^2 = 4\pi\alpha_s T^2$ is used in LBT simulation. Event selectors

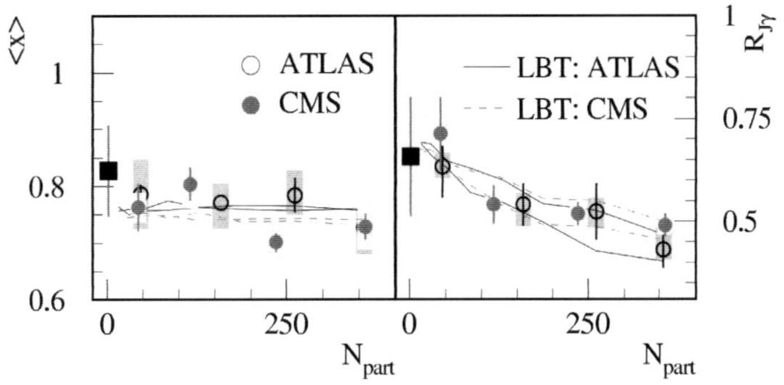

Figure 2. Averaged γ-jet asymmetry $\langle x\rangle = \langle p_T^{\rm jet}/p_T^{\gamma}\rangle$ (left) and jet survival rate (right) as functions of the number of participant nucleons in Pb+Pb collisions at $\sqrt{s} = 2.76$ TeV from LBT as compared to CMS data [8] with $\alpha_s = 0.15$–0.23 in LBT(dashed line), and ATLAS data[9] with 0.2–0.27 (solid line), respectively.

are used for CMS and ATLAS accordingly: For CMS data [8], $p_T^{\gamma} > 60$ GeV, $|\eta^{\gamma}| < 1.44$, $p_T^{\rm jet} > 30$ GeV, $|\eta^{\rm jet}| < 1.6$, and $\Delta\phi = |\phi^{\rm jet} - \phi^{\gamma}| > 7\pi/8$, and for ATLAS data, [9], $60 < p_T^{\gamma} < 90$ GeV, $|\eta^{\gamma}| < 1.3$, $p_T^{\rm jet} > 25$ GeV, $|\eta^{\rm jet}| < 2.1$, and $\Delta\phi > 7\pi/8$. As background has been subtracted in the experimental data, we refrain to add and subtract that from underlying events in our simulation. Jets are reconstructed with all partons from the simulation before hadronization. We have verified that the difference between jets reconstructed from partons and hadrons are rather small. In Fig. 1, we show the average energy loss as a function of time (left), and γ-jet azimuthal angle distribution (right) with $\alpha_s = 0.2$. The energy loss rises up initially, and saturates at $t \approx 10$ fm/c because of the decrease with time of the jet transport parameter $\hat{q}_a = \hat{q}_a^0(\tau_0/\tau)^{1+\alpha}$ with $\alpha \geq 0$ in an expanding medium, which eventually terminates parton energy loss. The analysis of the jet energy loss from different contributions, such as shower parton (p_3), recoiled thermal parton (p_4), as well radiated parton sampled from eq. (2), tells us that recoiled and radiated partons have non-negligible contributions to the jet energy inside the jet cone. On the right panel of Fig. 1, we found that the γ-jet azimuthal angle distribution (solid histograms) has a very good agreement with the data (points), which clearly shows a small broadening relative to the p+p (dotted line) reference in the large angle, while errors in the data are too big to confirm this. Therefore, we expect that experimental measurements are able to verify the small broadening when the errors are reduced significantly. In Fig. 2, we show the averaged γ-jet asymmetry $\langle x\rangle = \langle p_T^{\rm jet}/p_T^{\gamma}\rangle$ and jet survival rate $R_{J\gamma}$ from LBT (lines) and experimental data from CMS [8] and ATLAS [9]. With fixed values of $\alpha_s = 0.15$–0.23 for CMS and 0.2–0.27 for ATLAS, LBT results are consistent with CMS and ATLAS data, yet have small dependence on the experimental cutoff. We also notice that $R_{J\gamma}$ is more sensitive to jet quenching than $\langle x\rangle$.

An enhancement at both small and large $z_{\rm jet} = p_L/E_{\rm jet}$ in jet fragmentation function was observed at LHC[10]. The phenomena can be explained as a result of the large energy fraction of leading particles in a reconstructed jet. To verify this, we make a comparison study of reconstructed (left) and γ-tagged (right) jet fragmentation function in Fig. 3. On the left panel, the reconstructed jet fragmentation function shows a similar property as the experimental observation. However, the γ-tagged jet fragmentation function on the right panel indeed shows an enhancement at small z_{γ} but a significant reduction at large z_{γ}. This γ-tagged fragmentation function is more sensitive to jet medium interaction, and therefore a better probe of medium properties and jet transport parameter.

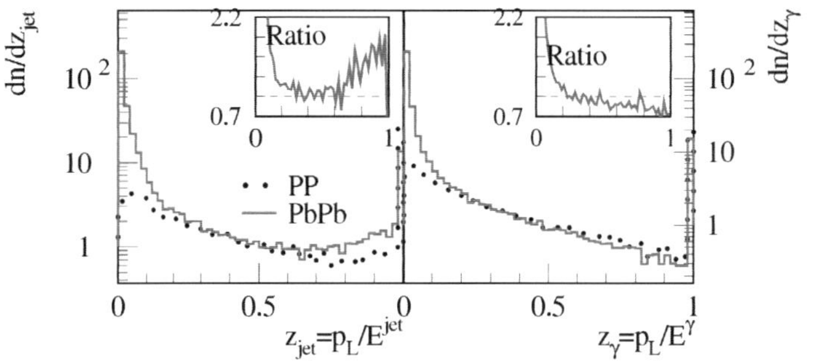

Figure 3. The reconstructed (left) and γ-tagged jet fragmentation functions (right) within a jet cone $R = 0.3$ of γ-tagged jets in central (0%–10%) Pb+Pb collisions at $\sqrt{s} = 2.76$ TeV.

3. Conclusion

In this talk, we have discussed jet quenching and γ-jet correlations within a linearized Boltzmann transport model at LHC. With the value of $\alpha_s \approx 0.2$, we have observed that the 15% energy fraction of reconstructed jet is redistributed out side the jet-cone due to multiple elastic scattering and induced radiation in a hot QCD medium. The recoiled and radiated partons are observed to have important contributions to the energy of the reconstructed jet. By comparing γ-jet azimuthal angle distribution with CMS data, we observed a nice agreement between them. In addition, a noticeable broadening on the tail of the γ−jet azimuthal angle distribution in Pb+Pb as compared to p+p collisions is observed, which, hopefully, is going to be observable in more precise experimental measurements with reduced errors in the future. A nice agreement between γ-jet asymmetry and jet survival rate with fixed α_s in LBT with CMS and ATLAS data shown that jet survival rate is a better probe for jet quenching and medium properties. The reconstructed jet fragmentation function from LBT exhibits an enhancement at both small and large $z_{\rm jet}$ due to the dominance of leading particle in a reconstructed jet. We have shown that the γ-tagged jet fragmentation function is a better measurement of parton energy loss and a sensitive probe of the jet transport parameters as it's more sensible to jet quenching.

Acknowledgments

This work is supported by the NSFC under Grant No. 11221504, China MOST under Grant No. 2014DFG02050, the Major State Basic Research Development Program in China (No. 2014CB845404), U.S. DOE under Contract No. DE-AC02-05CH11231 and within the framework of the JET Collaboration. Y.Z. was also supported by the Sofja Kovalevskaja program of the Alexander von Humboldt foundation, the DFG graduate school *Quantum Fields and Strongly Interacting Matter*, and European Research Council Grant No. HotLHC ERC-2001-StG-279579.

References

[1] Wang X N and Gyulassy M 1992 *Phys. Rev. Lett.* **68** 1480
[2] Li H, Liu F, Ma G L, Wang X N, and Zhu Y 2011 *Phys. Rev. Lett.* **106** 012301
[3] Wang X N and Zhu Y 2013 *Phys. Rev. Lett.* **111** 062301
[4] Wang X N and Guo X F 2011 *Nucl. Phys.* **A696**, 788
[5] Cacciari M, Salam G P and Soyez G 2012 *Eur. Phys. J. C* **72**, 1896
[6] Hirano T, Heinz U W, Kharzeev D, Lacey R, and Nara Y 2006 *Phys. Lett. B* **636**, 299
[7] Wang X N and Gyulassy M 1991 *Phys. Rev. D* **44**, 3501
[8] CMS Collaboration 2012 *Phys. Lett. B* **718**, 773
[9] ATLAS Collaboration 2012 Report No. ATLAS-CONF-2012-121.
[10] CMS Collaboration 2012 Report No. CMS-PAS-HIN-12-013.

Hot Quarks 2014 IOP Publishing
Journal of Physics: Conference Series **612** (2015) 012008 doi:10.1088/1742-6596/612/1/012008

Jet Measurements in Heavy-ion Collisions with CMS

Raghav Kunnawalkam Elayavalli on behalf of the CMS collaboration
Rutgers University, Piscataway, USA 08854
E-mail: raghav.k.e@cern.ch

Abstract. The production of a strongly interacting medium in heavy-ion collisions is identified through suppression of high transverse momentum jets leading to an effect known as jet quenching. Detailed measurements of nuclear modification factors and energy flow of the aforementioned quenched jets are performed with the CMS detector. These new results extend previous measurements to large angles with respect to the quenched jets and increase reach in the transverse momentum range. High statistics pp, pPb and PbPb data taken in 2011-13 are used in the following analysis. A new data-driven method to estimate the underlying event contribution from the forward calorimeter energy distribution is employed which takes into account possible event by event flow modulation. The flavor dependence of energy loss of jets is studied by tagging heavy flavor b-jets in these systems.

Presented at *Hot Quarks 2014*

Keywords: quark gluon plasma, nuclear modification factor, jet quenching, b jets

1. Introduction

Relativistic heavy-ions with sufficient energy density are expected to produce a strongly interacting medium known as the quark gluon plasma (QGP) [1]. Large transverse momentum (p_T) jets are of a special importance because their medium interactions lead to a modification in their yield, i.e. they are quenched [2] when compared to pp collisions at the same energy. This was seen at RHIC in central AuAu collisions [3, 4, 5, 6] and at the LHC [7]. Recent datasets with integrated luminosity 150 μb^{-1} in PbPb collisions at $\sqrt{s_{NN}} = 2.76$ TeV and 35 nb^{-1} in pPb collisions at $\sqrt{s_{NN}} = 5.02$ TeV. A reference pp dataset with integrated luminosity (5.3 pb^{-1} at $\sqrt{s} = 2.76$ TeV) is used for the calculation of the nuclear modification factor.

The jet quenching effect is assumed to be dependent on the flavor of the fragmenting quark. For example when the parton p_T is comparable with the parton mass, the radiation is expected to be less due to the dead-cone effect [8]. Heavy flavor quark jets are tagged using a discriminant based on the displaced vertex and the b-jet yields are compared between the three systems to study the flavor dependence of quenching. A novel method of studying the distribution of track p_T around the jet axis for a particular jet (in a dijet system) is performed to recover the energy lost due to quenching by going to large jet radii in the ϕ, η plane. The detector coordinate system is defined with ϕ as the azimuthal angle and η, the pseudorapidity. The differences in charged particle multiplicity and momentum flow carried by charged particles near the leading and the sub-leading jets in dijet events are measured in PbPb and pp systems.

Events in heavy-ion collisions are also classified according to their centrality or the degree of overlap of the two colliding nuclei. Events with a very large impact parameter correspond to a small overlap region and hence are termed as peripheral events. A small impact parameter pictorially represents two Lorentz contracted nuclei with a head-on collision and these are termed as most central collisions. The centrality classes are separated based on the energy deposited in the Hadronic Forward (HF) calorimeters for the PbPb events and by looking at the number of reconstructed charged particle tracks (event multiplicity) in the pPb system.

2. Jet reconstruction

A full description of the CMS experiment is available elsewhere [9]. Jets are reconstructed using the CMS particle flow algorithm [10] which combines all stable particle signatures like calorimeter deposits, track information and muon chamber hits. These particle flow objects are then clustered using the anti-k_T algorithm [11] encoded in the FastJet [12] framework. The clustering algorithms use a resolution parameter R, which is a p_T weighted distance between objects in the η and ϕ space to specify the "size" of the reconstructed jet. To study the momentum flow in dijet measurements, calorimetric jets that are reconstructed from energy deposits in the calorimeters are used. A discriminator called the Simple Secondary Vertex tagger (SSV) [13] which is based on the flight distance significance of the reconstructed secondary vertices is used to tag heavy flavor jets. The Underlying Event (UE) contribution or the background energy is removed using an iterative-pileup subtraction technique [14]. The HF/Voronoi algorithm [15] is a new background subtraction method which models the central underlying event density as a nonlinear response of the HF azimuthal profile. This improves the estimation of the central UE by incorporating several non-zero components of the anisotropic flow of the QGP. The estimated energy is subtracted on particle level and is later equalized to even out non physical negative energies (which arise due to the subtraction).

3. Nuclear modification factors

One of the important jet tomography variables is the nuclear modification factor (R_{AA} or R_{pA} depending on the collision system) defined as

$$R_{AA} = \frac{dN^{AA}_{jets}/dp_T d\eta}{T_{AA} \cdot d\sigma^{pp}_{jets}/dp_T d\eta}.$$

Here $T_{AA} = N_{coll}/\sigma^{NN}_{inel}$ is the nuclear overlap function defined in terms of the inelastic nucleon-nucleon cross section (σ^{NN}_{inel}) and the effective number of binary nucleon nucleon collisions (N_{coll}) calculated from a Glauber model [16].

The nuclear modification factor for fully reconstructed anti-k_T jets in PbPb collisions (closed squares) [7] shown in Fig. 1 (left) is compared with the bottom quark tagged jets [17] (b-jets) on the right. Jet suppression is found to be flavor independent at high p_T and in central PbPb events. The jets are also suppressed by a factor of about 0.5 independent of the jet transverse momentum. The R_{pA} is shown in open squares for inclusive jets on the left [18] and b-jets on the right [19]. Within the uncertainties no deviation from unity is observed. The large uncertainties in the R_{pA} stem from the pp reference extrapolated from published inclusive jet pp spectra at $\sqrt{s} = 7$ TeV using the anti-k_T algorithm with the resolution parameter of $R = 0.5, 0.7$ [20, 21]. The center of mass energy dependence is calculated by the ratio of the simulated spectra at $\sqrt{s} = 5.02$ to $\sqrt{s} = 7$ TeV and was used to scale down the measured pp spectra. A detailed description of the jet measurement and the reference extrapolation can be found in [18].

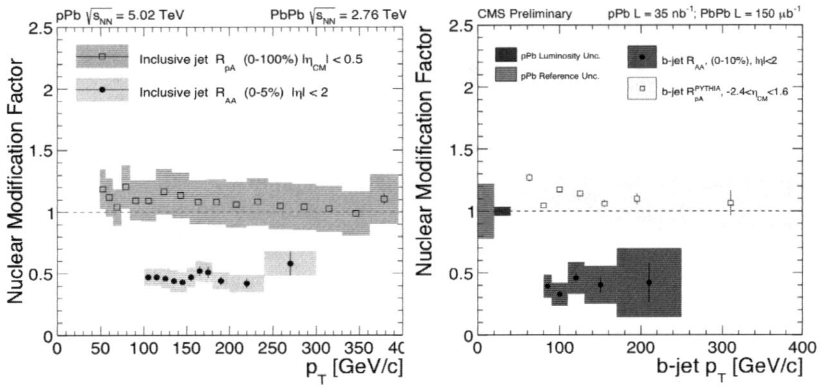

Figure 1. Left: Inclusive jet ($R = 0.3$, anti-k_T algorithm) nuclear modification factor R_{AA} (closed circles) in the most central PbPb events and R_{pPb} (open squares). Right: b-jet nuclear modification factor measured in pPb (open squares) and central PbPb (closed circles) collisions. Statistical uncertainty is represented by bars and systematics by boxes. Luminosity uncertainty is indicated by the boxes at b-jet $p_T = 0$.

4. Regaining lost energy

To study dijets where one of the jets is quenched, the dijet asymmetry ratio $A_J = (p_{T,1} - p_{T,2})/(p_{T,1} + p_{T,2})$ is calculated and the most unbalanced dijets (with large $A_J > 0.22$) are selected [22]. The direction of the dijet is defined as $\phi_{Dijet} = \frac{1}{2}(\phi_1 + (\pi - \phi_2))$. The missing p_T, defined as

$$\not{p}_T^{||} = \sum_i -p_T^i \cos(\phi_i - \phi_{Dijet})$$

is used to study the angular flow in rings of $\Delta R = \sqrt{\Delta\phi^2_{trk,Jet} + \Delta\eta^2_{trk,Jet}}$. The top panels of Fig. 2 show the missing p_T measurement in central PbPb on the right, peripheral PbPb collisions in the middle and pp collisions on the left. The bottom panels show the difference between PbPb and pp for the central and peripheral collisions on the right and center respectively. The missing p_T is shown as stacked colored histograms (color online) differentiated based on the p_T and the lines, dotted and solid represent the cumulative distribution for pp and PbPb collisions. For large $\Delta R \approx 2$, the missing p_T gets balanced by low p_T particles which is found to be similar for PbPb and pp collisions.

5. Conclusions

Nuclear modification factors measured for inclusive and b-jets in central PbPb collisions show evidence of strong suppression and are jet flavor independent. Extending this measurement to lower p_T for the heavy flavor jets might highlight the dependence (if any exists) since mass becomes an important factor in that kinematic regime. In contrast to the PbPb measurement, the pPb modification factor does not show any enhancement or suppression in the p_T range reconstructed for inclusive jets and b-jets. To further quantify the energy loss in the quenched jets, the momentum flow of tracks in unbalanced dijet pairs in both PbPb and pp collisions is studied. There is about 35 GeV/c of p_T missing from the most quenched jet in these asymmetric back to back dijet pairs. This loss in p_T is recovered by collecting low $p_T \leq 2$ GeV/c tracks up to $\Delta R \approx 2$ around the dijet axis. The angular pattern of the momentum flow is found to be

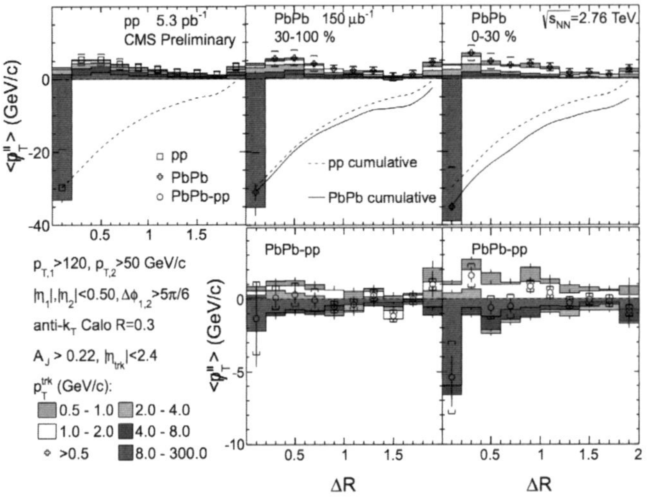

Figure 2. Upper row: Differential missing p_T distributions for pp, 30-100% and 0-30% PbPb data for various p_T ranges (colored boxes), as a function of ΔR. Also shown is the total $< \not{p}_T^{||} >$ for pp (open boxes), PbPb data (cross). Dashed lines indicate the cumulative $< \not{p}_T^{||} >$ for events with large dijet asymmetry. Bottom row: Difference between PbPb and pp where the bars and brackets represent statistical and systematic uncertainties respectively.

similar between PbPb and pp while the fragmentation is inherently different between pp and central PbPb collisions.

References

[1] Harris J.W. and Muller B. *Annu. Rev. Nucl. Part. Sci.* **46**, 71 (1996)
[2] Bjorken J.D. FERMILAB-PUB-82-059-THY.
[3] Arsene I. *et al.* [BRAHMS Collaboration] *Nucl. Phys. A* **757**, 1 (2005)
[4] Adcox K. *et al.* [PHENIX Collaboration] *Nucl. Phys. A* **757**, 184 (2005)
[5] Back B.B.*et al.* [PHOBOS Collaboration] *Nucl. Phys. A* **757**, 28 (2005)
[6] Adams J. *et al.* [STAR Collaboration] *Nucl. Phys. A* **757**, 102 (2005)
[7] [CMS Collaboration] CMS-PAS-HIN-12-004
[8] Dokshitzer Y.L. and Kharzeev D.E. *Phys. Lett. B* **519**, 199. (2001)
[9] Chatrchyan S. et al. [CMS Collaboration] *JINST* **3**, S08004 (2008)
[10] [CMS Collaboration] CMS-PAS-PFT-09-001
[11] Cacciari M, Salam G. P and Soyez G *Eur. Phys.* J. C **72**, 1896 (2012)
[12] Cacciari M, Salam G. P and Soyez G *JHEP* **04**, 063, (2008)
[13] Chatrchyan S. et al. [CMS Collaboration] *JINST* **8** P04013 (2013)
[14] Kodolova O, Vardanian I, Nikitenko A *et al. Eur. Phys. J.* C **50**, 117 (2007)
[15] [CMS Collaboration] CMS-DP-2013-018
[16] Miller M. L, Reygers K, Sanders S. J and Steinberg P *Ann. Rev. Nucl. Part. Sci.* **57**, 205 (2007)
[17] Chatrchyan S. et al. [CMS Collaboration] *Phys. Rev. Lett.* **113**, 132301 (2014)
[18] [CMS Collaboration] CMS-PAS-HIN-14-001
[19] [CMS Collaboration] CMS-PAS-HIN-14-000
[20] Chatrchyan S. et al. [CMS Collaboration] *Phys.Rev.Lett.* **107** 132001 (2013)
[21] Chatrchyan S. et al. [CMS Collaboration] *Phys.Rev.D* **87** 112002 (2012)
[22] [CMS Collaboration] CMS-PAS-HIN-14-010

Measurement of W-boson production in p–Pb collisions at $\sqrt{s_{\mathrm{NN}}} = 5.02$ TeV with ALICE at the LHC

Jianhui Zhu[1,2]**, on behalf of the ALICE collaboration**
[1]Key Laboratory of Quark and Lepton Physics (MOE) and Institute of Particle Physics, Central China Normal University, Wuhan, China
[2]SUBATECH, Ecole des Mines de Nantes, Université de Nantes, CNRS-IN2P3, Nantes, France
E-mail: Jianhui.Zhu@cern.ch

Abstract. In hadronic collisions, electroweak bosons are produced in initial hard scattering processes and they are not affected by the strong interaction. In proton–proton collisions, they have been suggested as standard candles for luminosity monitoring and their measurement can improve the evaluation of detector performances. In nucleus-nucleus and proton–nucleus collisions, W-bosons allow one to check at first order the validity of binary collision scaling, while small deviations allow to study the nuclear modifications of parton distribution functions. The W-boson production in p–Pb collisions at $\sqrt{s_{\mathrm{NN}}} = 5.02$ TeV is measured via the contribution of W-boson decays to the inclusive p_T-differential muon yield reconstructed with the ALICE muon spectrometer at forward ($2.03 < y^{\mu}_{\mathrm{cms}} < 3.53$) and backward ($-4.46 < y^{\mu}_{\mathrm{cms}} < -2.96$) rapidity. This paper reports the production cross section of muons from W-boson decays for $p^{\mu}_{\mathrm{T}} > 10$ GeV/c and the yields normalised to the average number of binary nucleon-nucleon collisions as a function of the event activity.

1. Introduction

The high collision energies available at the Large Hadron Collider (LHC) at CERN allow for an abundant production of hard probes, such as quarkonia, high p_{T} jets and intermediate vector bosons ($W^{\pm}$, Z^0). The W bosons are produced in initial hard parton scattering processes, with a formation time of the order of $1/M_{\mathrm{W}} \sim 0.003$ fm/c and, having a lifetime of 0.09 fm/c, they decay before the formation of the quark-gluon plasma (QGP), which is a deconfined phase of QCD matter that is formed in high-energy heavy-ion collisions. Furthermore, their leptonic decay products do not interact strongly with the hot and dense QCD medium. In proton-proton collisions, due to precise theoretical predictions, one can use W-boson production to cross check the luminosity. In proton-nucleus and nucleus-nucleus collisions, precise measurements of W-boson and Z-boson production can constrain the modification of the Parton Distribution Functions (PDFs) in the nucleus [1] and they can be used to test the scaling of hard particle production with the number of binary nucleon-nucleon collisions at first order [2].

2. Experimental setup and data taking

The ALICE experiment is equipped with various detectors for triggering, tracking and particle identification [3]. Muons are reconstructed in the muon spectrometer, which covers the pseudo-rapidity range $-4 < \eta_{\mathrm{lab}} < -2.5$, consists of a thick front absorber, a dipole magnet, five tracking stations, a muon filter and two trigger stations. Other detectors are used for event selection and characterization. The inner tracking system consists of six layers of silicon detectors covering the range $|\eta_{\rm lab}| < 0.9$. In this analysis, the interaction vertex is measured with the two innermost layers, which are equipped with Silicon Pixel Detectors (SPD). The number of clusters in the outer layer of the SPD (CL1) can be used to characterize the event activity. An array of scintillators placed at both sides of the interaction point, the VZERO detector (V0A, V0C), provides the trigger and event activity estimation. A third estimation of the event activity is obtained with the Zero Degree Calorimeters (ZNA, ZNC) located at both sides of the detector at 112.5 m distance from the interaction point along the beam line. Due to the energy asymmetry of the two colliding beams, the centre-of-mass system of nucleon-nucleon collisions is shifted by $\Delta y = 0.465$ in the direction of the proton beam, with respect to the laboratory frame. Experimental data have been collected with two beam configurations (p-Pb and Pb-p) by inverting the orbits of the two particle species. The rapidity regions covered by the muon spectrometer in the two cases are $2.03 < y^{\mu}_{\rm cms} < 3.53$ (forward, p-Pb) and $-4.46 < y^{\mu}_{\rm cms} < -2.96$ (backward, Pb-p), where positive rapidities refer to the situation where the proton beam is moving towards the muon spectrometer. The data have been collected at $\sqrt{s_{\rm NN}} = 5.02$ TeV with minimum bias (MB) and single muon triggers. The former is defined requiring hits in both sides of the VZERO detector in coincidence with the beam counters. The latter is defined by asking, in addition to the MB condition, for a high transverse momentum muon in the trigger system of the spectrometer. The trigger selection is not sharp in momentum, but roughly corresponds to $p_{\rm T} > 4$ GeV/c. The integrated luminosities for the forward and backward rapidity measurements are 4.9 $\rm nb^{-1}$ and 5.8 $\rm nb^{-1}$, respectively. Events are selected by requiring that the timing in VZERO and ZDC detectors are compatible with a nominal p-Pb interaction and that the interaction vertex is reconstructed. The track in the muon tracking system is required to be in the geometrical acceptance ($-4 < \eta_{\rm lab} < -2.5$), to have a polar angle measured at the end of the absorber of $170° < \theta_{\rm abs} < 178°$ and match a tracklet in the trigger system in order to reject the remaining contamination from hadrons. Furthermore, the correlation between momentum and distance of closest approach (DCA) to the interaction vertex was used to remove beam-gas collisions and secondary particles produced in the absorber.

3. W-boson signal extraction

The present analysis is based on the extraction of the W-boson contribution from the transverse momentum distribution of inclusive muons. At high $p_{\rm T}$, the main contributions to the yield of inclusive muons come from the muonic decays of W-bosons, the di-muon decays of $\rm Z^0/\gamma^*$ bosons and the semi-muonic decays of beauty hadrons. The yield of muons from W-boson decays can be obtained through a fit based on suitable parameterizations of the different components. The distribution of muons from intermediate vector boson decays is described by MC templates obtained from POWHEG simulations [4]. The isospin dependence of the W-boson differential cross-section [5] is accounted for by simulating separately p-p and p-n collisions and then summing the results together according to:

$$\frac{1}{N_{\rm p\text{-}Pb}} \cdot \frac{{\rm d}N_{\rm p\text{-}Pb}}{{\rm d}p_{\rm T}} = \frac{Z_{\rm Pb}}{A_{\rm Pb}} \cdot \frac{1}{N_{\rm p\text{-}p}} \cdot \frac{{\rm d}N_{\rm p\text{-}p}}{{\rm d}p_{\rm T}} + \frac{A_{\rm Pb} - Z_{\rm Pb}}{A_{\rm Pb}} \cdot \frac{1}{N_{\rm p\text{-}n}} \cdot \frac{{\rm d}N_{\rm p\text{-}n}}{{\rm d}p_{\rm T}} \tag{1}$$

where $A_{\rm Pb} = 208$ and $Z_{\rm Pb} = 82$. The generation of muons from $\rm Z^0/\gamma^*$ decays is performed in an equivalent way. For the description of the background at low $p_{\rm T}$ ($10 < p_{\rm T} < 40$ GeV/c), which originates from the decays of beauty hadrons, three approaches are adopted: MC templates based on pQCD calculations with FONLL [6], functional form which was already successfully adopted for analogous measurements by the ATLAS experiment at the LHC [7] and another function adapted from the second term of the function used by ATLAS, which can be written as $f(p_{\rm T}) = a \cdot \frac{e^{b\cdot\sqrt{p_{\rm T}}}}{p_{\rm T}^c}$. The different contributions are summed together in the final fit function:

$$f(p_{\rm T}) = N_{\rm bkg} \cdot f_{\rm bkg}(p_{\rm T}) + N_{\mu\leftarrow {\rm W}} \cdot f_{\mu\leftarrow {\rm W}}(p_{\rm T}) + N_{\mu\leftarrow {\rm Z}^0/\gamma^*} \cdot f_{\mu\leftarrow {\rm Z}^0/\gamma^*}(p_{\rm T}) \quad (2)$$

where $f_{\rm bkg}$ can be either a functional form or the MC template for muons from heavy-flavour decays, and $f_{\mu\leftarrow {\rm W}}$ and $f_{\mu\leftarrow {\rm Z}^0/\gamma^*}$ are the MC templates for muons from W and Z^0/γ^* boson decays, respectively. The number of muons from W decays ($N_{\mu\leftarrow {\rm W}}$) is a free parameter, while the ratio $N_{\mu\leftarrow {\rm Z}^0/\gamma^*}/N_{\mu\leftarrow {\rm W}}$ is fixed to the value obtained with POWHEG. Two examples of combined fit are shown in figure 1.

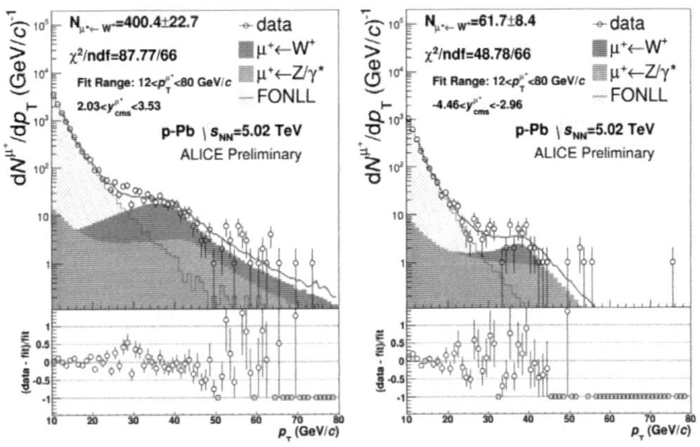

Figure 1. Examples of signal extraction at forward (left) and backward (right) rapidity using a FONLL-based MC template to describe the background

The signal extraction is affected by the uncertainties in the description of the shapes of signal and background. The effect of the detector alignment is taken into account by producing the templates in simulations using different residual alignment assumptions. The uncertainty on the background description is determined by using either FONLL-based MC templates or functional forms and by varying the $p_{\rm T}$ range in the fit. The relative contribution of Z^0/γ^* is varied based on POWHEG and PYTHIA [8] predictions. The signal was extracted many times by varying the above conditions and the final result is chosen as the weighted average of all of the trials. The systematic uncertainties is given by the RMS of the trials. It is worth noting that POWHEG does not account for the nuclear modification of the PDFs. Hence, the signal extraction using PYTHIA with EPS09 parameterization of nuclear PDFs [9] was included in the calculation of the systematic uncertainties.

4. Cross section of muons from W-boson decays in p-Pb collisions

After signal extraction, the yields of muons from W-boson decays are corrected for acceptance and efficiency and converted into cross sections. The measured production cross section of muons from W-boson decays at forward and backward rapidity in p-Pb collisions is shown in figure 2 and compared with theoretical predictions based on NLO pQCD calculations with CT10 [10] and EPS09 PDFs. The yield of muons from W-boson decays, corrected for acceptance and efficiency and normalised to the number of equivalent MB events can be divided by the average number of binary collisions $\langle N_{\rm coll}\rangle$ in each event activity bin, in order to obtain the yield per binary collision. $\langle N_{\rm coll}\rangle$ is determined from a Glauber-model based analysis in the case of V0A and CL1 estimators, while for the ZN estimator it is computed assuming that the particle density at mid-rapidity is proportional to $N_{\rm part}$. All about $\langle N_{\rm coll}\rangle$ measurement are described in [2]. The measured yield/$\langle N_{\rm coll}\rangle$ as a function of event activity at forward and backward rapidity in p-Pb collisions are shown in figure 3.

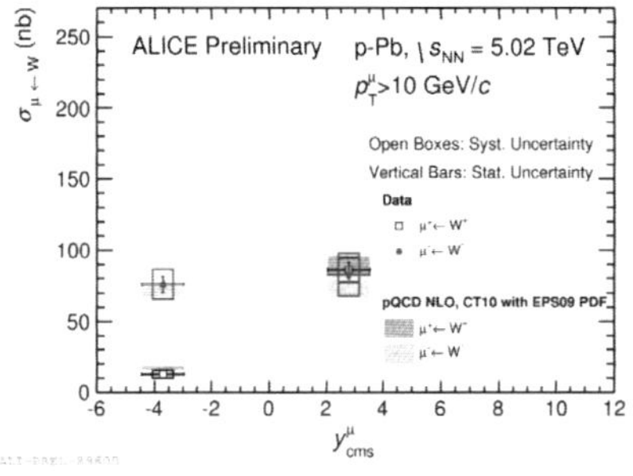

Figure 2. Cross section of muons from W-boson decays as a function of rapidity

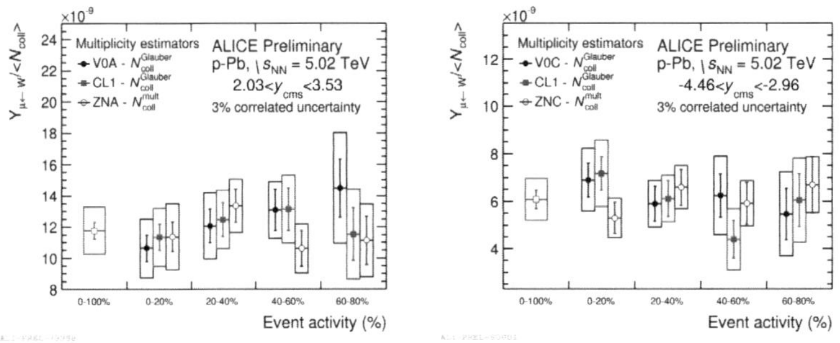

Figure 3. Yield/$\langle N_{\rm coll}\rangle$ as a function of event activity at forward and backward rapidity

5. Conclusion

The production cross section of muons from W-boson decays at forward and backward rapidity in the range $p_T^\mu > 10$ GeV/c has been measured in p-Pb collisions at $\sqrt{s_{\rm NN}} = 5.02$ TeV with the ALICE muon spectrometer. Theoretical predictions based on NLO pQCD calculations with CT10 and EPS09 PDFs agree with the data within uncertainties. The yield of muons from W-boson decays in different event activity bins scales with $\langle N_{\rm coll}\rangle$ within substantial uncertainties.

Acknowledgments

This work is supported partly by the Chinese Ministry of Science and Technology 973 grant 2013CB837803, NSFC Grant 11375071 and IRG11221504, CCNU13F026 and CSC grant No. 201306770033.

References

[1] Paukkunen H and Salgado C A 2011 *J. High Energy Phys.* **03** 071
[2] Toia A 2014 *Nuclear Physics A* **926** 78-84
[3] Aamodt K *et al.* (ALICE Collaboration) 2008 *JINST* **3** S08002
[4] Alioli S, Nason P, Oleari C and Re E 2008 *J. High Energy Phys.* **07** 060
[5] Conesa del Valle Z, CERN-THESIS-2007-102
[6] Cacciari M, Greco M and Nason P 1998 *J. High Energy Phys.* **05** 007
[7] ATLAS Collaboration, ATLAS-CONF-2011-078
[8] Sjostrand T, Mrenna S and Skands P 2006 *J. High Energy Phys.* **05** 026
[9] Eskola K J, Paukkunen H and Salgado C A 2009 *J. High Energy Phys.* **04** 065
[10] Lai H L *et al.* 2010 *Phys.Rev. D* **82** 074024

Hot Quarks 2014 IOP Publishing
Journal of Physics: Conference Series **612** (2015) 012010 doi:10.1088/1742-6596/612/1/012010

Decoding the quark-gluon plasma with jet quenching observables

Konrad Tywoniuk

Departament d'Estructura i Constituents de la Matèria and Institut de Ciències del Cosmos (IC-CUB),
Universitat de Barcelona, Martí i Franquès 1, 08028 Barcelona, Spain

E-mail: konrad@ecm.ub.edu

Abstract. We briefly describe two interesting aspects of heavy-ion collisions related to jets and medium dynamics. First, we discuss how the interaction with the medium change the energy and internal distributions of a QCD jet. Secondly, we discuss a possible manifestation of effects from the confining (chromo)magnetic scale for quark excitations in a realistic plasma.

1. Jet modifications in heavy-ion collisions

Jets emerging from heavy-ion collisions are unique probes of the underlying dynamics [1]. Being multi-particle configurations, their detailed substructure is very sensitive to interactions with the dynamical medium, characterised by the transport coefficient $\hat{q}$, in course of the branching process. The complexity of the resulting multi-scale evolution is reduced by the presence of a hard scale, related to the jet energy and opening angle $Q = E\Theta_{\rm jet}$, which is much larger than any characteristic momentum scale of the medium, see below [2]. This scale separation guarantees the validity of a perturbative framework. Since any coloured high-energy particle propagating through the medium (color current) perturbs the background field and leads to stimulated radiation, former studies mainly addressed the jet modifications in medium arising from induced radiation [3]. The dynamical interplay of jet and medium scales in course of the evolution is, however, to a large extent unknown, although models based on perturbative arguments exist [4].

The problem has been worked out in detail on the level of a time-like single emission off a dipole [5, 6]. It was shown that the decoherence of the dipole is governed by the "decoherence parameter," defined as $\Delta_{\rm med} = 1 - \exp[-(r_\perp Q_s)^2/12]$, which marks out two characteristic medium scales with simple geometric interpretations. On the one hand, $r_\perp = \theta_0 L$ is the transverse size of a dipole with opening angle θ_0 traversing a medium of length L. On the other hand, $Q_s = \sqrt{\hat{q}L}$ determines the accumulated transverse momentum (via Brownian motion) of a probe passing through a medium characterized by the diffusion coefficient $\hat{q}$. Whenever the dipole is small enough, such that $r_\perp^{-1} \gg Q_s$, the decoherence parameter is vanishing. This means that the medium only resolves the total charge of the dipole. In this case, it is also the total charge that gives rise to medium-induced radiation. In the opposite case, whenever the medium scale is large, $r_\perp^{-1} \ll Q_s$, the individual charges of the dipole are resolved and can induce radiation.

Journal of Physics: Conference Series **612** (2015) 012010 doi:10.1088/1742-6596/612/1/012010

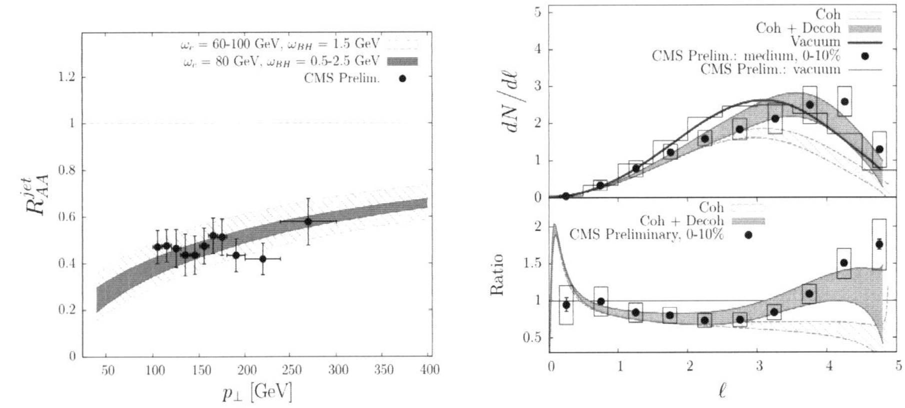

Figure 1. Left panel: The nuclear modification factor calculated with ω_c = 80 GeV as a function of jet $p_\perp$ for central Pb-Pb collisions. The dark (red) band includes the variation of $\omega_{\rm BH}$ around a central value of 1.5 GeV. The light (grey) band includes, in addition, a variation of $\omega_c \in$ [60, 100] GeV. The experimental data are taken from [13]. Right, upper panel: the longitudinal fragmentation function plotted as a function of $\ell = \ln 1/x$. Right, lower panel: the ratio of medium-modified and vacuum fragmentation functions for a jet with E = 100 GeV and $\Theta_{\rm jet}$ = 0.3. The experimental data are taken from [14]. The figures in both panels are taken from [12].

We therefore can think of the jet-medium interaction in a twofold way [2]. The interplay of scales at each step of the fragmentation determines simultaneously the number of independent (or color decoherent) sources of medium-induced radiation as well as the phase space for soft and collinear emissions [7, 8, 9]. The former process is crucial in redistributing the jet energy up to very large angles, mostly far out of the jet cone [10]. The latter modifies the familiar angular ordering features [11] and brings about modifications to the inter-jet distribution, mostly within the cone.

For a sample of the jets with the highest energy available in heavy-ion collision reconstructed at relatively small cone sizes, we expect to be dominated by collimated jets, i.e. situations where the major fraction of the jet energy is located close to the jet axis [2]. When calculating inclusive observables, such as the jet spectrum, one can therefore treat the whole jet as coherent system, interacting with the medium as a whole [12]. The inter-jet observables, e.g. the energy distribution of final-state hadrons in the jet, are however more sensitive to the effects of decoherence which can strongly alter the yield of soft particles inside the cone.

First let us address the modification to the inclusive jet spectrum, typically quantified through the nuclear modification factor defined as $R_{AA}^{\rm jet} \equiv {\rm d}^2 N^{\rm jet}_{\rm Pb\text{-}Pb}(p_\perp)/{\rm d}^2 p_\perp / (T_{\rm AA}\, {\rm d}^2 \sigma^{\rm jet}_{\rm p\text{-}p}(p_\perp)/{\rm d}^2 p_\perp)$, where $T_{\rm AA}$ is the nuclear overlap function. Thus, the spectrum after passing through the medium is computed by convoluting the jet cross-section in vacuum, proportional to the quark cross-section, with the distribution of quarks, $D_q^{\rm med}$, after passing through the medium, The distribution $D_q^{\rm med}$ was found from the rate equation described above with the parameters α_s = 0.5 and an energy cut-off $\omega_{\rm BH}$ = 1.5 GeV [12]. The central value obtained from the left panel in Fig. 1 results in $\omega_c \equiv Q_s^2 L/2$ = 80 GeV. To gauge the uncertainties related to the choice of our parameters, we have also allowed the medium parameters to vary giving rise to the shaded region in Fig. 1. To establish a closed set of physical energy and momentum scales, we have also chosen $L = 2.5$ fm resulting in the medium scale $Q_s = 3.6$ GeV, and $Q_s \ll Q$ as expected. This naturally leads to a small deflection of the jet while 14–19% of the energy flows out a cone of $\Theta_{\rm jet} = 0.3$ which agrees qualitatively with experimental measurements [15].

The modifications of the fragmentation functions of jets are more subtle. We will focus on the so called intra-jet energy distribution of hadrons $dN^{\rm vac}/d\ln(1/x) \equiv D^{\rm vac}(x;Q)$ which is typically plotted in terms of the variable $\ell = \ln(1/x_h)$ where x is fraction of the jet energy carried by the parton. The Q dependence of $D^{\rm vac}$ is governed by the MLLA evolution equations [11] which obey angular ordering. The thin, solid (grey) lines in the right panel of Fig. (1) represent the energy distribution within a completely coherent jet with reduced energy and fails to describe the data. The thick, solid (red) curves demonstrate the effect of decoherence, which plays a crucial role only for soft gluons. The bands represent uncertainties in the jet ($\sim 20\%$) and medium scales. See [12] for further details.

2. Novel excitations of hot quarks

In recent years, there has been compelling evidence that a strongly coupled system is created in ultrarelativistic heavy-ion collisions. We are tempted to ask: what are the fundamental degrees of freedom of such a system? And how can we use external probes, e.g. such as jets, to precisely pin down these features? The underlying dynamics is governed by QCD, the theory of strong interactions, but a perturbative expansion in the ordinary setup is tractable only for asymptotically large temperatures. There are nevertheless strong indications that a confinement mechanism has to be incorporated within perturbative resummation even when dealing with the QGP phase.

When addressing these questions, one has to take the problem of quantizing the theory seriously. A formalism to tackle this issue is the Gribov-Zwanziger (GZ) action, which is well known from studies of color confinement [16, 17]. In general covariant gauges, the ghost and gluon propagators are modified. Additionally, a new scale, the Gribov parameter γ_G, is generated from the fixing of the infrared residue gauge transformations after the Faddeev-Popov procedure. This scale is determined by a self-consistent gap equation, which at one-loop order gives $\gamma_G = \frac{3N_c}{16\sqrt{2}\pi} g^2 T$ at asymptotically high temperatures, where N_c is the number of colors [18, 19]. The Gribov parameter provides a fundamental infrared cutoff at the confining (chromo)magnetic scale for the GZ action and the resulting equations of state show stable and robust behavior that is consistent with lattice data down to nearly the deconfinement transition [19]. These results therefore highlight the significant role played by the magnetic scale in the phenomenologically relevant temperature regime.

In order to get a handle on what the relevant degrees of freedom are, we have recently studied the quasi-particle quark excitations of a finite temperature QCD plasma [20] using systematics of the hard-thermal-loop (HTL) effective theory [21]. Our results are plotted in Fig. 2 for a relevant range of couplings $g \lesssim 2$. As expected, we recover the screened quasi-particle $\omega = \omega_+(p;\gamma_G)$ and hole $\omega = \omega_-(p;\gamma_G)$ excitations of the plasma, depicted by dashed-dotted and dashed (blue) curves in Fig. 2. The low-momentum limit of dispersion relation corresponds to the screening mass, $\omega_\pm(0;\gamma_G) = m_q(\gamma_G)$, which in our calculation incorporates effects from the magnetic scale and reduces to the well know fermionic Debye mass, $m_q(0) = \sqrt{C_F/8}gT$.

In stark contrast to the conventional HTL expectation, we also find a third pole in the quark propagator with $\omega = \omega_G(\omega;\gamma_G)$. It describes massless fermionic excitations in the plasma which are governed by the disperstion relation $\omega = v_s p$ at small momenta, see solid (red) curves in Fig. 2. We have numerically established that $v_s \approx 1/\sqrt{3}$ (speed of sound) independent of the coupling for the studied range. Due to its masslessness, the new mode incorporate long-range correlations of the quark-gluon plasma rendering it strongly correlated. The range in momentum where this pole can exist scales with the magnetic scale through γ_G, and we have plotted it in Fig. 2 for two values of the coupling. Moreover, the pole goes along with a negative weight to

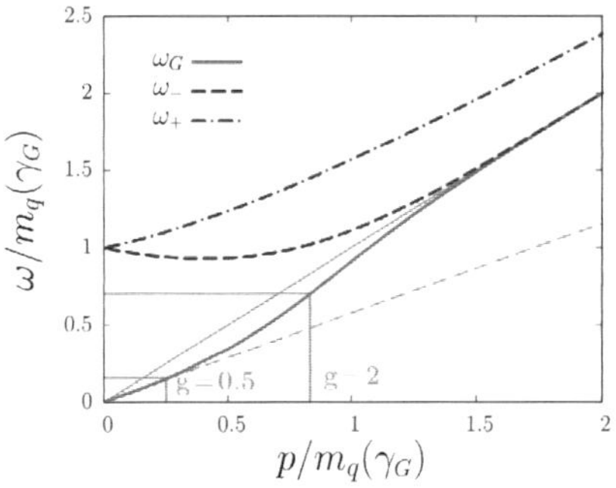

Figure 2. Dispersion relations (upper panel) and the corresponding residues (lower panel) for the plasmino (ω_+), anti-plasmino (ω_-) and Gribov (ω_G) poles. The scaling with $m_q(\gamma_G)$ has been numerically established for $g \lesssim 2$. Figure taken from [20].

the spectral function which signals effects of confinement [22], see [20] for a further discussion. It therefore does not represent a physical quark excitation but should rather be a resonance-like mode.

We conclude that these novel features, where the magnetic scale plays a prominent role, are characteristic of the non-Abelian nature of QCD. Their manifestation emphasizes the importance of the long-range confinement effects surviving at finite temperatures in the plasma and will play a crucial role for phenomenology in a realistic range of temperatures.

Acknowledgements. The author would like to thank Y. Mehtar-Tani and N. Su for fruitful collaboration. The author is supported by a Juan de la Cierva fellowship and by the research grants FPA2010-20807, 2009SGR502 and by the Consolider CPAN project and by FEDER.

References

[1] Mehtar-Tani Y, Milhano J G and Tywoniuk K 2013 *Int.J.Mod.Phys.* **A28** 1340013 (*Preprint* 1302.2579)
[2] Casalderrey-Solana J, Mehtar-Tani Y, Salgado C A and Tywoniuk K 2013 *Phys.Lett.* **B725** 357–360 (*Preprint* 1210.7765)
[3] Salgado C A and Wiedemann U A 2002 *Phys.Rev.Lett.* **89** 092303 (*Preprint* hep-ph/0204221)
[4] Zapp K, Ingelman G, Rathsman J, Stachel J and Wiedemann U A 2009 *Eur.Phys.J.* **C60** 617–632 (*Preprint* 0804.3568)
[5] Mehtar-Tani Y, Salgado C and Tywoniuk K 2012 *Phys.Lett.* **B707** 156–159 (*Preprint* 1102.4317)
[6] Casalderrey-Solana J and Iancu E 2011 *JHEP* **1108** 015 (*Preprint* 1105.1760)
[7] Mehtar-Tani Y, Salgado C A and Tywoniuk K 2011 *Phys.Rev.Lett.* **106** 122002 (*Preprint* 1009.2965)
[8] Mehtar-Tani Y and Tywoniuk K 2013 *JHEP* **1301** 031 (*Preprint* 1105.1346)
[9] Mehtar-Tani Y, Salgado C A and Tywoniuk K 2012 *JHEP* **1210** 197 (*Preprint* 1205.5739)
[10] Blaizot J P, Fister L and Mehtar-Tani Y 2014 (*Preprint* 1409.6202)
[11] Khoze V A and Ochs W 1997 *Int.J.Mod.Phys.* **A12** 2949–3120 (*Preprint* hep-ph/9701421)
[12] Mehtar-Tani Y and Tywoniuk K 2014 (*Preprint* 1401.8293)
[13] CMS Collaboration 2012 (*Preprint* CMS-PAS-HIN-12-004)
[14] CMS Collaboration 2012 (*Preprint* CMS-PAS-HIN-12-013)
[15] Chatrchyan S *et al.* (CMS Collaboration) 2011 *Phys.Rev.* **C84** 024906 (*Preprint* 1102.1957)
[16] Gribov V 1978 *Nucl.Phys.* **B139** 1
[17] Zwanziger D 1989 *Nucl.Phys.* **B323** 513–544
[18] Zwanziger D 2007 *Phys.Rev.* **D76** 125014 (*Preprint* hep-ph/0610021)
[19] Fukushima K and Su N 2013 *Phys.Rev.* **D88** 076008 (*Preprint* 1304.8004)
[20] Su N and Tywoniuk K 2014 (*Preprint* 1409.3203)
[21] Braaten E and Pisarski R D 1990 *Nucl.Phys.* **B337** 569
[22] Alkofer R and von Smekal L 2001 *Phys.Rept.* **353** 281 (*Preprint* hep-ph/0007355)

Study of Z boson production in pp, pPb and PbPb collisions in CMS

Anna Julia Zsigmond (for the CMS Collaboration)
Wigner Research Center for Physics, Budapest, Hungary
E-mail: anna.julia.zsigmond@cern.ch

Abstract. The Z boson production is studied in PbPb and pp collisions at $\sqrt{s_{NN}} = 2.76$ TeV and in pPb collisions at $\sqrt{s_{NN}} = 5.02$ TeV using data collected by the CMS experiment at the LHC. The Z boson production is observed to be proportional to the number of elementary nucleon-nucleon collisions in both PbPb and pPb collisions. The measured differential cross sections as a function of transverse momentum and rapidity are in agreement with NLO pQCD calculations using different parton distribution functions with and without nuclear effects. Using the PbPb and pp data, the nuclear modification factor is presented as well as the forward-backward ratio measured in pPb collisions.

1. Introduction

Thanks to its high center-of-mass energy and high luminosity, the LHC allows the study of Z and W boson production in heavy ion collisions for the first time. Electroweak bosons are essentially unmodified by the hot and dense medium created in nucleus-nucleus collisions, and their leptonic decays are of particular interest since the leptons pass through the medium without interacting strongly. However, in heavy ion collisions, the production of electroweak bosons can be affected by initial-state effects. The parton distribution functions (PDFs) can be modified in nuclei, which together with the fact that the nucleus contains neutrons besides protons, can modify the observed cross sections compared to pp collisions.

Based on the first PbPb collisions at the LHC in 2010, the CMS collaboration has reported results on Z and W boson production [1, 2]. In this paper, the study of Z boson production at the center-of-mass energy per nucleon pair $\sqrt{s_{\mathrm{NN}}} = 2.76$ TeV is presented using PbPb collision data collected in 2011 corresponding to an integrated luminosity of 150 μb^{-1} and using pp collision data recorded in 2013 with an integrated luminosity of 5.4 pb^{-1} [3]. These data samples allow for more precise measurement of the Z boson nuclear modification factor (R_{AA}) and its dependence on transverse momentum (p_{T}), rapidity (y) and collision centrality. The pPb collision data taken in 2013 is currently the best sample to constrain nuclear PDFs (nPDFs) in a previously unexplored region of phase space. The results on Z boson production cross section in pPb collisions at $\sqrt{s_{\mathrm{NN}}} = 5.02$ TeV corresponding to an integrated luminosity of 34.6 nb^{-1} are also presented [4].

2. Z bosons in PbPb and pp collisions at $\sqrt{s_{NN}} = 2.76$ TeV

The Z bosons are identified through their leptonic decays to muon or electron pairs. Signal candidates are selected by requiring an opposite charge lepton pair in the 60–120 GeV/c^2

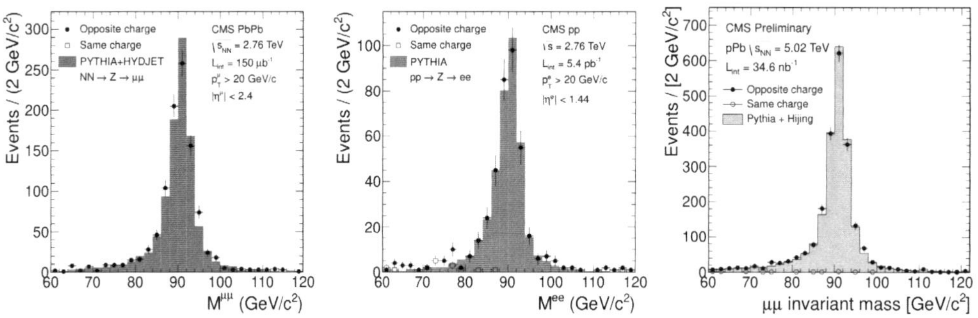

Figure 1. Invariant mass of the selected lepton pairs from PbPb (*left*), pp (*middle*) and pPb (*right*) collisions compared to the corresponding simulations.

invariant mass range with both leptons having $p_T > 20$ GeV/c and found to be in $|\eta^\mu| < 2.4$ for muons or $|\eta^e| < 1.44$ for electrons. The dimuon rapidity is limited to $|y| < 2$, while the dielectron rapidity is kept in the $|y| < 1.44$ range. Figure 1 shows examples of the invariant mass distribution of Z boson candidates in PbPb and pp collisions. Overlaid histograms are from PYTHIA+HYDJET [5] and PYTHIA [6] simulations that are used to determine the acceptance and efficiency corrections for the PbPb and pp sample, respectively.

The corrected Z boson yield is normalized by the number of minimum bias (MB) events or by the integrated luminosity in the two datasets. The results from the muon and electron decay channels are in agreement within the statistical uncertainties and a combination is performed in the common kinematic region $|y| < 1.44$. Figure 2 shows the normalized Z boson yield in PbPb collisions for the combined dilepton channel as a function of centrality, rapidity and transverse momentum. The results are in agreement with predictions from the POWHEG [7] generator in case of p_T dependence or with predictions from [8] in case of y dependence using CT10 PDF set [9] with and without the nuclear effects from EPS09 nPDF set [10].

The R_{AA} is calculated by dividing the corrected yield from PbPb collisions by the cross section measured in pp collisions and scaled by the nuclear overlap function (T_{AA}), which is proportional

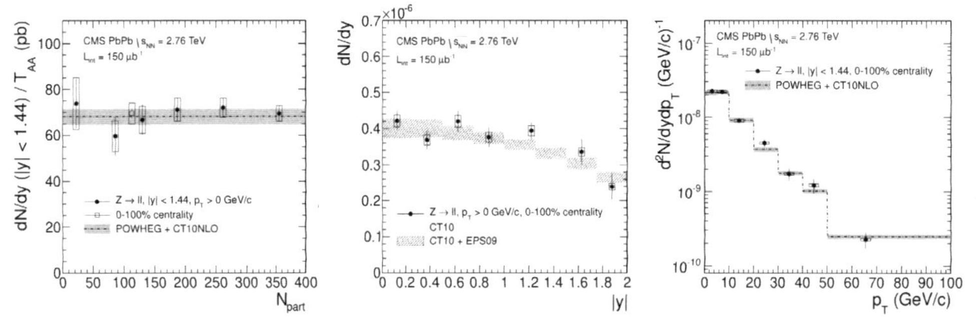

Figure 2. The Z boson yield per MB event measured in PbPb collisions shown for the combined dilepton channel as a function of event centrality (*left*), Z boson y (*middle*) and p_T (*right*). The yields are compared to predictions with (green) and without (yellow) nuclear modification effects vs. y, and to POWHEG (gray) vs. centrality and p_T.

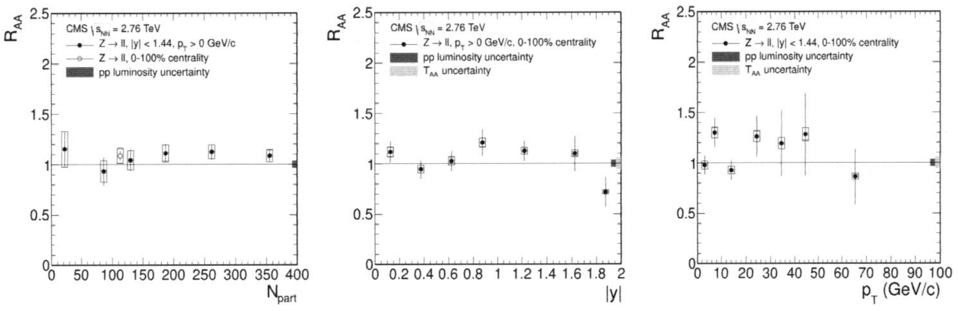

Figure 3. Nuclear modification factor of Z boson production for the combined dilepton channel as a function of N_{part} (*left*), y (*middle*) and p_{T} (*right*). Gray (green) box at $R_{\text{AA}} = 1$ corresponds to the pp luminosity (T_{AA}) uncertainty.

to the average number of binary nucleon-nucleon collisions (N_{coll}). Figure 3 shows the R_{AA} of Z boson production as a function of centrality, rapidity and transverse momentum. In every event class, it is compatible with unity and the integrated values are 1.06 ± 0.05 (stat.) $\pm$ 0.08 (syst.) in the dimuon and 1.02 ± 0.08 (stat.) ± 0.15 (syst.) in the dielectron channel. From the centrality dependence, one can confirm that the Z boson production in PbPb collisions scales with N_{coll}. The results as a function of y and p_{T} are compatible with unity within uncertainties and no initial state nuclear effects are visible with the current amount of data.

3. Z bosons in pPb collisions at $\sqrt{s_{NN}} = 5.02$ TeV

Due to the LHC magnet system, the p and Pb beam energies were different that resulted in a $\Delta y = 0.465$ shift of the nucleon-nucleon center-of-mass frame compared to the laboratory frame. The results are presented in the center-of-mass frame with positive rapidity values corresponding to the proton fragmentation region. The Z boson candidates are selected as an opposite-charge muon pair in the 60–120 GeV/c^2 mass range where both muons have $p_{\text{T}} > 20$ GeV/c and are within the $|\eta_{\text{lab}}| < 2.4$ muon detector coverage. The invariant mass of muon pairs is shown in the right-hand side of Figure 1 and compared to PYTHIA+HIJING [11] simulation that is used to determine the efficiency correction factors. The acceptance correction is calculated from POWHEG simulation with CT10 PDF set interfaced with PYTHIA parton shower.

The measured inclusive Z boson production cross section in pPb collisions is $\sigma_{\text{pPb}\to\text{Z}\to\mu\mu}(-2.5 < y_{\text{c.m.}} < 1.5) = 94.1 \pm 2.1$ (stat.) $\pm$ 2.4 (syst.) $\pm$ 3.3 (lumi.) nb using the calibrated integrated luminosity. For the same restricted rapidity range, the POWHEG simulation predicts 94.0 ± 4.7 nb after multiplying by the number of nucleons in the Pb nucleus ($A = 208$), which corresponds to the hypothesis of binary collision scaling in pPb. The inclusive result is in a good agreement with the NLO pQCD prediction.

The measured differential cross section as a function of rapidity in the center-of-mass frame is compared to predictions from MCFM generator [12] in the left-hand side of Figure 4. The MCFM predictions are calculated with MSTW2008NLO free proton PDF set [13] with and without the nuclear modification from EPS09 or DSSZ nPDF sets [10, 14] and multiplied by A. The measured differential cross section is consistent with the theory predictions within uncertainties that are dominated by the statistical uncertainties.

The forward-backward ratio, defined as $d\sigma(+y)/d\sigma(-y)$, is expected to be more sensitive to nuclear effects [8], because normalization uncertainties cancel both in theory and in experiment. The right-hand side of Figure 4 shows the measured forward-backward ratio as a function

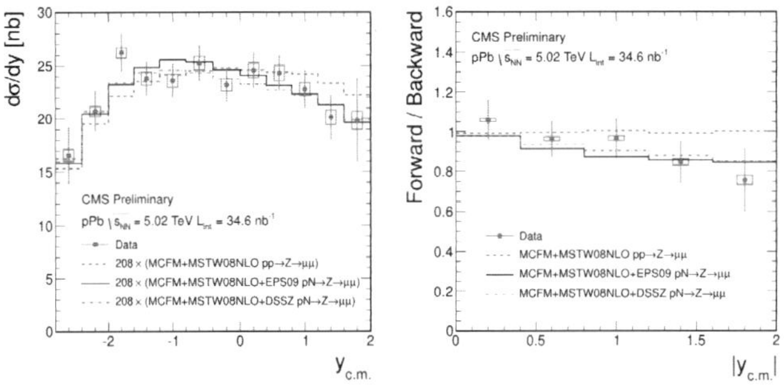

Figure 4. Rapidity differential cross section (*left*) and forward-backward ratio (*right*) of Z boson production in pPb collisions compared to predictions from MCFM with MSTW2008NLO PDF with and without the nuclear modification from EPS09 or DSSZ nPDFs.

of $|y_{\mathrm{c.m.}}|$ compared to the MCFM predictions. Due to the large statistical uncertainties, this measurement is not able to distinguish between different nPDF sets but it can constrain their uncertainties by adding new data points to the global fits in a previously unexplored region of the $Q^2 - x$ phase space.

4. Conclusions

CMS showed that Z boson production is unmodified by the hot and dense QCD medium produced in heavy ion collisions, and its yield scales with the number of binary nucleon-nucleon collisions. The nuclear modification factor does not exhibit large deviations from unity showing that nuclear effects are small with respect to the uncertainties of the PbPb measurements. The measurement of Z bosons in pPb collisions demonstrate that at first order, the cross section scales with number of binary nucleon-nucleon collisions. The results were compared to NLO theory predictions with and without nuclear modification, that show hints of nuclear effects but more luminosity is needed to distinguish between different nPDF sets. These measurements set constraints for the global fits of nPDFs in a previously unexplored region of phase space.

Acknowledgements

The author wishes to thank to the Hungarian Scientific Research Fund (K 109703) and the Swiss National Science Foundation (152601) for support.

References

[1] Chatrchyan S *et al.* (CMS Collaboration) 2011 *Phys.Rev.Lett.* **106** 212301 (*Preprint* 1102.5435)
[2] Chatrchyan S *et al.* (CMS Collaboration) 2012 *Phys.Lett.* **B715** 66 (*Preprint* 1205.6334)
[3] Chatrchyan S *et al.* (CMS Collaboration) 2014 (*Preprint* 1410.4825)
[4] CMS Collaboration 2014 (*Preprint* CMS-PAS-HIN-14-003) URL http://cds.cern.ch/record/1703012
[5] Lokhtin I P and Snigirev A M 2006 *Eur. Phys. J. C* **45** 211 (*Preprint* hep-ph/0506189)
[6] Sjöstrand T, Mrenna S and Skands P 2006 *JHEP* **05** 026 (*Preprint* hep-ph/0603175)
[7] Alioli S, Nason P, Oleari C and Re E 2008 *JHEP* **0807** 060 (*Preprint* 0805.4802)
[8] Paukkunen H and Salgado C A 2011 *JHEP* **1103** 071 (*Preprint* 1010.5392)
[9] Lai H L, Guzzi M, Huston J, Li Z, Nadolsky P M *et al.* 2010 *Phys.Rev.* **D82** 074024 (*Preprint* 1007.2241)
[10] Eskola K, Paukkunen H and Salgado C 2009 *JHEP* **0904** 065 (*Preprint* 0902.4154)
[11] Gyulassy M and Wang X N 1994 *Comput.Phys.Commun.* **83** 307 (*Preprint* nucl-th/9502021)
[12] Campbell J M, Ellis R K and Williams C 2011 *JHEP* **1107** 018 (*Preprint* 1105.0020)
[13] Martin A, Stirling W, Thorne R and Watt G 2009 *Eur.Phys.J.* **C63** 189 (*Preprint* 0901.0002)
[14] de Florian D, Sassot R, Zurita P and Stratmann M 2012 *Phys.Rev.* **D85** 074028 (*Preprint* 1112.6324)

High-p_{T} probes of p+Pb collisions with ATLAS

Dennis V. Perepelitsa, for the ATLAS Collaboration
Physics Department, Brookhaven National Laboratory, Upton, NY 11973
E-mail: dvp@bnl.gov

Abstract. Measurements of high p_{T} processes in ultrarelativistic proton–nucleus collisions are sensitive to changes in the partonic densities arising from the presence of the nuclear environment. Additionally, such measurements benchmark the so called "cold nuclear matter" effects and provide the reference for understanding the large suppression of high-p_{T} processes observed in nucleus–nucleus collisions. Furthermore, measurements of the centrality dependence of jet production at forward (proton-going) rapidities may shed light on the behavior of the proton wavefunction at large Bjorken-x. The latest ATLAS results for inclusive jets and charged particles in 28.9 nb^{-1} of 5.02 TeV proton–lead collisions at the LHC are presented. The centrality in these collisions is characterized through the sum of the total transverse energy in the lead-going forward calorimeter. The nuclear modification factors $R_{p\mathrm{Pb}}$ and R_{CP} are presented as a function of transverse momentum, rapidity and centrality. The jet R_{CP} in a large rapidity region is found to be modified in a way that depends only on the total jet energy.

1. Introduction

Measurements of high transverse momentum (p_{T}) processes of proton–lead (p+Pb) collisions probe the partonic structure of the nucleus and its possible modification [1], and provide constraints on how initial state effects may influence the observed suppression of these processes in lead–lead collisions. More generally, they test the relationship between hard process rates and the soft particle activity used to classify the geometry of nuclear collisions and, at high-x, may reveal novel features of the proton wave function [2]. In these proceedings, inclusive jet and charged hadron measurements in $\sqrt{s_{\mathrm{NN}}} = 5.02$ TeV p+Pb collisions by the ATLAS detector [3] at the LHC are presented as a function of jet or hadron p_{T} and center of mass rapidity, y^*. The data are described in more detail in Refs. [4, 5].

Modifications to inclusive jet or hadron production rates were investigated through the nuclear modification factor $R_{p\mathrm{Pb}} = (1/N_{\mathrm{evt}})(\mathrm{d}^2N/\mathrm{d}p_{\mathrm{T}}\mathrm{d}y^*)/(T_{\mathrm{pA}}\mathrm{d}^2\sigma/\mathrm{d}p_{\mathrm{T}}\mathrm{d}y^*)$, where the numerator is the per-event jet or hadron yield, T_{pA} is the mean nuclear thickness seen by the proton in events of the given centrality interval and $\mathrm{d}^2\sigma/\mathrm{d}p_{\mathrm{T}}\mathrm{d}y^*$ is the jet or hadron cross-section in pp collisions evaluated at the same $\sqrt{s}$. The modifications were also explored through the central-to-peripheral ratio, R_{CP}, which is the ratio of the $1/T_{\mathrm{pA}}$-scaled yield in a given centrality interval with respect to the yield in the 60–90% interval. The R_{CP} and $R_{p\mathrm{Pb}}$ measure deviations in the yield from the geometric expectation of an incoherent superposition of an equivalent number of nucleon-nucleon collisions (for which $R_{\mathrm{CP}} = 1$ and $R_{p\mathrm{Pb}} = 1$).

2. Data selection and centrality definition

The data used in this work primarily comprise 28.9 nb^{-1} of p+Pb collisions at $\sqrt{s_{\mathrm{NN}}} = 5.02$ TeV, and also 4.0 pb^{-1} and 156 pb^{-1} of pp data at $\sqrt{s} = 2.76$ TeV and 7 TeV, respectively, to

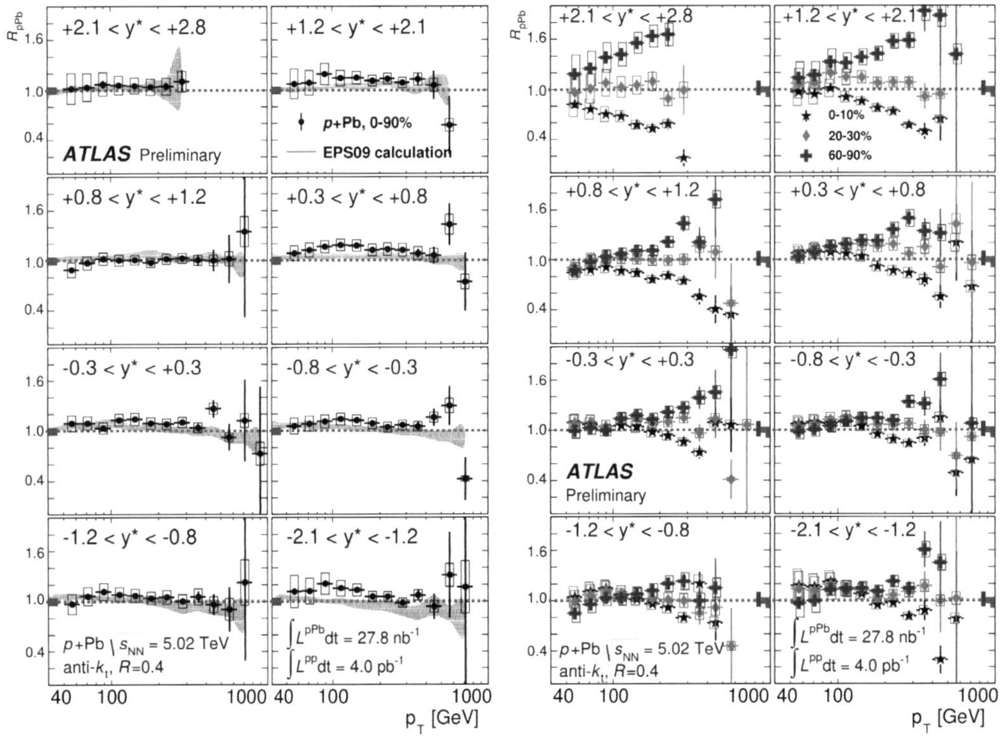

Figure 1. Jet $R_{p\mathrm{Pb}}$ as a function of p_T, with each panel a different selection on the center of mass rapidity y^* for (left) 0–90% events and (right) centrality-selected events. From Ref. [4].

construct the reference jet and hadron spectra at 5.02 TeV. All data sets were recorded using a combination of minimum bias and high level jet triggers. The minimum bias trigger condition was the presence of at least one hit in each side of the minimum bias trigger scintillator (MBTS) detector in p+Pb collisions, and the presence of a track reconstructed by the high-level trigger in pp collisions. Offline, a reconstructed vertex was required for all datasets. Additionally, p+Pb events were required to pass a number of additional cuts designed to select a set of events for which a centrality analysis was applicable and to which geometric parameters could be assigned. The centrality of p+Pb events was characterized through the sum of the total transverse energy, $\Sigma E_\mathrm{T}^\mathrm{Pb}$, in the Pb-going forward calorimeter $-4.9 < \eta < -3.1$ (in this work $y^*, \eta > 0$ denote the proton-going direction). A Monte Carlo (MC) Glauber-based analysis with nucleon–nucleon inelastic cross-section $\sigma_\mathrm{NN} = 70$ mb along with a modified wounded nucleon (WN) model was used to estimate T_pA in each centrality interval. More details are given in Ref. [6].

3. Jet and charged particle reconstruction and corrections

The jet reconstruction procedure closely follows that used in Pb+Pb collisions, which utilizes an iterative estimation and subtraction of the underlying event (UE) pedestal. Jets were reconstructed by applying the anti-k_t algorithm with $R = 0.4$ to calorimeter cells collected into $\Delta\eta \times \Delta\phi = 0.1 \times 0.1$ towers, and their kinematics updated to account for the UE subtraction procedure. The final set of jets was first corrected for any inadvertent inclusion of jets into the UE estimate, then corrected to the full hadronic scale through a calibration derived from MC

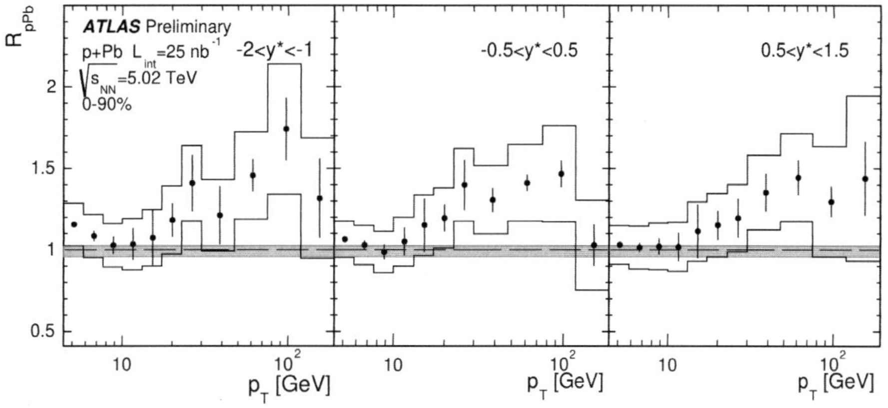

Figure 2. Charged particle $R_{p\mathrm{Pb}}$ vs. p_T, shown at three different rapidities. From Ref. [5].

simulation and finally received a small *in situ* energy scale correction. To avoid any corrections for the trigger efficiency, each p_T interval was populated exclusively by offline jets matched to the highest-luminosity trigger which was $> 99\%$ efficient in that interval. The performance of the jet reconstruction was evaluated with MC simulations of PYTHIA jet events overlaid onto minimum bias p+Pb events. The reconstruction efficiency was determined to be $> 99\%$ for $p_\mathrm{T} > 25$ GeV jets. The jet energy scale closure was found to be better than 2% (1%) for all $p_\mathrm{T} > 25$ GeV (> 100 GeV) jets. The detector-level jet spectra were corrected for p_T interval migration effects resulting from defects in the energy scale and finite energy resolution through a bin-by-bin correction procedure.

Charged tracks were reconstructed in the inner detector (ID) following an algorithm optimized for measurements in pp collisions, and selected for use in the analysis after the requirements of p_T-dependent quality cuts, such as a minimum number of hits in the pixel and silicon micro-strip components of the ID system. To further reduce the contribution of poorly reconstructed tracks or tracks from secondary particles, tracks were required to leave eight hits in the transition radiation tracker, and restricted to $|\eta| < 2.0$. To extend the p_T-reach of the measurement, tracks associated with jets matched to high p_T threshold jet triggers were used. The full charged particle spectrum was constructed by combining charged particles measured in minimum bias events with those associated with a number of jet triggers, with each trigger used exclusively in the p_T region where it was $> 99\%$ efficient. The raw spectrum was corrected for the presence of tracks associated with secondary particles, finite p_T resolution and reconstruction efficiency, using the same MC samples used above to determine the jet performance.

4. Nuclear modification factors for high-p_T jets and charged particles

The left panel of Figure 1 shows the $R_{p\mathrm{Pb}}$ for 0–90% events and a pQCD-based calculation incorporating the EPS09 nPDF set. The pp reference was derived from the measured jet cross-section at 2.76 TeV, interpolated using x_T scaling (derived from ATLAS measurements of jet spectra at 2.76 and 7 TeV) to 5.02 TeV. At all rapidities, the $R_{p\mathrm{Pb}}$ is consistent with a slight p_T-dependent enhancement above unity, in line with the expectations from EPS09. However, the centrality-dependent $R_{p\mathrm{Pb}}$, shown in the right panel of Figure 1, features strong deviations in the jet yield from the geometric expectation. The jet yield is suppressed in central collisions and enhanced in peripheral collisions. Furthermore, the modifications are systematically larger

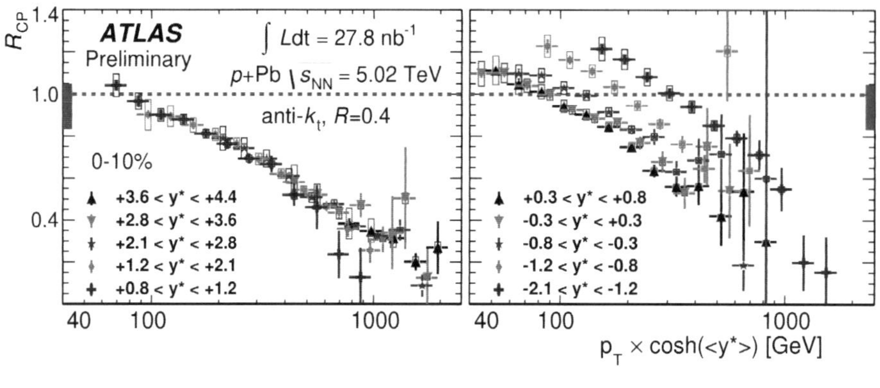

Figure 3. Jet R_{CP} as a function of $p_{\mathrm{T}}\cosh(y^*)$, plotted for different y^* values. From Ref. [4].

at higher p_{T} and at more forward rapidities.

The charged particle $R_{p\mathrm{Pb}}$ is shown in Figure 2. The pp reference was derived from ATLAS charged particle cross-sections in pp collisions at 2.76 TeV and 7 TeV, interpolated to 5.02 TeV using a linear dependence in $\sqrt{s}$. The variation in the result when the interpolation was performed as a function of $\log(s)$ or using x_{T} scaling were included in the systematic uncertainties. The $R_{p\mathrm{Pb}}$ is consistent with unity for $p_{\mathrm{T}} < 20$ GeV, but systematically increases at high-p_{T}, reaching a 30–40% excess at $p_{\mathrm{T}} \approx 60$ GeV in all rapidity intervals studied.

Finally, to explore the kinematic dependence of the jet modifications, the data were presented as a function of $p_{\mathrm{T}} \times \cosh(y^*) \approx p$, where p is the approximate total jet energy. Figure 3 shows the R_{CP} at all rapidities studied as a function of p. When plotted this way, the R_{CP} values for rapidities in the range $0.8 < y^* < 4.4$ follow a single trend that is a function of p alone.

5. Conclusion

In these proceedings, measurements of the nuclear modification factor for jets and charged hadrons at high-p_{T} by ATLAS are presented. In minimum bias p+Pb collisions, the jet rate is in line with expectations while the particle rate has a systematic enhancement. Together, these results are challenging to explain within standard nPDF-based frameworks and the assumption of minimal jet modification. Additional information, such as from measurements of other hard probe rates or of the fragmentation function in p+Pb collisions may be needed for a more complete understanding of these effects. Furthermore, the centrality-dependent jet rates feature strong deviations from expectations. At forward rapidities, the modifications are consistent with a function of the approximate total jet energy p alone. However, forward jet production at fixed p arises dominantly from parton–parton configurations in which the partonic x in the proton equals $2p/\sqrt{s}$. Therefore, the modifications are controlled by the proton x only and indicate the influence of an initial state proton, rather than nuclear, effect [7].

References

[1] C. Salgado, et al., J. Phys. G39 (2012) 015010; I. Helenius, et al., JHEP 1207 (2012) 073
[2] C. E. Coleman-Smith, et al., Phys. Rev. D89 (2014) 025019; M. Alvioli, et al., Phys. Rev. C 90 (2014) 034914
[3] ATLAS Collaboration, JINST 3 (2008) S08003
[4] ATLAS Collaboration, ATLAS-CONF-2014-024, http://cds.cern.ch/record/1702986
[5] ATLAS Collaboration, ATLAS-CONF-2014-029, http://cds.cern.ch/record/1704978
[6] ATLAS Collaboration, ATLAS-CONF-2013-096, http://cds.cern.ch/record/1599773
[7] A. Bzdak, et al., hep-ph/1408.3156; M. Alvioli, et al., hep-ph/1409.7381

Production of strange particles in charged jets in Pb–Pb and p–Pb collisions measured with ALICE

Vít Kučera[1,2] **(for the ALICE Collaboration)**
[1] Nuclear Physics Institute ASCR, Na Truhlářce 39/64, 180 86 PRAHA 8, Czech Republic
[2] IPHC, 23 rue du Loess - BP 28, F-67037 STRASBOURG CEDEX 2, France
E-mail: vit.kucera@cern.ch

Abstract. Measurements of spectra of identified particles produced in jets in heavy-ion collisions can provide further important insights into the interplay of various hadronisation processes which participate in the particle production in the hot and dense medium. In this contribution, we present the measurements of the p_{T} spectra of Λ baryons and $\mathrm{K}^0_{\mathrm{S}}$ mesons produced in association with charged jets in Pb–Pb collisions at $\sqrt{s_{\mathrm{NN}}} = 2.76\,\mathrm{TeV}$ and in p–Pb collisions at $\sqrt{s_{\mathrm{NN}}} = 5.02\,\mathrm{TeV}$. The results are obtained with the ALICE detector at the LHC, exploiting the excellent particle identification capabilities of this experiment. Baryon-to-meson ratios of the spectra of strange particles associated with jets are studied in central collisions and are compared to the ratios obtained for inclusive particles and for particles coming from the underlying event.

1. Introduction

An enhancement of the baryon-to-meson ratio is observed for inclusive spectra of light-flavour particles produced at intermediate transverse momenta (p_{T}) ($2\ \mathrm{GeV}/c < p_{\mathrm{T}} < 6\ \mathrm{GeV}/c$) in heavy-ion collisions when compared to the ratio in proton–proton (pp) collisions. It was first observed at the Relativistic Heavy Ion Collider [1, 2] and later at the Large Hadron Collider (LHC) as well [3]. The ratio of the p_{T} spectrum of Λ baryons to the spectrum of $\mathrm{K}^0_{\mathrm{S}}$ mesons measured in lead–lead (Pb–Pb) collisions with the ALICE experiment is increasing dramatically with centrality and reaching values three times greater than the ratio in pp collisions. This phenomenon also manifests in proton–lead (p–Pb) collisions and depends on particle multiplicity [4]. This enhancement is still not fully understood.

Parton fragmentation and hadronisation is a well-known concept of hadron production in perturbative QCD describing how a high-p_{T} parton, produced in a hard scattering process, fragments in vacuum into a collimated spray of hadrons called a jet. Fragmentation in vacuum cannot explain the observed enhancement. However, this process might be modified by interaction of the fragmenting partons with constituents of the hot and dense strongly interacting matter created in ultra-relativistic heavy-ion collisions.

Another frequently discussed scenario proposed to explain the shape of the observed baryon-to-meson ratio is parton recombination which assumes that multiple partons (close to each other in phase space) cluster together to form a hadron. In such case, a quark clusters with an anti-quark to form a meson and a baryon is produced when three quarks cluster which gives rise to a harder p_{T} spectrum of baryons with respect to that of mesons.

We aim to understand the origin of the $\Lambda/\mathrm{K}^0_\mathrm{S}$ enhancement by separating hadrons produced in jets from hadrons produced in the thermalised bulk. Results of this analysis will allow us to tell whether the baryon-to-meson ratio is enhanced due to the bulk effects in the plasma, like for example parton recombination or radial flow, or whether it is a phenomenon caused (also) by a modification of the jet fragmentation in the medium.

2. Analysis

The analysis is performed on data recorded in the ALICE detector at the LHC during the Pb–Pb and p–Pb runs with collisions at the centre-of-mass energies of $\sqrt{s_\mathrm{NN}} = 2.76$ TeV and 5.02 TeV respectively. Tracking of charged particles in the central barrel is provided by the Inner Tracking System and the Time Projection Chamber, both placed in a magnetic field of 0.5 T. Centrality in Pb–Pb collisions and multiplicity classes in case of p–Pb collisions are estimated from the multiplicity of charged particles measured in the V0 detectors at forward rapidities [5].

The analysis of neutral strange particles in charged jets is based on two main parts: reconstruction of charged jets and reconstruction of neutral strange particles.

The charged jets are reconstructed using tracks of charged primary particles having $p_\mathrm{T}^\mathrm{track} > 150\,\mathrm{MeV}/c$ and uniform distribution in azimuth and in pseudorapidity of range $|\eta_\mathrm{track}| < 0.9$. These tracks are used as input for the anti-k_t algorithm (implemented in the FastJet package [6]). Values used for the resolution parameter of the jet algorithm are $R = 0.2, 0.3$ in case of the Pb–Pb collisions whereas for the p–Pb collisions, it is feasible to use also $R = 0.4$ since the background from soft processes is not as large as in the Pb–Pb case. The definition of charged jets implies that neutral primary particles are not included in the jet constituents and do not contribute to the reconstructed jet momenta. Clusters reconstructed by the k_t algorithm are used to estimate the average level of background in each event which is then subtracted from the momenta of reconstructed jets [7]. In order to further reduce the contribution of fake jets, cuts are applied on jet p_T ($p_\mathrm{T}^\mathrm{jet,ch}$), p_T of the leading jet constituent and jet area ($A_\mathrm{jet,ch}$).

Neutral strange particles (V^0) are measured by reconstructing their weak decays into two charged daughter particles, namely $\mathrm{K}^0_\mathrm{S} \to \pi^+ + \pi^-$ and $\Lambda \to \mathrm{p} + \pi^-$. Cuts on parameters of the decay topology are used to suppress combinatorial background in the sample of V^0 candidates and the signal yield is extracted from the resulting invariant-mass distribution.

If a V^0 candidate is accepted, its distances from the axes of selected jets in the event are calculated to check whether it is in a jet cone. A jet is selected only if its cone is fully contained within the acceptance region of V^0 particles: $|\eta_\mathrm{jet,ch}| < |\eta_{\mathrm{V}^0}|^\mathrm{max} - R$. The candidate is considered to be inside a jet cone if the distance, calculated in azimuth and pseudorapidity, is less than the resolution parameter of the jet finder:

$$\sqrt{(\varphi_{\mathrm{V}^0} - \varphi_\mathrm{jet,ch})^2 + (\eta_{\mathrm{V}^0} - \eta_\mathrm{jet,ch})^2} < R. \tag{1}$$

Spectra of associated particles are studied for two intervals of jet momentum: $p_\mathrm{T}^\mathrm{jet,ch} > 10\,\mathrm{GeV}/c$ and $p_\mathrm{T}^\mathrm{jet,ch} > 20\,\mathrm{GeV}/c$. Raw spectra of V^0 particles associated with cones of charged jets in central Pb–Pb collisions, normalised per unit area in the $(\varphi \times \eta)$-space, are presented in Fig. 1. Area used for normalisation is the area of the region of associated particles.

Particles in jet cones do not originate only from the jet fragmentation but also from underlying event (UE). In order to estimate this contribution, the V^0 particles are also associated with regions where only underlying event is expected to be present. In case of Pb–Pb collisions, these regions are defined by five different methods (e.g. perpendicular or randomly placed cones).

Prior to subtraction of the contribution from UE, we apply the efficiency correction of the V^0 reconstruction, using the inclusive simulated particles. The pseudorapidity distribution of the simulated jets in the Monte Carlo sample does not correspond to the distribution of real selected

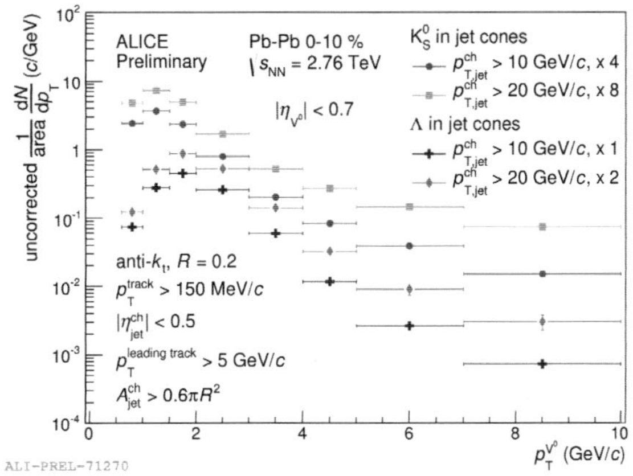

Figure 1. Raw spectra of V^0 particles associated with charged jets with $R = 0.2$ above a given $p_T^{jet,ch}$ threshold in central Pb–Pb collisions at $\sqrt{s_{NN}} = 2.76$ TeV.

jets. Since the efficiency correction is applied only as a function of $p_T^{V^0}$, the η distributions of simulated V^0 particles associated with jet cones and with the UE regions have to be reweighted in order to match the shapes observed in real data.

After the efficiency correction, spectra of V^0 particles in jets are obtained by subtracting the UE contribution from the spectra of particles in jet cones.

Strange particles in jets can be produced also by decays of jet constituents, namely in this analysis, we aim to account for the contribution of the Λ production from decays of $\Xi^{0,-}$ baryons. The corresponding feed-down fraction can only be roughly estimated since spectra of Ξ in jets have not been measured yet. We consider two scenarios: feed-down fraction calculated for spectra of inclusive Λ baryons and fraction obtained from Λ baryons in jets simulated in pp at $\sqrt{s} = 2.76$ TeV or 5.02 TeV with PYTHIA 8 tune 4C [8, 9]. These scenarios are taken as an upper and a lower limit of the systematic uncertainty on the feed-down contribution.

3. Results

Figure 2 shows the p_T dependence of the ratio of the corrected yields of Λ baryons to that of K^0_S mesons in charged jets in high-multiplicity p–Pb collisions for two $p_T^{jet,ch}$ thresholds, compared with the ratios for measured inclusive particles and for particles simulated in PYTHIA 8. The measured baryon-to-meson ratio in jets in p–Pb collisions is below the measured ratio of inclusive particles in p–Pb collisions, below the measured ratio of inclusive particles in pp collisions [10] (not shown) and is also below the ratio of inclusive particles obtained with PYTHIA simulation. Moreover, the measured ratio exhibits a surprising similarity with ratios of particles in jets simulated in PYTHIA and does not evince any significant dependence on R or $p_T^{jet,ch}$. The results indicate no visible modification of strangeness production in charged jets in p–Pb and the enhancement of the baryon-to-meson ratio thus seems to be due to bulk effects.

4. Summary

We presented the motivation for studying spectra of strange particles in jets in heavy-ion collisions, techniques used in the analysis of the Λ/K^0_S ratio in charged jets in Pb–Pb and p–Pb collisions were described and we showed the first results of such measurement, performed by ALICE in p–Pb collisions. The measured Λ/K^0_S ratio indicates that the enhancement originates

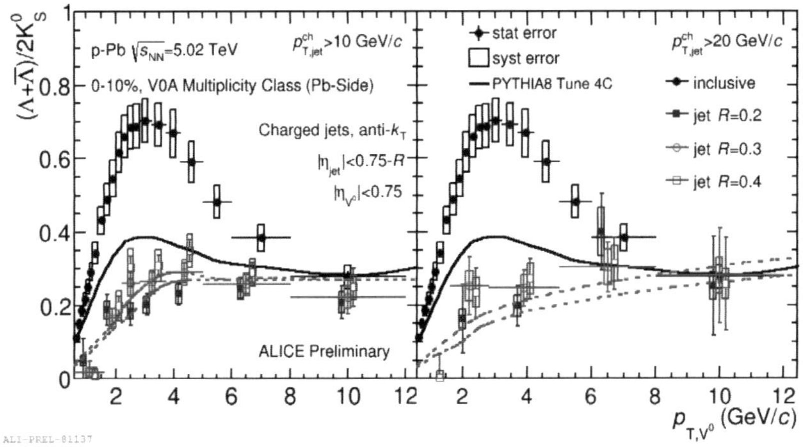

Figure 2. Λ/K^0_S ratio in charged jets with $R = 0.2, 0.3, 0.4$ in p–Pb collisions at $\sqrt{s_{NN}} = 5.02$ TeV for $p_T^{jet,ch} > 10$ GeV/c (left) and $p_T^{jet,ch} > 20$ GeV/c (right), compared with the inclusive ratio and simulation results. Black solid line indicates the inclusive ratio from PYTHIA 8. Red dashed lines denote the spread of ratios in PYTHIA jets for all used values of R.

from bulk effects. The determination of final corrections and uncertainties of the results in case of Pb–Pb collisions is under ongoing investigation.

Acknowledgments

The work has been supported by the grant LG 13031 of the Ministry of Education of the Czech Republic (MEYS) and by the MEYS grant CZ.1.07/2.3.00/20.0207 of the European Social Fund in the Czech Republic: "Education for Competitiveness Operational Programme".

References

[1] Abelev B *et al.* (STAR Collaboration) 2007 *Phys. Lett.* B **655** 104–113 (*Preprint* nucl-ex/0703040)
[2] Lamont M (STAR Collaboration) 2004 *J. Phys.* G **30** S963–S968 (*Preprint* nucl-ex/0403059)
[3] Abelev B B *et al.* (ALICE Collaboration) 2013 *Phys. Rev. Lett.* **111** 222301 (*Preprint* 1307.5530)
[4] Abelev B B *et al.* (ALICE Collaboration) 2014 *Phys. Lett.* B **728** 25–38 (*Preprint* 1307.6796)
[5] Carminati F *et al.* (ALICE Collaboration) 2004 *J. Phys.* G **30** 1517–1763
[6] Cacciari M, Salam G P and Soyez G 2012 *Eur. Phys. J.* C **72** 1896 (*Preprint* 1111.6097)
[7] Abelev B *et al.* (ALICE Collaboration) 2012 *JHEP* **1203** 053 (*Preprint* 1201.2423)
[8] Sjöstrand T, Mrenna S and Skands P Z 2006 *JHEP* **0605** 026 (*Preprint* hep-ph/0603175)
[9] Sjöstrand T, Mrenna S and Skands P Z 2008 *Comput. Phys. Commun.* **178** 852–867 (*Preprint* 0710.3820)
[10] Aamodt K *et al.* (ALICE Collaboration) 2011 *Eur. Phys. J.* C **71** 1594 (*Preprint* 1012.3257)

Multiplicity Dependence of (Anti-)Deuteron Production in p–Pb Collisions at $\sqrt{s_{\rm NN}} = 5.02$ TeV Measured with ALICE

J. Anielski for the ALICE Collaboration

Institut für Kernphysik, Universität Münster, Wilhelm-Klemm-Str. 9, 48149 Münster, Germany

E-mail: `j.anielski@wwu.de`

Abstract. The measurement of the (anti-)deuteron production as a function of transverse momentum and event multiplicity in p–Pb collisions at $\sqrt{s_{\rm NN}} = 5.02$ TeV is presented here. (Anti-)deuterons are identified via their energy loss $\mathrm{d}E/\mathrm{d}x$ in the time projection chamber and via their time-of-flight. Their production will be discussed within the context of thermal and coalescence models.

1. Measurement of (Anti-)Deuterons

The (anti-)deuteron production has been measured with the minimum-bias trigger data recorded by ALICE [1] during the 2013 p–Pb run at $\sqrt{s_{\rm NN}} = 5.02$ TeV at the LHC. The event selection is done as in [2] and the sample is divided in five multiplicity classes according to the multiplicity of the VZERO-A detector, which is located in the Pb-going direction at $2.8 < \eta < 5.1$.

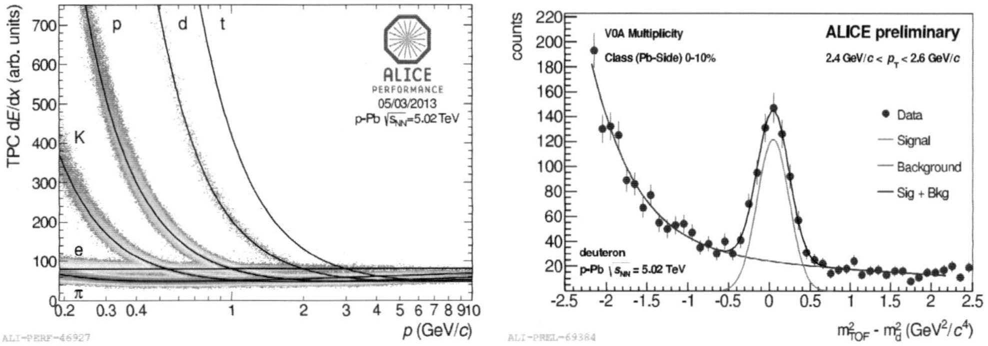

Figure 1: Deuteron identification with the TPC and TOF. On the left the energy loss per path-length as a function of momentum is shown. On the right an example of the fit to the squared TOF mass to reduce the remaining background is shown.

Fig. 1 on the left shows the specific energy loss of particles in the time projection chamber (TPC) [3] for p–Pb data. At low momenta (< 1 GeV/c) a clean identification of (anti-)deuterons is achieved by a symmetric 3σ cut around the expected energy loss. At higher momenta the particle bands cross and the mass measured with the time-of-flight detector (TOF) [4] is used in addition. Here, the cut of 3σ on the TPC $\mathrm{d}E/\mathrm{d}x$ signal reduces the amount of the TPC-TOF mismatched tracks and the remaining background is removed with a fit to the squared mass. This is shown for deuterons with 2.4 GeV/c $< p_{\mathrm{T}} <$ 2.6 GeV/c in Fig. 1 on the right.

The tracking efficiency and the TOF matching efficiency for deuterons and anti-deuterons is shown in Fig. 2. The efficiencies also account for particles that are absorbed by the detector material, which is negligible for deuterons, but sizeable for anti-deuterons. In addition, only very little data is available for the hadronic cross-section of anti-deuterons and different transport models give different results. For this work the efficiencies have been calculated with Geant 4 [5] and have been compared to results from Geant 3 [6] to estimate the systematic uncertainty. It amounts to about 15%.

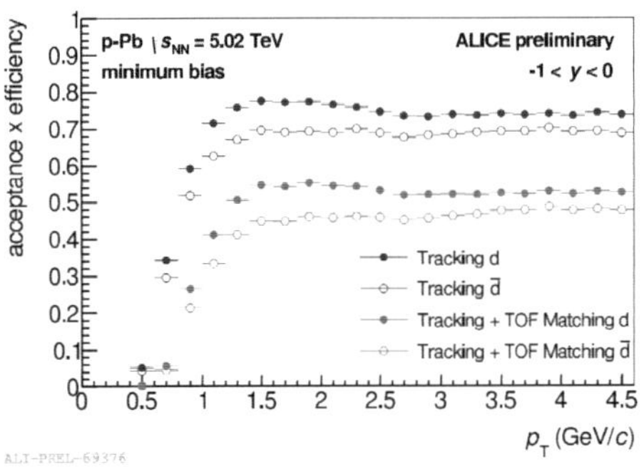

Figure 2: Tracking and TOF matching efficiencies for deuterons and anti-deuterons.

2. Results

The (anti-)deuteron transverse momentum distributions for different charged-particle multiplicities are shown in Fig. 3. Also shown are individual Blast-Wave fits (black lines) used to extrapolate to low and high p_{T} and to calculate the integrated yields $\mathrm{d}N/\mathrm{d}y$. A hardening of the spectra from low to high multiplicity is visible, which is also observed for light-flavor hadrons [7]. In heavy-ion collisions this hardening is commonly attributed to radial flow [8].

A central result of the deuteron measurement is the deuteron to proton ratio. It is shown in Fig. 4 for p–Pb and Pb–Pb collisions as a function of multiplicity and for minimum bias pp collisions. At low multiplicity the d/p ratio in p–Pb is consistent with the pp data. The ratio rises with multiplicity reaching the value of peripheral Pb–Pb collisions, where it saturates. The smooth transition between the different systems suggests that the d/p ratio is goverend by the event multiplicity. Please note, that the uncertainties for p–Pb are largely correlated across multiplicity bins.

The deuteron production in ultra-relativistic particle collisions is commonly discussed within coalescence [9, 10] and thermal models [11, 12, 13]. Thermal models describe the particle abundance with the chemical freeze-out temperature T_{ch}, the baryo-chemical potential μ_B and the source volume V. The chemical freeze-out marks the point in the evolution of particle collisions, where all inelastic processes cease and the chemical composition is fixed. They are very successful in describing not only the hadron production, but also the light nuclei production of heavy-ion collisions [14, 15, 16]. This is unexpected, because in thermal models the nuclei yield is fixed at $T_{\mathrm{ch}} \approx 150$ MeV and deuterons can hardly remain intact in this hot environment,

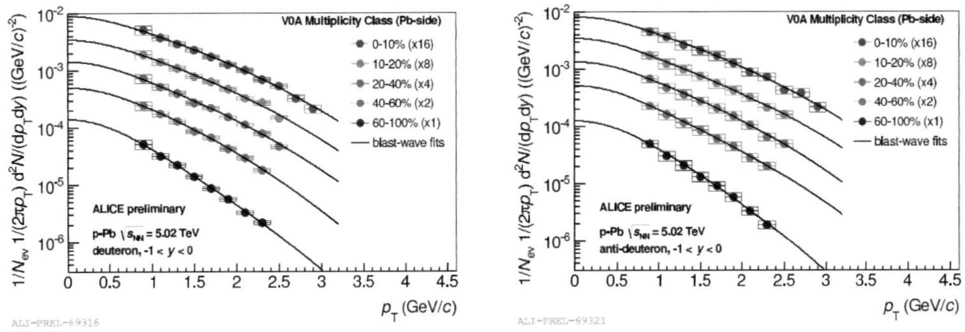

Figure 3: The deuteron (left) and anti-deuteron (right) transverse momentum distributions as a function of multiplicity.

because of their low binding energy $E_B = 2.2$ MeV [17]. On the other hand, the quality of current thermal model fits to high multiplicity p–Pb collisions is poor [18].

In coalescence models deuterons (and other light nuclei) are produced by protons and neutrons, that are close in phase space. In the most naive way, this would lead to an increased deuteron production for higher nucleon multiplicities. However, the d/p ratio for Pb–Pb collisions is constant with increasing centrality, although the nucleon multiplicity increases. Models, which take the source volume into account, i.e. are sensitive to the nucleon density, are therefore clearly favored. Here, the increasing nucleon multiplicity is cancelled by the increasing source volume, leading to a constant nucleon density. This is also consistent with the rising d/p ratio with multiplicity in p–Pb, where the effect of the increasing nucleon multiplicity probably dominates over the effect of the increasing source volume.

Within coalescence models the coalescence parameter for deuterons B_2 is defined as

$$E_d \frac{\mathrm{d}^3 N_\mathrm{d}}{(\mathrm{d}p_\mathrm{d})^3} = B_2 \left(E_\mathrm{p} \frac{\mathrm{d}^3 N_\mathrm{p}}{(\mathrm{d}p_\mathrm{p})^3} \right)^2 , \tag{1}$$

where $p_\mathrm{d} = 2p_\mathrm{p}$. Fig. 5 shows the resulting B_2 as a function of p_T and multiplicity for p–Pb and Pb–Pb collisions. Within the systematic uncertainties it is flat over the measured p_T-range in p–Pb collisions and peripheral Pb–Pb collisions, which is in agreement with the

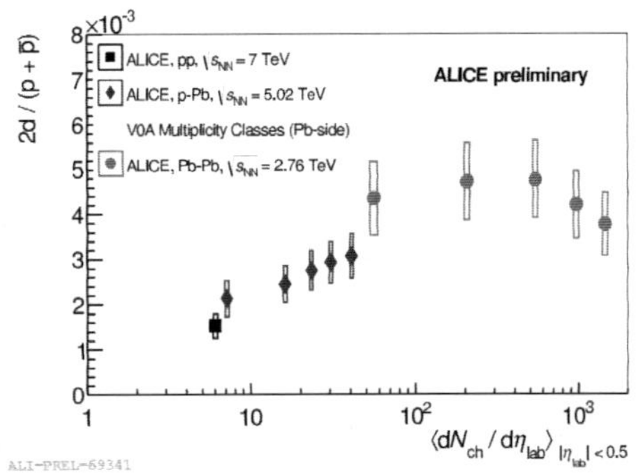

Figure 4: Deuteron over proton ratio as a function of charged-particle multiplicity at mid-rapidity ($|\eta| < 0.5$) for pp, p–Pb and Pb–Pb collisions at the LHC.

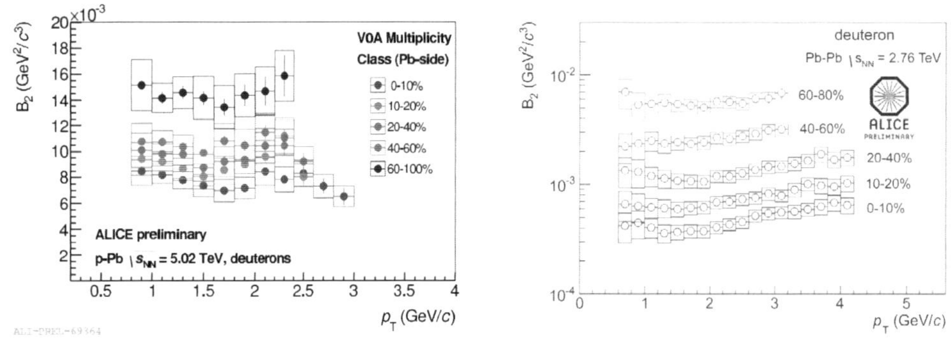

Figure 5: Coalescence parameter B_2 as a function of p_T and multiplicity for p–Pb and Pb–Pb. Please note the logarithmic scale on the right plot.

coalescence picture. In addition, a decreasing B_2 for increasing multiplicity is observed, which can be attributed to the increasing source volume. In fact, B_2 could be governed by the same homogeneity volume as HBT [19]. For $p_T > 2.0$ GeV/c and centralities > 40% of Pb–Pb collisions B_2 increases with the transverse momentum, which could be due to the collectivity of the emerging particles.

3. Summary

The transverse momentum distributions of (anti-)deuterons as a function of multiplicity in p–Pb collisions at $\sqrt{s_{NN}} = 5.02$ TeV have been measured with ALICE. The deuteron over proton ratio rises with multiplicity and is consistent with pp and Pb–Pb collisions at low and high multiplicities, respectively. This favors coalescence models, that are based on the nucleon density. The coalescence parameter B_2 is flat over the measured p_T-range, which is in agreement with the expectation. Thermal models successfully describe the particle production including the light nuclei production in Pb–Pb collisions, but current fits to high multiplicity p–Pb collisions [18] are less successful.

References

[1] Aamodt K *et al.* (ALICE) 2008 *JINST* **3** S08002
[2] Abelev B *et al.* (ALICE) 2013 *Phys. Rev. Lett.* **110** 032301
[3] Alme J, Andres Y, Appelshauser H, Bablok S, Bialas N *et al.* 2010 *Nucl. Instrum. Meth.* **A622** 316–367
[4] Akindinov A, Alici A, Agostinelli A, Antonioli P, Arcelli S *et al.* 2013 *Eur. Phys. J. Plus* **128** 44
[5] Agostinelli S *et al.* (GEANT) 2003 *Nucl. Instrum. Meth.* **A506** 250–303
[6] Brun R, Carminati F and Giani S 1994 CERN-W5013
[7] Abelev B *et al.* (ALICE) 2014 *Phys. Lett.* **B728** 25 – 38
[8] Abelev B *et al.* (ALICE) 2013 *Phys. Rev.* **C88** 044910
[9] Butler S T and Pearson C A 1963 *Phys. Rev.* **129**(2) 836–842
[10] Kapusta J I 1980 *Phys. Rev.* **C21**(4) 1301–1310
[11] Wheaton S, Cleymans J and Hauer M 2009 *Computer Phys. Comm.* **180** 84 – 106
[12] Andronic A, Braun-Munzinger P and Stachel J 2009 *Phys. Lett.* **B673** 142 – 145
[13] Petran M, Letessier J, Rafelski J and Torrieri G 2014 *Comput. Phys. Comm.* **185** 2056–2079
[14] Andronic A, Braun-Munzinger P, Stachel J and Stocker H 2011 *Phys. Lett.* **B697** 203–207
[15] Cleymans J, Kabana S, Kraus I, Oeschler H, Redlich K *et al.* 2011 *Phys. Rev.* **C84** 054916
[16] Andronic A, Braun-Munzinger P, Redlich K and Stachel J 2011 *J. Phys.* **G38** 124081
[17] Beringer J *et al.* (PDG) 2012 *Phys. Rev.* **D86** 010001
[18] Floris M 2014 (*Preprint* 1408.6403)
[19] Scheibl R and Heinz U W 1999 *Phys. Rev.* **C59** 1585–1602

Hot Quarks 2014 IOP Publishing
Journal of Physics: Conference Series **612** (2015) 012015 doi:10.1088/1742-6596/612/1/012015

Chromo-Weibel instabilities in classical Yang-Mills evolution

Maximilian Attems

Departament d'Estructura i Constituents de la Matèria and Institut de Ciències del Cosmos (ICCUB), Universitat de Barcelona, Martí i Franquès 1, 08028 Barcelona, Spain
Frankfurt Institute for Advanced Studies - Johann Wolfgang Goethe-Universität, Ruth-Moufang-Str. 1, 60438 Frankfurt am Main, Germany
Yukawa Institute for Theoretical Physics, Kyoto University
Kitashirakawa Oiwakecho, Sakyo-ku, Kyoto 606-8502, Japan

E-mail: attems@icc.ub.edu

Abstract. The standard model of relativistic heavy-ion collisions describes the hadronic collision in four distinct stages: the initial Color glass condensate, the glasma, the quark gluon plasma and finally the hot hadron gas. Exploring the transition from at first Lorentz contracted sheets of hadrons, the color glass condensate, to thermal QCD matter is the core of this work. The study of anisotropically expanding plasmas shows that non-Abelian plasmas isotropize fast enough thanks to the chromo-Weibel instability. In recent work we find filamentation instabilities in SU(3) Yang-Mills plasma. Real time simulations are a highly promising approach to shed light on the fast thermalization of non-equilibrium matter. This involves to further develop both the theoretical foundations of the description of initial-state fluctuations and the computational framework for the turbulent dynamics of nuclear matter at extreme energy densities.

1. Introduction

A considerable amount of work on the study of the early soft fields [1, 2, 3, 4, 5, 6, 7, 8, 9, 10] that are generated by two ultra-relativistic colliding color sheets has been done from a weakly coupled field-theory perspective in order to better understand the early fast thermalization. In the context of collective excitations Chromo-Weibel instabilities [11, 12, 13, 14, 15, 16, 17, 18, 19, 20] are a subject of intense research.

2. Expanding Vlasov-Yang-Mills

The dynamics of the color glass condensate, the initial effective field theory description, gives rise to very interesting dynamics. In particular there are strong chromo-electric and chromo-magnetic fields that shape longitudinal flux tubes. As a consequence, one has to deal with a very anisotropic momentum distribution in the plasma that leads to the chromo-Weibel instability also called filamentation instability. The instability causes soft modes, the expected chromo-magnetic and chromo-electric quantum fluctuations, to exponentially grow and is characterized by a final turbulent cascade.

Figure 1. Visualization of a single space-time development of color correlations in a non-Abelian plasma in Bjorken expansion.

Fig. 1 shows a single run real-time evolution depicting the Weibel instability in the forward lightcone. See [4] for details on the Vlasov-Yang-Mills setup. The SU(2) chromo-color is mapped to the RGB color space. Of course on average over several runs one would only see white noise, but the single run nicely depicts the filaments that can exponentially grow from small initial random fluctuations. If one looks carefully one can also find depictions of the three vertex splitting (which is the most efficient process in energy transport from small to higher modes). Note that the most unstable mode is responsible for the creation of the filaments. This mode with the highest growth rate has almost the same wavelength for different anisotropies. In the expanding case the anisotropy of this free streaming setup grows with proper time.

3. SU(3) Yang-Mills box

The transverse/longitudinal electric and magnetic components of the field energy density are given by

$$\begin{aligned} \mathcal{E} = \mathcal{E}_T + \mathcal{E}_L &= \mathcal{H}_T^B + \mathcal{H}_T^E + \mathcal{H}_L^B + \mathcal{H}_L^E \\ &= \operatorname{tr}\left[\left(F_{zx}^2 + F_{zy}^2\right) + \left(\Pi_x^2 + \Pi_y^2\right) + F_{xy}^2 + \Pi_z^2\right], \end{aligned} \tag{1}$$

where the gauge-covariant field strength tensor is defined as $F_{\alpha\beta} = \partial_\alpha A_\beta - \partial_\beta A_\alpha - ig[A_\alpha, A_\beta]$ and the canonical conjugate field momenta as $\Pi_\alpha(x) = \partial_0 A_\alpha(x)$.

Fig. 2 shows the local transversal energy densities averaged over 30 runs in a SU(3) Yang-Mills evolution [21]. It was found that the chromo-Weibel instability is at first stronger in chromo-magnetic fields. Hence we show its behavior in the xz plane. In Fig. 2 the perpendicular axis is in the longitudinal direction (along the beam axes) and the horizontal axis is in the transversal direction. Each box is for a given time step. The chosen runs have 120 sites in the longitudinal direction and 40 in the transverse. The saturation scale $Q_s = g^2\mu$ is set to $Q_s = 2GeV$ for typical values for ultrarelativistic collisions at the Large Hadron Collider (hence each upper time step value is to be divided by 100 to be expressed in fm/c).

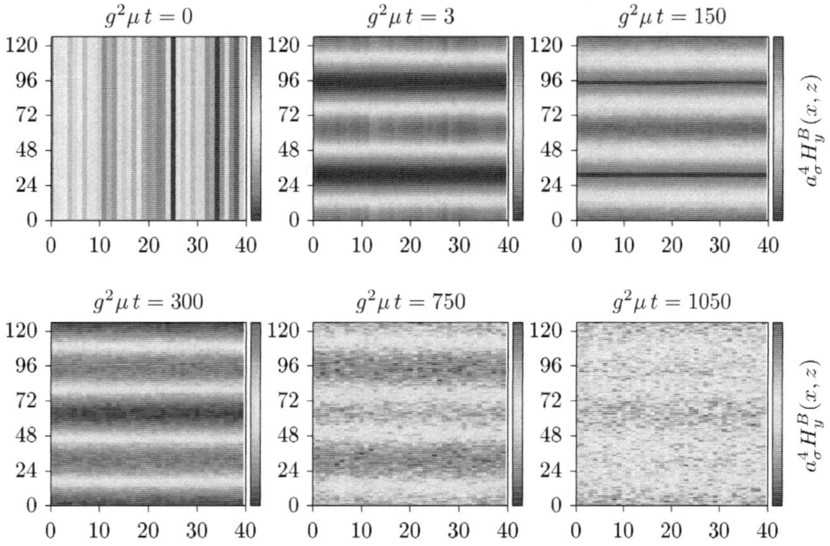

Figure 2. Visualization of the local transversal chromo-magnetic energy densities in a fixed box SU(3) Yang-Mills evolution in the yz plane for several timesteps.

At startup of the evolution there are only small random fluctuations in the chromo-magnetic fields that get rapidly ordered in the next timestep perpendicular to the beam direction. As the initial condition initializes the chromo-electric fields with the strong leading terms the Weibel instabilities are not immediately visible in its local transversal energy component, but they turn out equally strong after some evolution. The time it takes for this to happen of course depends on the initial amplitude of the initialized fluctuations.

As expected the check plots in Fig. 3 show that there are no filamentation instabilities happening in the transversal plane for both the local longitudinal chromo-electric and chromo-magnetic energy components. The chosen timesteps (startup plus two additional) for this test case do not matter, as any of them will show a random value outcome. At any timestep of the evolution the Gauss law constraint is respected, and the fields are Coulomb gauge fixed.

4. Conclusions

SU(3) Yang-Mills evolution with anisotropic initial conditions exhibits filamentation instabilities in the transverse components of the local energy densities. We found evidence for the emergence of the chromo-Weibel instability that isotropizes the soft field pressures. We are planning to improve our numerical results by adding fermionic degrees of freedom and also studying pure Yang-Mills dynamics in an longitudinally expanding geometry. Preliminary results of this upcoming work with fermionic degrees of freedom shown at the HotQuarks 2014 conference plot a fast fermionic energy rise, which would lead to a strong initial entropy production. In the future it would be interesting to compare the momentum-space anisotropy dependence of many important heavy-ion collision observables with experimental results.

Acknowledgments

M.A. thanks the O. Philipsen and C. Schäfer for collaboration on this project. M.A. thanks the Yukawa Institute for Theoretical Physics, Kyoto University, where this work was initiated during

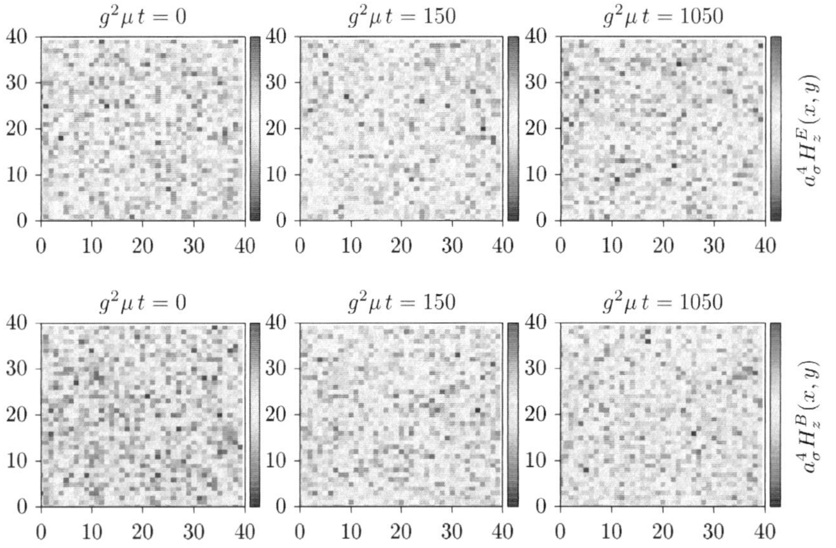

Figure 3. Visualization of the local longitudinal chromo-electric and chromo-magnetic energy density in a fixed box SU(3) Yang-Mills evolution in the transversal xy plane for three different timesteps.

the YITP-T-13-05 on New Frontiers in QCD". M.A. acknowledges support by the Helmholtz Association through the Helmholtz Young Investigator Grant No. VH-NG-822, by the Ministerio de Economìa y Competitividad for the grants FPA2013-40360-ERC and by Generalitat de Catalunya for 2014-SGR-104.

References

[1] Romatschke P and Venugopalan R 2006 *Phys.Rev.Lett.* **96** 062302 (*Preprint* hep-ph/0510121)
[2] Romatschke P and Venugopalan R 2006 *Phys.Rev.* **D74** 045011 (*Preprint* hep-ph/0605045)
[3] Romatschke P and Rebhan A 2006 *Phys.Rev.Lett.* **97** 252301 (*Preprint* hep-ph/0605064)
[4] Rebhan A, Strickland M and Attems M 2008 *Phys.Rev.* **D78** 045023 (*Preprint* 0802.1714)
[5] Fukushima K and Gelis F 2012 *Nucl.Phys.* **A874** 108–129 (*Preprint* 1106.1396)
[6] Berges J, Boguslavski K and Schlichting S 2012 *Phys.Rev.* **D85** 076005 (*Preprint* 1201.3582)
[7] Attems M, Rebhan A and Strickland M 2013 *Phys.Rev.* **D87** 025010 (*Preprint* 1207.5795)
[8] Berges J, Boguslavski K, Schlichting S and Venugopalan R 2014 *Phys.Rev.* **D89** 074011 (*Preprint* 1303.5650)
[9] Fukushima K 2014 *Phys.Rev.* **C89** 024907 (*Preprint* 1307.1046)
[10] Epelbaum T and Gelis F 2013 *Phys.Rev.Lett.* **111** 232301 (*Preprint* 1307.2214)
[11] Heinz U W 1984 *Nucl.Phys.* **A418** 603C–612C
[12] Mrowczynski S 1988 *Phys.Lett.* **B214** 587
[13] Pokrovsky Y and Selikhov A 1988 *JETP Lett.* **47** 12–14
[14] Mrowczynski S 1993 *Phys.Lett.* **B314** 118–121
[15] Blaizot J P and Iancu E 2002 *Phys.Rept.* **359** 355–528 (*Preprint* hep-ph/0101103)
[16] Romatschke P and Strickland M 2003 *Phys.Rev.* **D68** 036004 (*Preprint* hep-ph/0304092)
[17] Arnold P B, Lenaghan J and Moore G D 2003 *JHEP* **0308** 002 (*Preprint* hep-ph/0307325)
[18] Romatschke P and Strickland M 2004 *Phys.Rev.* **D70** 116006 (*Preprint* hep-ph/0406188)
[19] Rebhan A and Steineder D 2010 *Phys.Rev.* **D81** 085044 (*Preprint* 0912.5383)
[20] Carrington M E, Deja K and Mrowczynski S 2014 *Phys.Rev.* **C90** 034913 (*Preprint* 1407.2764)
[21] Attems M, Philipsen O and Christian S upcoming to appear

Probing novel long-range correlation phenomena in pPb collisions with identified particles at CMS

Zhenyu Chen, on behalf of the CMS Collaboration
Physics & Astronomy MS-61, Rice University, Houston, US
E-mail: zc11@rice.edu

Abstract. Observation of a long-range near-side two-particle correlation (known as the "Ridge") in high-multiplicity pp and pPb collisions opened up new opportunities of exploring novel QCD dynamics in small collision systems. To further investigate the origin of this phenomenon, studies of two-particle correlations with identified K_s^0 and $\Lambda/\overline{\Lambda}$ trigger particles in 5.02 TeV pPb collisions are presented, and compared to PbPb collisions over a similar multiplicity range. The K_s^0 and $\Lambda/\overline{\Lambda}$ are cleanly reconstructed via their secondary decay vertices over a wide pseudorapidity and transverse momentum range. The second-order anisotropy harmonics (v_2) of K_s^0 and $\Lambda/\overline{\Lambda}$ are extracted from long-range correlations as a function of particle multiplicity and p_T. The wide p_T coverage and rich sample of high multiplicity pPb events allow: (1) a precise examination of the mass ordering effect of v_n at low p_T as predicted by hydrodynamics for a collectively expanding medium; (2) exploration of possible constituent quark number scaling of v_2 between mesons and baryons as was observed in high-energy nucleus-nucleus collisions.

1. Introduction

Studies of multiparticle correlations provide important insights into the underlying mechanism of particle production in high-energy collisions of both protons and nuclei. The observed long-range (large relative pseudorapidity), two-particle correlations in AA collisions are suggested to mainly arise from the collective flow of a strongly interacting, expanding medium. In hydrodynamic models, the azimuthal correlation structure of emitted particle pairs is typically characterized by its Fourier components, such as the second Fourier components, v_2, known as elliptic flow.

Elliptic flow of identified particles measured in AA collisions exhibits a dependence on the particle species [1]. The mass ordering effect of v_2 in AA collisions is present in hydrodynamic models due to stronger influence of radial flow for heavier particles. Furthermore, a variety of constituent quark scaling features has been observed in AA collisions for v_2 at low and intermediate p_T range ($p_T < 6$ GeV). In particular, v_2 as a function of the hadron transverse kinetic energy, $\mathrm{KE_T} = \sqrt{m^2 + p_\mathrm{T}^2} - m$, scales with the number of constituent quarks n_q in the hadron [2].

Similar long-range correlations have been recently discovered by CMS in proton-proton (pp) [3] and proton-lead (pPb) [4] collisions with high final-state particle multiplicity at the CERN LHC. While it is widely accepted that this long-range structure in the large AA systems is a consequence of collective flow, its origin in small collision systems is still unclear.

An analysis of two-particle correlations with identified strange hadrons, $\mathrm{K_S^0}$ and $\Lambda/\overline{\Lambda}$, in pPb collisions at a center-of-mass energy per nucleon pair ($\sqrt{s_{NN}}$) of 5.02 TeV is presented [5].

By associating a K^0_S or $\Lambda/\overline{\Lambda}$ particle with a charged particle, the anisotropy harmonics v_2 are extracted from long-range two-particle correlations as a function of strange hadron p_T and event multiplicity. The ratios of v_2 to the number of constituent quarks are obtained as a function of hadron transverse kinetic energy per quark for both K^0_S and $\Lambda/\overline{\Lambda}$ to examine the validity of constituent quark number scaling. Results from pPb collisions are directly compared with those from PbPb collisions over a broad range of similar multiplicities.

2. Reconstruction of K^0_S and $\Lambda/\overline{\Lambda}$candidates

The reconstruction technique for K^0_S and $\Lambda/\overline{\Lambda}$ candidates (generally referred to as V^0s) at CMS is described in [5, 6]. The V^0 candidates are reconstructed by combining pairs of oppositely charged tracks that displaced from the primary vertex and form a good secondary vertex. Due to the long lifetime of K^0_S and $\Lambda/\overline{\Lambda}$ particles, requirements on the significance of the V^0 decay length to be greater than 5 and on the angle θ^{point} between the V^0 momentum vector and the vector connecting the primary and V^0 vertices to satisfy $\cos\theta^{point} > 0.999$ are applied to reduce background contributions.

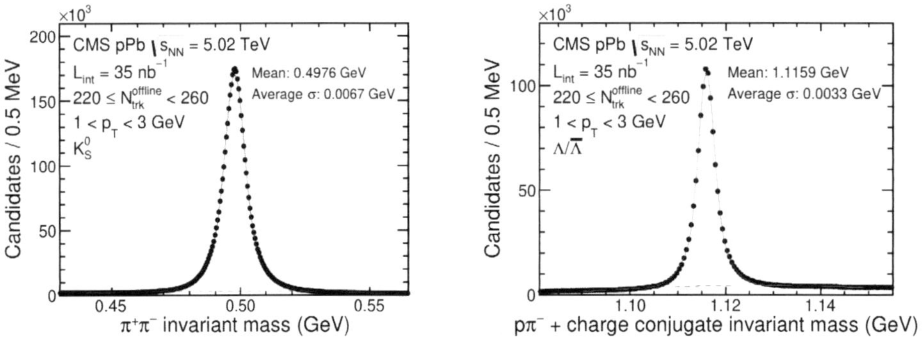

Figure 1. Invariant mass distribution of K^0_S (left) and $\Lambda/\overline{\Lambda}$ (right) candidates in the p_T range of 1–3 GeV for $220 \leq N_{trk}^{offline} < 260$ in pPb collisions at $\sqrt{s_{NN}} = 5.02$ TeV. The solid line shows the fit function of a double Gaussian plus a 4th-order polynomial (dashed lines).

Examples of invariant mass distributions of reconstructed K^0_S and $\Lambda/\overline{\Lambda}$ candidates are shown in Fig. 1 for pPb data. The V^0 peaks can be clearly identified with little background. The true V^0 signal peak is well described by a double Gaussian function (with a common mean), while the background is modeled by a 4th-order polynomial function. The mass window of $\pm2\sigma$ wide around the center of the peak is defined as the "peak region", where σ represents the root mean square of the two standard deviations of the double Gaussian functions weighted by the yields. A "sideband region" is chosen that includes V^0 candidates from outside the $\pm3\sigma$ mass range around the V^0 mass to the limit of the mass distributions shown in Fig. 1.

3. Results

The method of extracting v_n from the two-particle correlation functions is described in [4, 7]. The v_n values are first extracted for V^0 candidates from the peak region and sideband region, denoted as v_n^{obs} and v_n^{bkg}, respectively. The v_n signal of true V^0 particles is denoted by v_n^{sig} and is obtained by $v_n^{sig} = \frac{v_n^{obs}-(1-f^{sig})\times v_n^{bkg}}{f^{sig}}$. Here, f^{sig} represents the signal yield fraction in the peak region determined by the fits to the mass distribution shown in Fig. 1. This fraction exceeds 80% for $\Lambda/\overline{\Lambda}$ candidates at $p_T > 1$ GeV and is above 95% for K^0_S candidates over the entire p_T range.

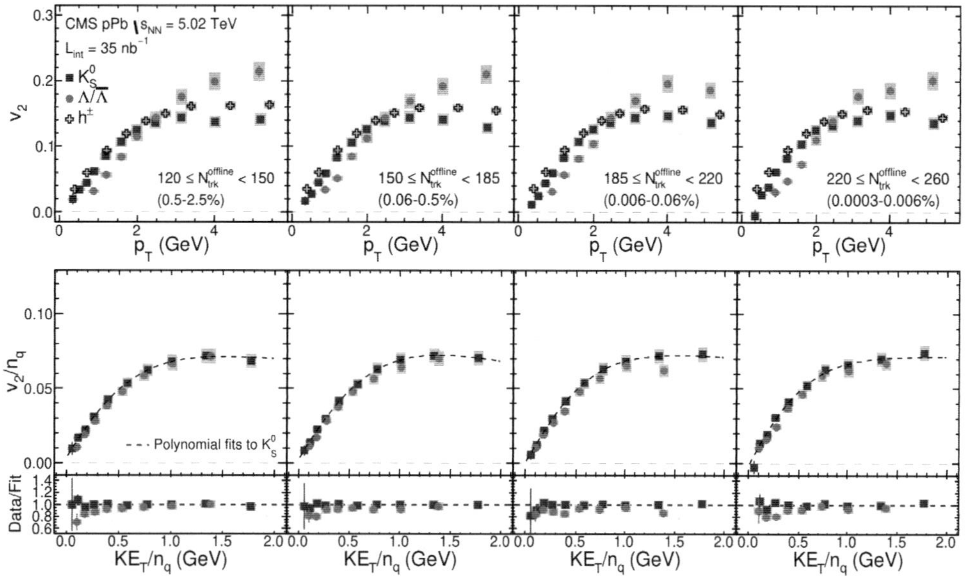

Figure 2. Top row: the v_2 results for K^0_S, $\Lambda/\overline{\Lambda}$, and inclusive charged particles as a function of p_T for four high multiplicity ranges. Middle row: the v_2/n_q ratios for K^0_S and $\Lambda/\overline{\Lambda}$ as a function of KE_T/n_q, along with a fit to the K^0_S results using a polynomial function. Bottom row: ratios of v_2/n_q for K^0_S and $\Lambda/\overline{\Lambda}$ to the fitted polynomial function as a function of KE_T/n_q. The error bars correspond to statistical uncertainties, while the shaded areas denote the systematic uncertainties.

The elliptic (v_2) flow harmonics of K^0_S and $\Lambda/\overline{\Lambda}$ particles are extracted in [5]. In Fig. 2, the $v_2^{sig}$of K^0_S and $\Lambda/\overline{\Lambda}$ particles are plotted as a function of p_T for four multiplicity ranges pPb collisions. A separation of the v_2 values between K^0_S and $\Lambda/\overline{\Lambda}$ particles as a function of p_T is observed. The previously CMS published v_2 results [8] for inclusive charged particles are shown as open cross markers for comparison. The v_2 values of K^0_S particles are larger than those for $\Lambda/\overline{\Lambda}$ particles in the $p_T < 2$ GeV region for all high-multiplicity ranges. Both of them are consistently below the v_2 values of inclusive charged particles (mostly pions). The data indicate a mass ordering behavior where lighter particle species exhibit a stronger azimuthal anisotropy signal, consistent with expectations in hydrodynamic models. At higher p_T the v_2 values of $\Lambda/\overline{\Lambda}$ particles are larger than those of K^0_S.

The scaling behavior of v_2 divided by the number of constituent quarks as a function of transverse kinetic energy per quark, KE_T/n_q, is investigated for high-multiplicity pPb events in the middle row of Fig. 2. The v_2 distributions for K^0_S and $\Lambda/\overline{\Lambda}$ particles are found to be in agreement after scaling by the number of quarks. The result of fitting a polynomial function to the K^0_S data is shown in the middle row of Fig. 2. The bottom row of Fig. 2 shows the n_q-scaled v_2 results for K^0_S and $\Lambda/\overline{\Lambda}$ particles divided by this polynomial function fit. The data/fit ratios indicate that the scaling is valid to better than 10% over most of the KE_T/n_q range. In AA collisions, this approximate scaling behavior is conjectured to be related to quark recombination which postulates that collective flow is developed among constituent quarks before they combine to form final-state hadrons.

The particle species dependence of v_2 and its scaling behavior is also studied in PbPb data over the same multiplicity ranges as for the pPb data, as shown in Fig. 3. For all multiplicity

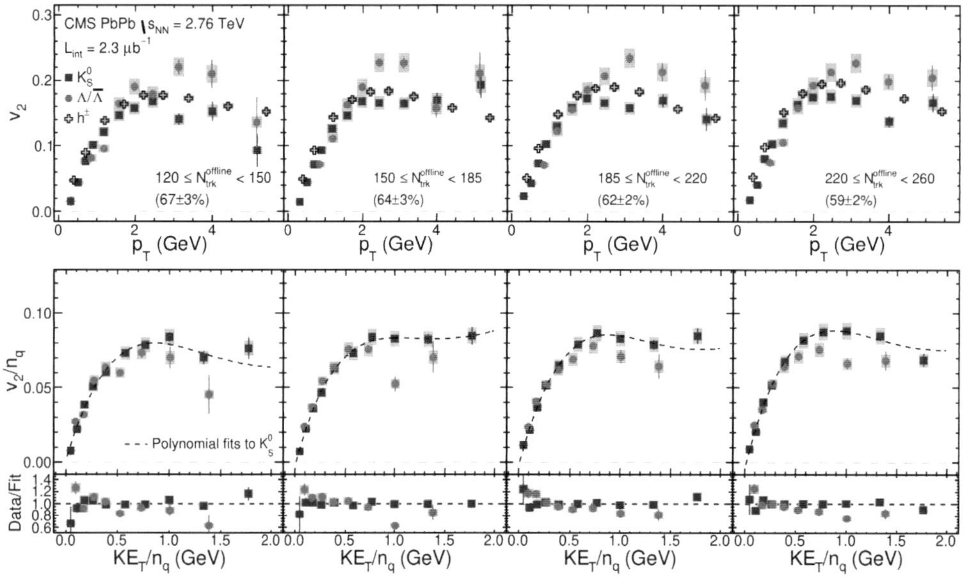

Figure 3. Top row: the v_2 results for K_S^0, $\Lambda/\overline{\Lambda}$, and inclusive charged particles (open crosses) as a function of p_T for four multiplicity ranges. Middle row: the v_2/n_q ratios for K_S^0 and $\Lambda/\overline{\Lambda}$ as a function of KE_T/n_q. Bottom row: ratios of v_2/n_q for K_S^0 and $\Lambda/\overline{\Lambda}$ to the fitted polynomial function as a function of KE_T/n_q. The error bars correspond to statistical uncertainties, while the shaded areas denote the systematic uncertainties.

ranges, the mass ordering effect of v_2 for different particle species is found to be less evident in PbPb data than in pPb data. This may indicate, in hydrodynamic models, a stronger radial flow is developed in the pPb system than that of a PbPb system at the same multiplicity. Moreover, compared to pPb data, the n_q-scaled v_2 data in PbPb at similar multiplicities suggest a stronger violation of constituent quark number scaling, up to 25%.

References

[1] STAR Collaboration, "Experimental and theoretical challenges in the search for the quark gluon plasma: The STAR Collaboration's critical assessment of the evidence from RHIC collisions," *Nucl. Phys. A*, vol. 757, p. 102, 2005.

[2] PHENIX Collaboration, "Scaling properties of azimuthal anisotropy in Au+Au and Cu+Cu collisions at $\sqrt{s_{NN}}$ = 200 GeV," *Phys. Rev. Lett.*, vol. 98, p. 162301, 2007.

[3] CMS Collaboration, "Observation of long-range near-side angular correlations in proton-proton collisions at the LHC," *JHEP*, vol. 09, p. 091, 2010.

[4] CMS Collaboration, "Observation of long-range near-side angular correlations in pPb collisions at the LHC," *Phys. Lett. B*, vol. 718, p. 795, 2013.

[5] CMS Collaboration, "Long-range two-particle correlations of strange hadrons with charged particles in pPb and PbPb collisions at LHC energies." Submitted to Phys. Lett. B, arXiv:1409.3392, 2014.

[6] CMS Collaboration, "Strange particle production in pp collisions at $\sqrt{s}$ = 0.9 and 7 TeV," *JHEP*, vol. 05, p. 064, 2011.

[7] CMS Collaboration, "Long-range and short-range dihadron angular correlations in central PbPb collisions at $\sqrt{s_{NN}}$= 2.76 TeV," *JHEP*, vol. 07, p. 076, 2011.

[8] CMS Collaboration, "Multiplicity and transverse momentum dependence of two- and four-particle correlations in pPb and PbPb collisions," *Phys. Lett. B*, vol. 724, p. 213, 2013.

Journal of Physics: Conference Series **612** (2015) 012017 doi:10.1088/1742-6596/612/1/012017

Experimental overview of p(d)–A physics

Roberto Preghenella

European Organisation for Nuclear Research (CERN), Geneva, Switzerland

Istituto Nazionale di Fisica Nucleare (INFN), Sezione di Bologna, Italy

E-mail: preghenella@bo.infn.it

Abstract. This document reports on the main topics discussed in the experimental overview talk on p(d)–A results given at the Hot Quarks 2014 Workshop. Recent measurements of the production of high transverse momentum hadrons and jets, inclusive identified particles and their correlations, quarkonia and vector bosons are presented. They represent a small set of the currently available results, which provide a deeper understanding of the hot QCD matter produced in A–A collisions and reveal the presence of unexpected phenomena in small systems.

1. Introduction

A wealth of physics results are reported by RHIC and LHC experiments, providing further understanding of the dynamics of the hot QCD matter produced in nucleus-nucleus (A–A) reactions and exciting insight for genuine proton(deuteron)-nucleus (p–A, d–A) physics. Measurements of the production of high transverse momentum hadrons and jets allow to conclude that the suppression observed in A–A collisions is not due to an initial-state effect, but it is rather the fingerprint of jet quenching in hot QCD matter. The study of inclusive identified particle spectra and their correlations revealed unexpected phenomena that promptly suggested the presence of collective behaviour in small systems and triggered further investigations. Results on quarkonia production also show that cold nuclear matter effects are important for the interpretation of heavy-ion results and the study of the production of the excited quarkonium states reveals sizeable and intriguing final-state effects as well. Electroweak boson production provide significant constraints to the nuclear modifications of parton distribution functions. These are the main topics discussed in the following sections of these proceedings, which report on a limited selection of experimental results. Given the large amount of new results, many other interesting topics related to p(d)–A physics, like centrality determination, detailed study of the production of jets and their modifications, strangeness and nuclei production and open heavy-flavour observables could not be discussed here.

2. Understanding initial-state and cold nuclear matter effects

High transverse momentum hadrons, arising from the fragmentation of partons scattered with large momentum transfer, provide an excellent probe of the high energy-density matter created in relativistic heavy-ion collisions. In the absence of medium effects, hard scattering yields should scale with the average number of inelastic nucleon-nucleon collisions (binary scaling). One of the main observations in central A–A collisions was a large suppression of high-p_T hadron yields with respect to binary-scaled pp results [1–3]. It is therefore of paramount interest to determine experimentally the modification of high-p_T hadron yields due to initial-state nuclear effects.

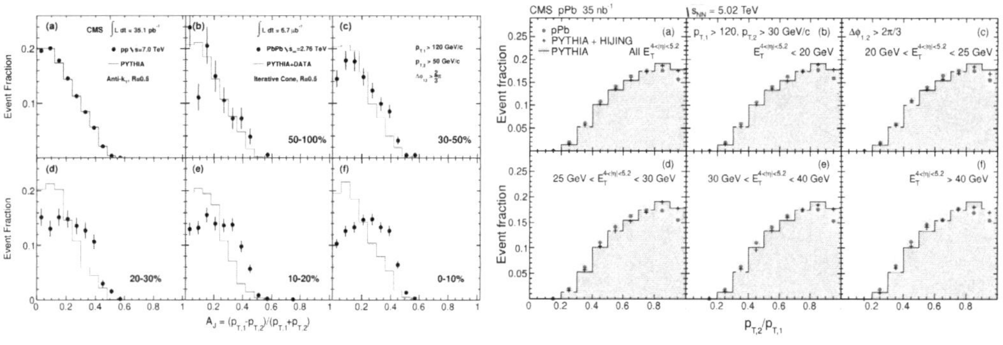

Figure 1. (left) Dijet asymmetry ratio for 7 TeV pp collisions and 2.76 TeV Pb–Pb collisions in several centrality bins [8]. (right) Dijet transverse momentum ratio distributions in p–Pb events without any selection on the HF transverse energy and for different classes [9].

Measurements by RHIC experiments reported on the absence of suppression of high-p_T hadrons in d–Au collisions at $\sqrt{s_{NN}}$ = 200 GeV [4, 5]. The results were based on measurements of the inclusive hadron production and two-particle azimuthal distributions at high p_T and provided evidence that the suppression phenomena seen in central Au–Au collisions at RHIC are due to final-state interactions with the dense system generated in the collision. Measurements of the charged-particle p_T spectra and the nuclear modification in inelastic p–Pb collisions at the $\sqrt{s_{NN}}$ = 5.02 TeV [6] show a nuclear modification factor consistent with unity at high momenta, indicating that also at the LHC the strong suppression of hadron production in Pb–Pb collisions is not due to initial-state effects, but it is rather the fingerprint of jet quenching in hot QCD matter.

Thanks to the wide coverage and high performance of the calorimetric systems installed in LHC experiments, it was possible to directly observe jet quenching phenomena by studying the centrality dependence of dijet production, in particular the dijet asymmetry in Pb–Pb collisions. The first observation of an enhancement of events with such large dijet asymmetries, not observed in proton-proton collisions, has been reported [7,8]. Results on dijet asymmetry in $\sqrt{s_{NN}}$ = 2.76 TeV Pb–Pb collisions are shown in several centrality bins and compared to $\sqrt{s}$ = 7 TeV pp collisions in Figure 1 (left). Dijet production has also been measured in p–Pb collisions at $\sqrt{s_{NN}}$ = 5.02 TeV [9] and the results are shown in Figure 1 (right): in contrast to what is seen in Pb–Pb collisions, no significant dijet transverse momentum imbalance is observed in p–Pb data. This further confirms that the imbalance observed in Pb–Pb collisions is not originating from initial-state effects.

It is worth mentioning that the surprising observation reported in recent measurements of inclusive charged hadron nuclear modification in minimum bias p–Pb collisions at the $\sqrt{s_{NN}}$ = 5.02 TeV. An excess compared to binary-scaled pp data is observed at high $p_T > 30$ GeV/c by ATLAS and CMS [10,11]. The excess is larger than expected from the EPS09 [12] nuclear parton distribution functions and it is even more surprisingly not observed in jet measurements [13,14]. ALICE data [15] do not show such enhancement, though the p_T reach is limited. Statistical and systematic uncertainties are large and in particular there are substantial uncertainties in the pp reference, which had to be interpolated from higher and lower energy results. It is therefore of greatest importance to improve the precision and establish the excess with proton-proton reference data measured at the same energy in the next LHC run.

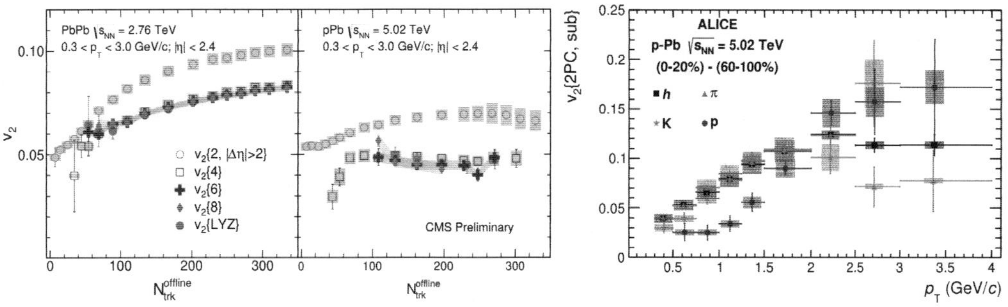

Figure 2. (left) Anisotropic flow v_2 results obtained from multi-particle cumulants and LYZ method in Pb–Pb at $\sqrt{s_{\rm NN}}$ = 2.76 TeV (left) and p–Pb at $\sqrt{s_{\rm NN}}$ = 5.02 TeV [30]. (right) Anisotropic flow v_2 results for hadrons, pions, kaons and protons in p–Pb at $\sqrt{s_{\rm NN}}$ = 5.02 TeV obtained in the 0–20% multiplicity class after subtraction of the correlation from the 60–100% [26].

3. Similarities to A–A and hints for collectivity

The measurements of inclusive charged-hadron nuclear modification in minimum bias p–Pb collisions [15] show a weak indication of a Cronin-like enhancement at intermediate $p_{\rm T}$, where a stronger enhancement is seen at lower energies [4, 5, 16]. Traditional explanations of the Cronin enhancement involve multiple soft scatterings in the initial state, prior to the hard scattering, and subsequent fragmentation of the hard scattered parton [17]. This process on the other hand does not account for the observed particle species dependence [18] that suggests final state effects. A strong baryon-to-meson enhancement is observed both at RHIC and LHC [15, 19, 20] with a clear evolution with multiplicity. This is similar to the pattern observed in high-energy heavy-ion collisions [21], where the effect is usually attributed to collective radial expansion. In contrast to the inclusive distribution, the ratio within jets in high-multiplicity p–Pb collisions does not exhibit such baryon enhancement [22].

Measurements at the LHC in high-multiplicity pp and p–Pb collisions have revealed a near-side long-range "ridge" structure in the two-particle correlation functions [23–25]. The observation of an unexpected "double-ridge" structure in the two-particle correlations in high-multiplicity p–Pb collisions has also been reported [26–29]. In A–A collisions, long-range correlations are interpreted as a consequence of the collective hydrodynamic expansion of the produced medium from the initial geometry: the presence of a ridge structure in p–A collisions promptly suggested collective behaviour and triggered further investigations.

The azimuthal particle distributions are characterised by their Fourier components. Of particular importance are the elliptic and triangular flow, namely the second and third Fourier coefficients. They most directly reflect the medium response to the initial collision geometry and fluctuations, allowing to shed light on the transport properties of the medium. Measurements of the second-order azimuthal anisotropy Fourier harmonic v_2 [30, 31] using multi-particle correlations are shown in Figure 2 (left). Significant multi-particle correlation signals are observed in both p–Pb and Pb–Pb collisions, which show in addition a remarkable agreement between the results derived from the four- or more particle-correlation methods. This strongly supports the collective nature of the observed signals.

Angular correlations between unidentified charged trigger particles and identified associated particles (pions, kaons, protons, K^0_S and Λ) are also measured in p(d)–A collisions [32–34]. The results show an elliptic-flow pattern qualitatively similar to the one observed in heavy-ion collisions [35]. The results for pion, kaon and proton v_2 are shown in Figure 2 (right). A mass

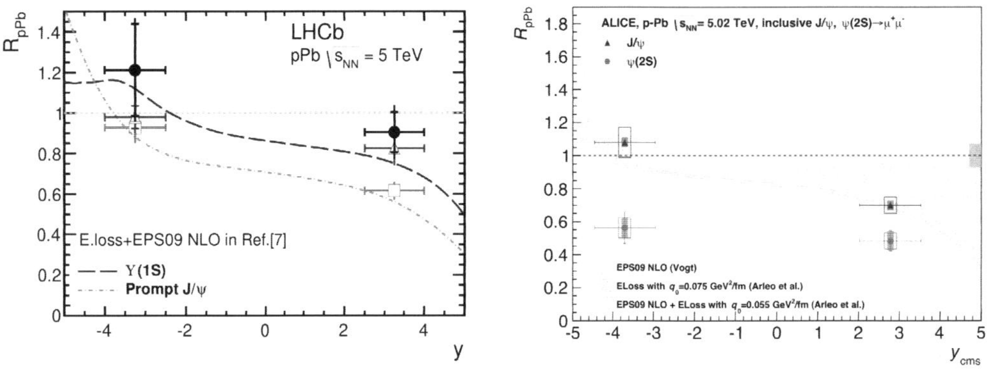

Figure 3. (left) Nuclear modification factor R_{pPb} for $\Upsilon(1S)$ (black circles), prompt J/ψ (red squares), and J/ψ from b-hadron decays (blue triangles) [39]. (right) Nuclear modification factor R_{pPb} for $\psi(2S)$ compared to the corresponding quantity for J/ψ [42].

ordering effect at low transverse momenta is consistent with expectations from hydrodynamic model calculations [36] assuming a collectively expanding system.

4. Quarkonia: further hints for final state effects

A clear suppression of J/ψ production with respect to binary-scaled pp collisions is observed in p–Pb collisions at $\sqrt{s_{\mathrm{NN}}} = 5.02$ TeV in the forward region, corresponding to the proton direction [37, 38]. No suppression is present in the backward region. The nuclear modification factor has been determined separately for prompt J/ψ mesons and J/ψ from b-hadron decays and is compared with results for $\Upsilon(1S)$ [39] in Figure 3 (left). The suppression of the production of J/ψ from b-hadron decays and of $\Upsilon(1S)$ is found to be smaller than in the case of prompt J/ψ mesons. Theoretical predictions based on nuclear shadowing, as well as on models including, in addition, a contribution from partonic energy loss, are in fair agreement with the experimental results [40, 41]. The measurements show that cold nuclear matter effects are important for interpretations of quarkonium-related quark-gluon plasma signatures in heavy-ion collisions.

The inclusive production of $\psi(2S)$ has also been measured [42] and its nuclear modification factor is compared to the measurement of the same quantity for J/ψ and to theoretical models in Figure 3 (right). The results show a significantly larger suppression of the $\psi(2S)$ compared to that measured for J/ψ and to models, both in the forward and in the backward region. A stronger suppression of $\psi(2S)$ relative to J/ψ was observed in previous fixed-target experiments at central rapidity, while at forward rapidity no difference was found within uncertainties [43, 44]. This feature was interpreted in terms of pair break-up: at central and backward rapidity the time spent by the $c\bar{c}$ state in the nuclear medium (crossing time) is typically larger than the formation time of the resonances, so that the loosely bound $\psi(2S)$ can be more easily dissociated than the J/ψ. Conversely, in forward production the crossing time is smaller than the formation time and the influence of the nucleus on the pre-hadronic state is the same, independent of the particular resonance being produced. No break-up effects depending on the final charmonium state should be expected at forward rapidity. As a consequence, the strong difference observed between $\psi(2S)$ and J/ψ suppression in the forward region represents a clear indication for sizeable final-state effects in $\psi(2S)$ production. In addition, the ratios of the excited-to-ground state cross sections $\psi(2S)/J/\psi$ and $\Upsilon(nS)/\Upsilon(1S)$ are found to decrease with the charged-particle multiplicity [45, 46]. Other final state effects should be considered to describe these features,

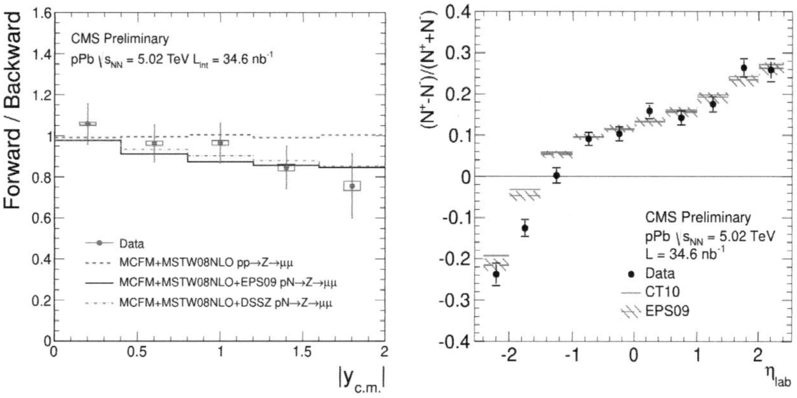

Figure 4. (left) Forward-backward ratio of Z boson cross section in p–Pb collisions as a function of rapidity compared to predictions [47]. (right) W boson lepton-charge asymmetry in p–Pb collisions computed as a function of the lepton pseudorapidity in the laboratory frame [49].

including the interaction of the $c\bar{c}$ pair with the final-state hadronic system created in the collision.

5. Constraints on nPDFs with vector bosons

Electroweak boson production is an important benchmark measurement that can provide constraints allowing to probe the nuclear modification of parton distribution functions in a phase-space region that was never studied before. The production of W and Z bosons in proton-nucleus collisions at the LHC has been reported [47–50]. Results on forward-backward ratio of Z boson cross section and of W boson lepton-charge asymmetry are shown in Figure 4. At large rapidity, the Z forward-backward ratio deviates from unity by an amount which is compatible with nuclear PDF (nPDF) modifications, though due to the large uncertainties, these measurements are not able to distinguish between different nuclear PDF sets. Nonetheless, these measurement provide significant input to nPDF fits with a significant constraining power in a previously unexplored region of the phase space. The W lepton-charge asymmetry, a sensitive probe of the down to up quark PDF, deviates from calculations for $\eta < -1$, suggesting that different nuclear modifications can occur for up and down quark PDFs.

6. Summary

In summary, p(d)–A collision results both at RHIC and LHC show no indication of quenching of high-p_T hadrons and jets, confirming that the strong suppression in A–A is a final-state hot QCD matter effect. However, a still unexplained enhancement of high-p_T hadrons is observed by CMS and ATLAS. Bulk identified-particle production reveals features that are reminiscent of A–A phenomena, triggering studies on genuine collective effects in small systems. This is strongly supported by the observation of non-zero elliptic flow from multi-particle correlations and by the mass-dependence of p_T spectra and v_2. The measurement of quarkonia production and of the suppression of quarkonia ground states is in line with the predictions from nuclear parton shadowing and/or initial-state energy-loss. The observation of a stronger suppression of quarkonia excited states provides a clear evidence for exotic final-state $c\bar{c}$ pair break-up phenomena. Furthermore, the measurement of vector boson production provides significant constraints for nPDF fits.

Acknowledgements

Thanks to the Hot Quarks 2014 organisers for the invitation to such a delightful and constructive Workshop, to the CERN ALICE group for financial support and advices, to the LHC and RHIC Collaborations for the large amount of high-quality results and to Maria Valentina Carlucci for proofreading this report.

References

[1] Adler C *et al.* (STAR Collaboration) 2002 *Phys.Rev.Lett.* **89** 202301 (*Preprint* nucl-ex/0206011)
[2] Adcox K *et al.* (PHENIX Collaboration) 2002 *Phys.Rev.Lett.* **88** 022301 (*Preprint* nucl-ex/0109003)
[3] Abelev B *et al.* (ALICE Collaboration) 2013 *Phys.Lett.* **B720** 52–62 (*Preprint* 1208.2711)
[4] Adams J *et al.* (STAR Collaboration) 2003 *Phys.Rev.Lett.* **91** 072304 (*Preprint* nucl-ex/0306024)
[5] Adler S *et al.* (PHENIX Collaboration) 2003 *Phys.Rev.Lett.* **91** 072303 (*Preprint* nucl-ex/0306021)
[6] Abelev B *et al.* (ALICE Collaboration) 2013 *Phys.Rev.Lett.* **110** 082302 (*Preprint* 1210.4520)
[7] Aad G *et al.* (ATLAS Collaboration) 2010 *Phys.Rev.Lett.* **105** 252303 (*Preprint* 1011.6182)
[8] Chatrchyan S *et al.* (CMS Collaboration) 2011 *Phys.Rev.* **C84** 024906 (*Preprint* 1102.1957)
[9] Chatrchyan S *et al.* (CMS Collaboration) 2014 *Eur.Phys.J.* **C74** 2951 (*Preprint* 1401.4433)
[10] ATLAS Collaboration 2014 ATLAS-CONF-2014-029
[11] CMS Collaboration 2013 CMS-PAS-HIN-12-017
[12] Eskola K, Paukkunen H and Salgado C 2009 *JHEP* **0904** 065 (*Preprint* 0902.4154)
[13] ATLAS Collaboration 2014 ATLAS-CONF-2014-024
[14] CMS Collaboration 2014 CMS-PAS-HIN-14-001
[15] Abelev B B *et al.* (ALICE Collaboration) 2014 *Eur.Phys.J.* **C74** 3054 (*Preprint* 1405.2737)
[16] Cronin J, Frisch H J, Shochet M, Boymond J, Mermod R *et al.* 1975 *Phys.Rev.* **D11** 3105
[17] Accardi A 2002 (*Preprint* hep-ph/0212148)
[18] Adare A *et al.* (PHENIX Collaboration) 2013 *Phys.Rev.* **C88** 024906 (*Preprint* 1304.3410)
[19] Abelev B B *et al.* (ALICE Collaboration) 2014 *Phys.Lett.* **B728** 25–38 (*Preprint* 1307.6796)
[20] Chatrchyan S *et al.* (CMS Collaboration) 2014 *Eur.Phys.J.* **C74** 2847 (*Preprint* 1307.3442)
[21] Abelev B *et al.* (ALICE Collaboration) 2013 *Phys.Rev.* **C88** 044910 (*Preprint* 1303.0737)
[22] Zhang X (ALICE Collaboration) 2014 (*Preprint* 1408.2672)
[23] Khachatryan V *et al.* (CMS Collaboration) 2010 *JHEP* **1009** 091 (*Preprint* 1009.4122)
[24] Chatrchyan S *et al.* (CMS Collaboration) 2013 *Phys.Lett.* **B718** 795–814 (*Preprint* 1210.5482)
[25] Abelev B *et al.* (STAR Collaboration) 2009 *Phys.Rev.* **C80** 064912 (*Preprint* 0909.0191)
[26] Abelev B *et al.* (ALICE Collaboration) 2013 *Phys.Lett.* **B719** 29–41 (*Preprint* 1212.2001)
[27] Aad G *et al.* (ATLAS Collaboration) 2013 *Phys.Rev.Lett.* **110** 182302 (*Preprint* 1212.5198)
[28] Aad G *et al.* (ATLAS Collaboration) 2013 *Phys.Lett.* **B725** 60–78 (*Preprint* 1303.2084)
[29] Chatrchyan S *et al.* (CMS Collaboration) 2013 *Phys.Lett.* **B724** 213–240 (*Preprint* 1305.0609)
[30] CMS Collaboration 2014 CMS-PAS-HIN-14-006
[31] Aad G *et al.* (ATLAS Collaboration) 2014 CERN-PH-EP-2014-201 (*Preprint* 1409.1792)
[32] Abelev B B *et al.* (ALICE Collaboration) 2013 *Phys.Lett.* **B726** 164–177 (*Preprint* 1307.3237)
[33] Khachatryan V *et al.* (CMS Collaboration) 2014 CMS-HIN-14-002 (*Preprint* 1409.3392)
[34] Adare A *et al.* (PHENIX Collaboration) 2014 (*Preprint* 1404.7461)
[35] Abelev B B *et al.* (ALICE Collaboration) 2014 (*Preprint* 1405.4632)
[36] Bozek P, Broniowski W and Torrieri G 2013 *Phys.Rev.Lett.* **111** 172303 (*Preprint* 1307.5060)
[37] Aaij R *et al.* (LHCb Collaboration) 2014 *JHEP* **1402** 072 (*Preprint* 1308.6729)
[38] Abelev B B *et al.* (ALICE Collaboration) 2014 *JHEP* **1402** 073 (*Preprint* 1308.6726)
[39] Aaij R *et al.* (LHCb Collaboration) 2014 *JHEP* **1407** 094 (*Preprint* 1405.5152)
[40] Albacete J o 2013 *Int.J.Mod.Phys.* **E22** 1330007 (*Preprint* 1301.3395)
[41] Arleo F and Peigne S 2013 *JHEP* **1303** 122 (*Preprint* 1212.0434)
[42] Abelev B B *et al.* (ALICE Collaboration) 2014 (*Preprint* 1405.3796)
[43] Leitch M *et al.* (NuSea Collaboration) 2000 *Phys.Rev.Lett.* **84** 3256–3260 (*Preprint* nucl-ex/9909007)
[44] Abt I *et al.* (HERA-B Collaboration) 2007 *Eur.Phys.J.* **C49** 545–558 (*Preprint* hep-ex/0607046)
[45] Adare A *et al.* (PHENIX Collaboration) 2013 *Phys.Rev.Lett.* **111** 202301 (*Preprint* 1305.5516)
[46] Chatrchyan S *et al.* (CMS Collaboration) 2014 *JHEP* **1404** 103 (*Preprint* 1312.6300)
[47] CMS Collaboration 2014 CMS-PAS-HIN-14-003
[48] Aaij R *et al.* (LHCb Collaboration) 2014 *JHEP* **1409** 030 (*Preprint* 1406.2885)
[49] CMS Collaboration 2014 CMS-PAS-HIN-13-007
[50] ATLAS Collaboration 2014 ATLAS-CONF-2014-020

The impact of the LHC nuclear program on nPDFs

H Paukkunen[a,b] **and P Zurita**[c]

[a]Department of Physics, University of Jyväskylä, P.O. Box 35, FI-40014 University of Jyväskylä, Finland
[b]Helsinki Institute of Physics, P.O. Box 64, FI-00014 University of Helsinki, Finland
[c]Departamento de Física de Partículas and IGFAE, Universidade de Santiago de Compostela, E-15782 Galicia, Spain

E-mail: hannu.t.paukkunen@jyu.fi, pia.zurita@usc.es

Abstract.
The proton-lead and lead-lead runs at the LHC are providing an enormous amount of data sensitive to the nuclear modifications of the initial state. The measurements explore a region of phase space not probed by previous experiments opening a possibility to test and, hopefully, also improve the current knowledge of nuclear parton densities. In this talk, we discuss to what extent the present quantitative results for the charge asymmetry in electroweak boson production show sensitivity to the nuclear parton distributions.

1. Introduction

A detailed description of the partonic structure of nucleons prior to a hard scattering is a key ingredient in precise description of high-energy collider data. As such structure cannot yet be determined from the first principles of QCD, sets of parton density functions (PDFs) are traditionally extracted from experimental data by means of global fits in the framework of perturbative QCD. The accuracy reached in the determination of free-proton parton distributions is astonishing, with sets available up to $\mathcal{O}(\alpha_s^2)$ (NNLO) [1, 2, 3, 4]. The situation is quite different when the free proton is substituted by a nucleus: as the amount of data is far more limited and the kinematical coverage rather restricted in comparison to the free-proton case, the nuclear PDFs (nPDFs) are still far less precise [5, 6, 7, 8]. Most of the available nuclear data come from deep-inelastic scattering (DIS) which is a process dominated by the valence and the sea quark distributions leaving especially the gluon densities unconstrained. To improve this state of affairs, results from the LHC proton-lead and lead-lead runs should provide with a completely novel information.

In this talk, we consider the electroweak boson production in both proton-lead and lead-lead collisions measured by ATLAS [9] and CMS [10] comparing these data with the theoretical predictions using nPDFs. All the data used here are preliminary and only limited conclusions can be drawn.

2. The LHC constraining potential

The kinematical range probed by the pre-LHC nuclear experiments is very limited, the momentum fraction x not reaching much below 10^{-2} at perturbative scales $Q \gtrsim 1\,\mathrm{GeV}$. This entails that predictions sensitive to lower values of x may depend strongly on the original

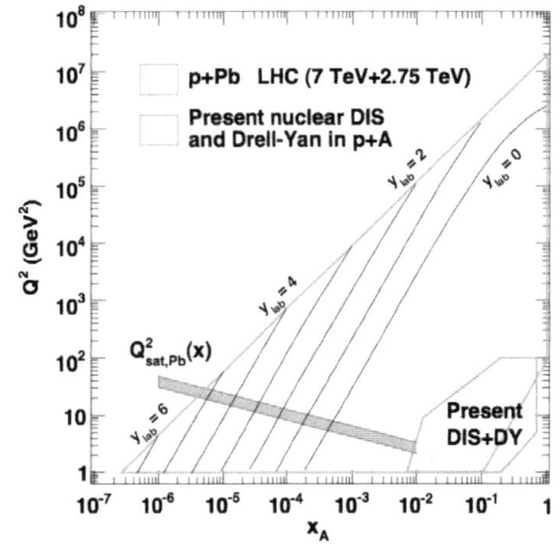

Figure 1. Kinematical coverage of the LHC proton-lead run compared with the one from earlier DIS and Drell-Yan (DY) experiments.

parameterization used in the individual nPDF fits. Especially the gluon distributions appear rather problematic: for example, in DSSZ [8] the gluon is fixed by hand to follow a certain shape while the gluon distribution of EPS09 [6] presents a much different behaviour with almost 100% uncertainty at $Q \approx 1.3\,\text{GeV}$. It is then clear that more data, both from other type of experiments and in new kinematical regions, are required in order to improve the nPDFs. In that aspect the LHC can be extremely useful, as it can reach down to $x \approx 10^{-6}$ — a four-order-of-magnitude extension to the currently explored region. As can be sen from Figure 1, also the Q^2 range achievable is quite wider. Therefore, the recently recorded data from the proton-lead and lead-lead runs at the LHC look a very promising source of new information which should be carefully analyzed and eventually also included into the future extractions of nPDFs.

3. Data and theory: $W^{\pm}$ charge asymmetries

From the wide variety of LHC data we shall focus on the charged electroweak boson ($W^{\pm}$) production in both proton-lead and lead-lead. These particular measurements are directly sensitive to the sea and valence quark distributions and should thereby be helpful e.g. to disentangle the strange-quark contribution from the other light flavors. However, due to the large mass of $W^{\pm}$, indirect constraints of gluon distributions emerging via the parton evolution may be obtainable.

Let us begin with the $W^{\pm}$ production measured by CMS in proton-lead collisions at $\sqrt{s} = 5.02$ TeV [10], available as a function of the pseudorapidity η of the charged lepton originating from the $W^{\pm}$ decay. The theoretical calculations have been performed to NLO accuracy using the CT10NLO [2] free-proton PDFs corrected for the nuclear effects with EPS09 [6]. The comparisons to the absolute cross-sections are presented in leftmost panels of Figure 2. The band enclosed by the blue lines is the errorband derived from the EPS09 error sets. In both cases the behaviour of the data is qualitatively similar as that of the theory predictions, but the correspondence is not perfect. We note that the uncertainties in the cross-section measurements are dominated by the systematics which can be suppressed by taking suitable ratios. The most common one is the charge asymmetry which is shown in the rightmost panel of Figure 2. Indeed, the "scatter" of the data gets attenuated in comparison to the individual absolute distributions and especially in the forward rapidities (the proton-going direction) the agreement

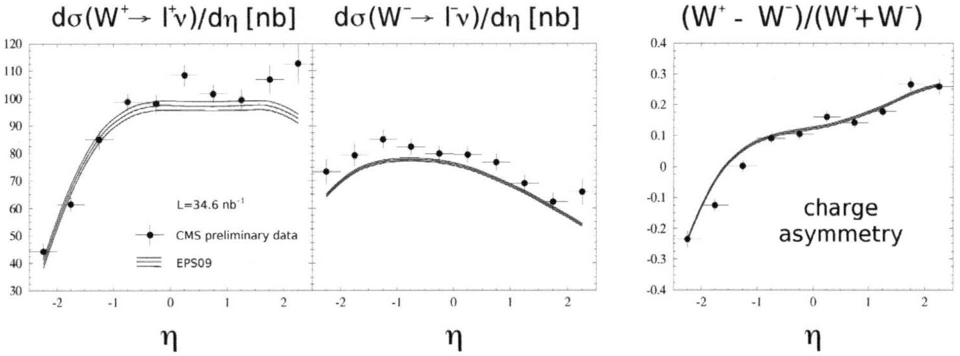

Figure 2. Left: W^+ differential cross-section as a function of η in p-Pb collisions measured by CMS [10] compared to the EPS09 [6] prediction (blue curve). **Center**: The same for W^-. **Right**: The charge asymmetry $(W^+ - W^-)/(W^+ + W^-)$. The proton PDF uncertainty has not been included.

with the NLO calculation seems almost perfect. In the backward region (the lead-going direction) there is a clear discrepancy in two rapidity bins[1]. At this stage it is impossible to draw strict conclusions concerning the observed discrepancy before the measurements by the ALICE and ATLAS collaborations become public.

Let us now turn to $W^\pm$ production in lead-lead collisions at $\sqrt{s} = 2.76$ TeV, measured by the ATLAS collaboration [9]. The charge asymmetry is shown in Figure 3 and compared with the predictions including EPS09 nuclear corrections (red) and with CT10NLO proton PDFs only (green). Both predictions are more or less compatible with these preliminary data. As the nuclear corrections induced by EPS09 are almost identical for W^+ anf W^- production, they tend to cancel when forming the charge asymmetry (as was discussed in Ref. [11]).

4. Summary

New data from the LHC has a potential to constrain nPDFs in ways that no other experiment has been able to do. In this short contribution we discussed how the measured charge asymmetries in $W^\pm$ production compare to the NLO pQCD predictions. In the case of lead-lead collisions we explicitly demonstrated that this particular observable appears not to undergo a very large nuclear modification and the predictions with and without nuclear corrections are compatible with the ATLAS data. The preliminary CMS data for the charge asymmetry in proton-lead collisions displays a certain discrepancy in comparison to the NLO predictions. Whether there is true physics behind this discrepancy remains to be confirmed by the other LHC experiments. However, we admit that our comparison here involved only a one particular set of nPDFs and the bias caused by this should be checked in a more systematic manner.

Acknowledgments

During this work H.P. was supported by the Academy of Finland, Project No. 133005. P.Z. was supported by European Research Council grant HotLHC ERC-2011-StG-279579; by Ministerio de Ciencia e Innovación of Spain under the Consolider-Ingenio 2010 Programme CPAN (CSD2007-00042); by Xunta de Galicia; and by FEDER.

[1] Even adding the uncertainty from the CT10NLO error sets does not seem to cure the observed difference [10]

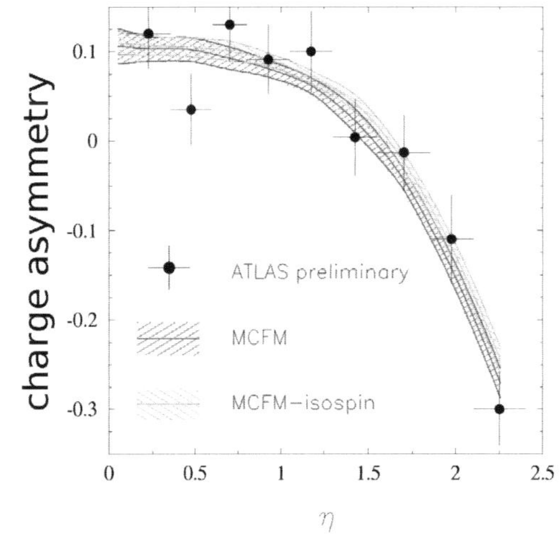

Figure 3. Charge asymmetry for $W^{\pm}$ production in Pb-Pb collisions measured by ATLAS [9]. The green band is the prediction assuming no nuclear effect other than different number of protons and neutrons in the lead nucleus (isospin effect). The red band, corresponds to the EPS09-corrected result with the CT10NLO and EPS09 uncertainties added in quadrature.

References

[1] S. Forte and G. Watt, Ann. Rev. Nucl. Part. Sci. **63** (2013) 291 [arXiv:1301.6754 [hep-ph]].
[2] H. -L. Lai, M. Guzzi, J. Huston, Z. Li, P. M. Nadolsky, J. Pumplin, C. -P. Yuan and , Phys. Rev. D **82** (2010) 074024 [arXiv:1007.2241 [hep-ph]].
[3] A. D. Martin, W. J. Stirling, R. S. Thorne and G. Watt, Eur. Phys. J. C **63** (2009) 189 [arXiv:0901.0002 [hep-ph]].
[4] R. D. Ball, V. Bertone, S. Carrazza, C. S. Deans, L. Del Debbio, S. Forte, A. Guffanti and N. P. Hartland *et al.*, Nucl. Phys. B **867** (2013) 244 [arXiv:1207.1303 [hep-ph]].
[5] M. Hirai, S. Kumano and T.-H. Nagai, Phys. Rev. C **76** (2007) 065207 [arXiv:0709.3038 [hep-ph]].
[6] K. J. Eskola, H. Paukkunen and C. A. Salgado, JHEP **0904** (2009) 065 [arXiv:0902.4154 [hep-ph]].
[7] I. Schienbein, J. Y. Yu, C. Keppel, J. G. Morfin, F. Olness and J. F. Owens, Phys. Rev. D **77** (2008) 054013 [arXiv:0710.4897 [hep-ph]].
[8] D. de Florian, R. Sassot, P. Zurita and M. Stratmann, Phys. Rev. D **85** (2012) 074028 [arXiv:1112.6324 [hep-ph]].
[9] G. Aad *et al.* [ATLAS Collaboration], arXiv:1408.4674 [hep-ex].
[10] CMS Collaboration [CMS Collaboration], CMS-PAS-HIN-13-007.
[11] H. Paukkunen and C. A. Salgado, JHEP **1103** (2011) 071 [arXiv:1010.5392 [hep-ph]].

The Impact of e+A Collisions on Nuclear PDFs

M R Moldes
Departamento de Física de Partículas and IGFAE, Universidade de Santiago de Compostela, E-15706 Santiago de Compostela, Galicia, Spain
E-mail: manoel.rodriguez-moldes@usc.es

Abstract. In this text we comment on the importance of measurements of electron-nucleus collisions and how that data would allow us to check the validity of the collinear factorization framework and constrain the existing set of nuclear parton distribution functions. We exemplify this using the EPS09 nPDF set and F_2 pseudodata generated from a saturation model.

1. Why e+A collisions?

One fundamental consequence of QCD dynamics is the emergence of parton saturation, a regime in which the evolution of the theory is non-linear due to how the parton densities grow at asymptotically small x. This parton saturation is characterized by the momentum $Q_s(x)$, the so-called saturation scale, and separates the dilute and dense regimes. The smaller x is, the higher $Q_s(x)$ gets. Data from e+A collisions can prove that the non-linearity of QCD evolution is truly essential or render it insignificant.

e+A collisions can also bring measurements with strong evidence of saturation, as well as some knowledge about the properties of this dense regime.

From a theoretical point of view, QCD is best approximated in the saturation regime by the Color Glass Condensate (CGC) effective theory. One of CGC weak points is that its results need, among other non-perturbative inputs, the impact parameter dependence of the gluon density. e+A measurements can give information about this and so help reduce the uncertainties of the CGC.

But one of the main reasons to study e+A collisions resides in the fact that they will provide precise measures of the longitudinal structure function F_L, that is related to the distribution of gluons. A linear combination of F_L and the structure function F_2 (related to the distribution of quarks) gives the total cross section of the process γ^*+A, measured in inclusive deep inelastic scattering (DIS). Having precise data of F_2 and F_L is of utmost importance to be able to answer a hugely interesting question: if linear DGLAP evolution models (such as EPS09 [1]) can simultaneously describe F_2 and F_L data coming from e+A collisions, taking into account that saturation behaves as the saturation models that are now in the market (such as the running-coupling Balitsky-Kovchegov (rcBK) equation [2]) predict.

More information about what e+A measurements can bring us is provided in [3].

2. What do we want to do?

The objective of this work is to constrain the nuclear parton distribution functions (nPDFs) given some e+A pseudodata of the structure function F_2. There is work in progress to do the same with F_L pseudodata.

3. How do we do it?

Our work consists of two different parts: how to generate and how to deal with the pseudodata. We consider an e+A DIS process in which an electron interacts with a nucleus via a virtual photon. This process can be looked at from two different frames: the Bjorken and the dipole frames. In the Bjorken one, we can express F_2 through the usual relation

$$F_2(x, Q^2) = \sum_q e_q^2 \left[xq(x, Q^2) + x\bar{q}(x, Q^2)\right] \tag{1}$$

via the nPDFs $xq(x, Q^2)$, while, in the dipole frame, F_2 takes the following form

$$F_2(x, Q^2) = \frac{Q^2}{4\pi^2\alpha_{em}}(\sigma_T + \sigma_L) \tag{2}$$

where σ_T and σ_L are the transverse (T) and longitudinal (L) cross section of the process, respectively [4]. We generate our pseudodata with expression (2), and given the equivalence between the two frames, we can use it to impose some constrains on the nPDFs via a reweighting process.

3.1. How do we generate the pseudodata?

In the dipole frame (valid in the regime $x \ll 1$), the cross section of the process γ^*+p is given by

$$\sigma_{T,L}^{\gamma^* h}(x, Q^2) = \sum_q \int_0^1 \mathrm{d}z \int \mathrm{d}^2 r \, |\Psi_{T,L}^q(e_q, m_q, z, Q^2, r)|^2 \, \sigma_{dip}^{ep}(r, x) \tag{3}$$

the convolution of $|\Psi_{T,L}^q|^2$, the wavefunction squared of a virtual photon splitting into a quark-antiquark dipole (the expressions to lowest order in α_{em} can be found in [5]) and $\sigma_{dip}^{ep}(r, x)$, the cross section of the quark-antiquark dipole scattering off the hadronic target [6]. The extension to γ^*+A is realized by the substitution $\sigma_{dip}^{ep}(r, x) \to \sigma_{dip}^{eA}(r, x)$. The optical theorem tell us that this new cross section can be calculated as the integral over impact parameter b of the (imaginary part of the) dipole-nucleus scattering amplitude:

$$\sigma_{dip}^{eA}(r, x) = \int \mathrm{d}^2 b \; \sigma_{dip}^{eA}(r, x, b) \tag{4}$$

All the information about the strong interactions and the x-dependence of the scattering process is encoded in this dipole amplitude.

The model of the nucleus we are using [7][8] has a Woods-Saxon density distribution $T_A(b)$ without fluctuations. The dipole-nucleus amplitude takes the following form:

$$\sigma_{dip}^{eA}(r, x, b) = 2\left[1 - \exp\left(-\frac{1}{2} A T_A(b) \sigma_{dip}^{ep}(r, x)\right)\right] \tag{5}$$

which is an average of the dipole cross section over the fluctuating positions of the nucleons in the nucleus, i.e., a mean field approach.

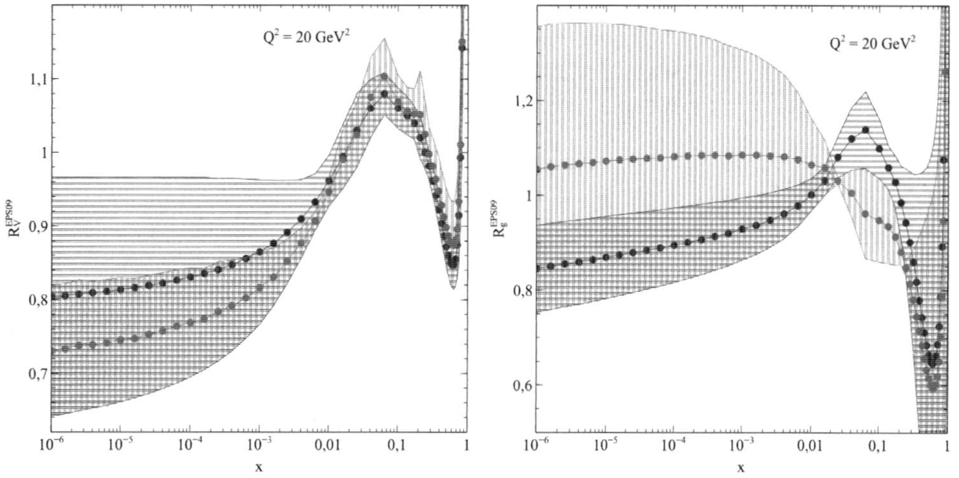

Figure 1. Preliminary results at $Q^2 = 20$ GeV^2 for the valence quark (left) and gluon (right) distributions. In blue, data of the original EPS09 fit and, in red, data of the reweighted EPS09.

3.2. How do we deal with the pseudodata?

Once the pseudodata is generated according to the model described above, we have to see how it affects the nPDFs. To do so, we perform a reweighting of the existing nPDFs that takes into account the new pseudodata. A reweighting is a process developed to save time, since it can give us a quantitative estimation of the degree of incompatibility between the data and the theory [9][10][11]. In case the new data and the theory are compatible, the nPDFs get *reweighted* and the resulting nPDFs are linear combinations of the original ones and include the new data. If data and theory are incompatible, a new global fit for the nPDFs is mandatory.

The performed reweighting is a Hessian method [12] that follows the steps indicated in [13]. The consistency of the new data with the original nPDFs is given by the relation between the so-called *penalty term* P and the tolerance $\Delta\chi^2$ of the original fit. If $P \ll \Delta\chi^2$, the existing nPDFs accepts the new data without causing trouble with the other data. However, if $P \gtrsim \Delta\chi^2$ the new data may be in conflict with the nPDFs considered, so it is not guaranteed that the result of the reweighting process would be reliable.

4. What do we get?

The set of nPDFs we have used is EPS09. The preliminary results of the reweighting process with the generated pseudodata for F_2 can be seen in Figure 1. For the valence quark, we can see that there is a suppression on the shadowing part with respect to the original fit, as well as an important constrain on the uncertainties. The reweighted nPDF follows the behaviour of the original fit. As for the gluon distribution, there is an enhancement in the small-x region, but the new uncertainties, far from being constrained, exceed the old ones, so we can't make any claim. Results for the sea quark distribution follow the same pattern as those for the gluon distribution. The trouble with the gluon and sea results can be understood if we pay attention to the penalty of the reweighting.

The values for the tolerance $\Delta\chi^2$ of the EPS09 nPDF set and the penalty term for the performed reweighting process are listed in Table 1. As we can observe, $P \gg \Delta\chi^2$, so the reweighted EPS09 with the F_2 pseudodata might not be reliable.

One possibility for the discrepancy we see between the pseudodata and theory is that the

Table 1. Reweighting results

nPDF	$\Delta\chi^2$	Penalty
EPS09	50	817,48

collinear factorization framework is not right as it is for e+A and it should be improved adding more QCD dynamics in the formalism. But we cannot claim this hypothesis is true until we have pseudodata from a more realistic saturation model.

5. Are there any conclusions?

We have seen that e+A measurements can give a lot of useful information as well as shed some light on important topics such as saturation. We have commented on how to use F_2 data from e+A collisions to enforce some contrains, if possible, into the existing nPDFs. We have done it for the EPS09 set and see that the pseudodata might not be compatible with the rest of the data in the nPDF. We wonder if this incompatibility comes from the collinear factorization framework, but given that the results are not final we cannot venture to take any further conclusion yet.

Acknowledgments

The author wants to thank Cyrille Marquet and Pía Zurita, collaborators in this work. This work is supported by European Research Council grant HotLHC ERC-2011-StG-279579, by Xunta de Galicia (Consellería de Educación); by the Spanish Consolider-Ingenio 2010 Programme CPAN and by FEDER.

References

[1] K. J. Eskola, H. Paukkunen and C. A. Salgado, JHEP **0904**, 065 (2009), [arXiv:0902.4154 [hep-ph]]
[2] J. L. Albacete and Y. V. Kovchegov, Phys. Rev. D **75**,125021(2007).
[3] C. Marquet, these proceedings.
[4] J. L. Albacete, N. Armesto, J. G. Milhano and C. A. Salgado, Phys. Rev. D **80**, 034031(2009), [arXiv:0902.1112v2 [hep-ph]]
[5] K. Golec-Biernat and M. Wüsthoff, Phys. Rev. D**59** (1999) 014017, [hep-ph/9807513].
[6] J. L. Albacete and C. Marquet, [arXiv:1401.4866 [hep-ph]].
[7] N. Armesto, Eur.Phys.J. **C26** (2002) 35 [arXiv:hep-ph/0206017 [hep-ph]].
[8] T. Lappi and H. Mäntysaari, [arXiv:1309.6963 [hep-ph]].
[9] W. T. Giele and S. Keller, Phys. Rev. D **58** (1998) 094023 [hep-ph/9803393].
[10] R. D. Ball *et al.* [NNPDF Collaboration], Nucl. Phys. B **849** (2011) 112 [Erratum-ibid. B **854** (2012) 926] [Erratum-ibid. B **855** (2012) 927] [arXiv:1012.0836 [hep-ph]].
[11] R. D. Ball, V. Bertone, F. Cerutti, L. Del Debbio, S. Forte, A. Guffanti, N. P. Hartland and J. I. Latorre *et al.*, Nucl. Phys. B **855** (2012) 608 [arXiv:1108.1758 [hep-ph]].
[12] J. Pumplin, D. Stump, R. Brock, D. Casey, J. Huston, J. Kalk, H. L. Lai, W. K. Tung, Phys. Rev. D **65** (2001) 014013 [hep-ph/0101032].
[13] H. Paukkunen and P. Zurita, [arXiv:1402.6623 [hep-ph]].

Hot Quarks 2014 IOP Publishing
Journal of Physics: Conference Series **612** (2015) 012020 doi:10.1088/1742-6596/612/1/012020

From light- to heavy-quark fragmentation in jets: current measurements in pp collisions with the ALICE detector and perspectives

C Bianchin for the ALICE Collaboration

Utrecht University, Princetonplein, 5 3584 CC Utrecht, The Netherlands

E-mail: chiara.bianchin@cern.ch

Abstract. The goal of the ALICE experiment at the LHC is to investigate the properties of the high-density state of strongly-interacting matter produced in Pb–Pb collisions. Partons propagating in a coloured medium interact with it leading to a softening of the parton p_{T} spectrum. The parton fragmentation function can be influenced by in-medium interactions leading to particle dependent modifications. Fragmentation function measurements in pp collisions are fundamental as reference for the Pb–Pb measurements and per se in order to constrain fragmentation models. The jet fragmentation into charged particles, pions, kaons, and protons measured by ALICE in pp collisions are presented. Different PYTHIA tunes are challenged in reproducing jet fragmentation into light particles. The measurement of the fragmentation of charm into heavier hadrons like the D mesons is being explored as well. The first studies and the perspectives for the LHC-Run2 are reported.

1. Introduction

Jets are collimated sprays of particles associated with hard-scattered partons. The study of jet production and fragmentation allows us to test our understanding of perturbative and non-perturbative aspects of QCD. Heavy quarks (Q), charm and beauty, are formed in initial hard scattering processes, hence on a shorter time scale with respect to the early hot and dense quark-gluon plasma phase. They interact with the medium via both inelastic (medium-induced gluon radiation) and elastic (collisional energy loss) processes [1]. Quarks are expected to lose less energy with respect to gluons (g), due to their smaller colour coupling factor. The small-angle gluon radiation is reduced for heavier quarks with moderate energy-over-mass values (so called "dead-cone effect") [2], with respect to light quarks (q). A hierarchy $E_{\mathrm{loss}}(Q) < E_{\mathrm{loss}}(q) < E_{\mathrm{loss}}(g)$ is hence expected. The fragmentation process inside a dense medium could thus be modified with respect to the vacuum in different ways for different partons.

2. Light quark and gluon fragmentation

At the LHC gluons are the most abundant, but the modelling of their fragmentation is poorly constrained by data, since the e^+e^- experiments mostly constrain quarks. Charged jets are reconstructed in the ALICE central barrel from primary charged particle tracks measured in the Inner Tracking System (ITS) and the Time Projection Chamber (TPC). Tracks with transverse momentum $p_{\mathrm{T}} > 0.15$ MeV/c in the pseudo-rapidity interval $|\eta| < 0.9$ are clustered with the

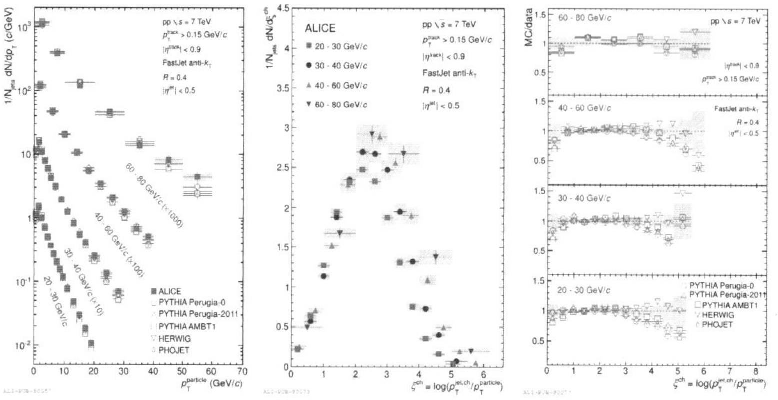

Figure 1. Transverse momentum distribution of charged tracks inside the jet (left) and their momentum fraction ξ (middle) in $p_{\mathrm{T}}^{\mathrm{jet,ch}}$ intervals from 20 to 80 GeV/c. The particle-p_{T} spectra are compared with Monte Carlo simulations (see legend) in the left panel, while the right panel shows the ratios of the ξ distribution in data and Monte Carlo simulations [8].

FastJet [3] anti-k_T algorithm with a resolution parameter $R = 0.4$ using the boost-invariant p_{T} recombination scheme. Jets which are fully contained within the detector acceptance are selected. The jet constituents' transverse momentum spectra (jet fragmentation) are measured in the leading (highest p_{T}) jets (Fig. 1, left panel). The distribution is corrected for detector effects bin-by-bin using simulations.

The middle panel of Fig. 1 shows the distribution of $\xi = \log(1/z^{\mathrm{ch}}) = \log(p_{\mathrm{T}}^{\mathrm{jet,ch}}/p_{\mathrm{T}}^{\mathrm{particle}})$. The subtracted underlying contribution was measured inside perpendicular cones transverse to the jet axis. The data are compared with different PYTHIA 6 tunes [4], HERWIG [5], and PHOJET [6]. The ratio of Monte Carlo to data in the right panel shows a general agreement at $\xi < 3$ and a slightly better description of the data by HERWIG at larger ξ.

The jet fragmentation measurement of identified particles (π, K, p) is shown in Fig. 2. The measurement utilizes the TPC for the dE/dx information. At high (> 4 GeV/c) p_{T} the dE/dx is very similar for the three species, hence advanced techniques based on fits to the TPC signal [9, 10] were used to identify particles. The results are corrected for detector effects, including tracking efficiency, track p_{T} resolution, jet p_{T} resolution, secondary particle and muon (fake pion) contamination. The fragmentation function distribution peaks at about $z^{\mathrm{ch}} \sim 0.1$ and a scaling is visible in the two higher $p_{\mathrm{T}}^{\mathrm{jet,ch}}$ bins.

Figure 3 shows the ratio of the measured p_{T} spectra of identified particles in jets over simulations with different PYTHIA 6 tunes (Perugia0 with and without colour reconnection and Perugia2011 [7]). The simulations reproduce fairly well the distribution of kaons and of pions at $p_{\mathrm{T}}^{\mathrm{track}} \gtrsim 400$ MeV/c. Protons are more challenging for models to be reproduced.

3. Heavy quark fragmentation

Measurements of the D$^{*\pm}$ meson momentum fraction ($X_E = E/E_{beam}$, where E is the D$^{*\pm}$ meson energy) from e^+e^- and ep experiments [11, 12] showed that charm quarks fragment harder than light quarks, namely with $z \approx 0.6$. Direct charm production via pair creation, gluon splitting into a charm pair, and beauty feed-down contribute to the D$^{*\pm}$ meson yields. The D$^{*\pm}$ mesons originating from gluon splitting can be isolated via topological selection of the region of phase space where the largest invariant mass is concentrated. The fragmentation is softer than that of charm from pair creation and it is well reproduced by JETSET Monte Carlo simulations [11].

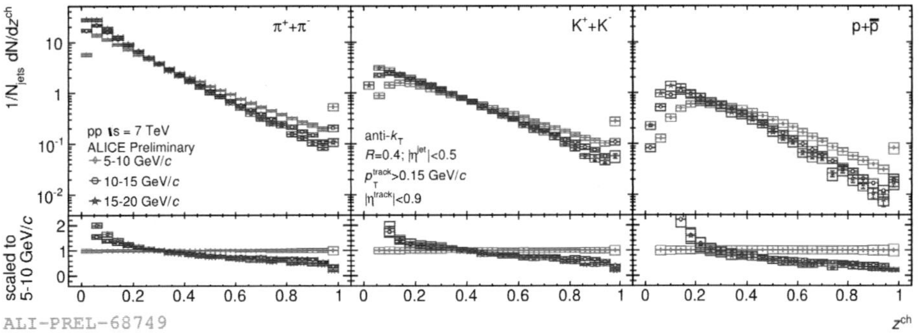

Figure 2. Jet momentum fraction $z^{\rm ch}$ of $\pi^+ + \pi^-$ (left), $\rm K^+ + K^-$ (middle), and $\rm p + \bar{p}$ (right) in jets for the ranges $p_{\rm T}^{\rm jet,ch}$ 5-10 GeV/c (red), 10-15 GeV/c (blue), 15-20 GeV/c (green).

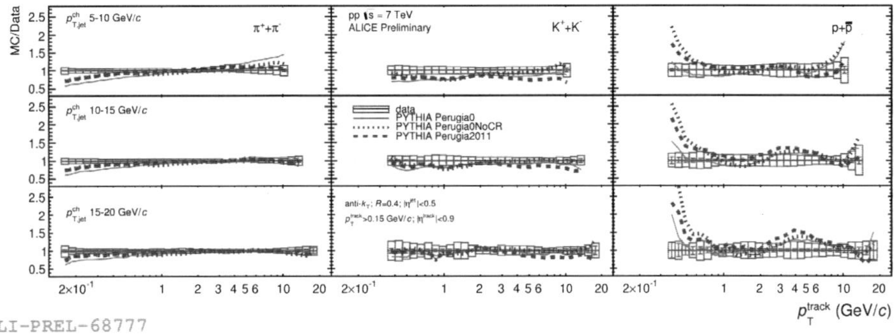

Figure 3. Ratio of the measured $p_{\rm T}$ spectrum of identified particles (left to right $\pi^+ + \pi^-$, $\rm K^+ + K^-$, and $\rm p + \bar{p}$) in charged jets and different PYTHIA tunes (see legend).

In hadronic collisions the initial parton energy is not known and the momentum fraction is defined as $z = p_{||}^{\rm D}/p^{\rm jet}$, where $p_{||}^{\rm D}$ is the component of the D meson momentum parallel to the jet axis and $p^{\rm jet}$ is the jet momentum magnitude. The contribution from gluon splitting dominates at azimuthal angles close to the jet axis. At LHC energies gluons dominate, hence the charm production via gluon splitting is expected to become more important.

The ATLAS Collaboration published a measurement of the $\rm D^{*\pm}$ meson production in jets in pp collisions at $\sqrt{s} = 7$ TeV and reported a large discrepancy between data and Monte Carlo simulations, both leading and next-to-leading order calculations [13]. Monte Carlo predictions fail to describe the data at small values of z, and this is most marked at low jet transverse momentum ($20 < p_{\rm T} < 60$ GeV/c). The ALICE experiment can play an important role in the investigation of the low-$p_{\rm T}$, low-z region. The first studies were performed with the pp sample at $\sqrt{s} = 8$ TeV using events triggered by the ElectroMagnetic Calorimeter (EMCal). The reconstruction of the $\rm D^{*\pm}$ candidates is performed with topological selections and particle identification as in [14]. Charged tracks are clustered in jets with the FastJet algorithm as (cf. Sec. 2) and finally $\rm D^{*\pm}$ mesons in a radius[1] of 0.4 around their axis are selected as $\rm D^{*\pm}$ mesons in jets. An invariant mass analysis is applied to the $\rm D^{*\pm}$ mesons in jets. As an example, the extracted signal via a fits are shown in the left panel of Fig. 4 for 3 bins of uncorrected

[1] The radial distance is defined as $R(\rm D, jet) = \sqrt{(\phi_D - \phi_{jet})^2 + (\eta_D - \eta_{jet})^2}$

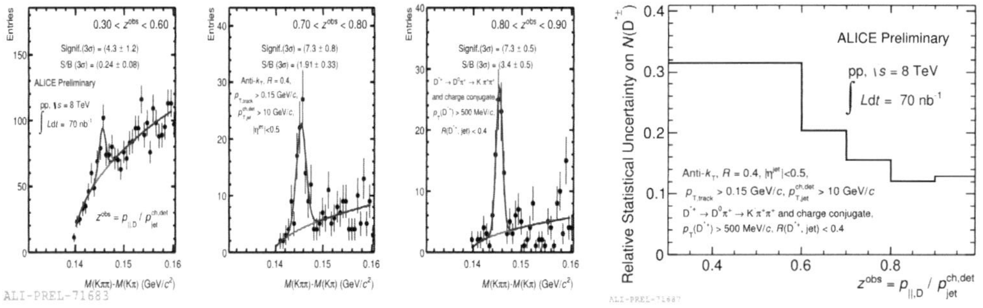

Figure 4. D$^{*\pm}$ meson invariant mass in 3 bins of uncorrected charged jet momentum fraction z^{obs} (left) and relative uncertainties on the signal extraction (right) in pp collisions at $\sqrt{s} = 8$ TeV.

charged jet momentum fraction. The signal is extracted in bins of z^{obs}, where the subscript *obs* indicates that the $p_{\rm T}$ spectrum of the jets is not yet corrected for detector effects. In the right panel of Fig. 4 the relative uncertainty on the signal as a function of z^{obs} is reported. The large magnitude of the uncertainties does not allow to proceed to the corrections needed (the already mentioned $p_{\rm T}^{\rm jet,ch}$ correction, but also D meson efficiency and the feed-down subtraction). In order to explore the low-$p_{\rm T}$, low-z region a large minimum bias sample is needed. ALICE plan for LHC-Run2 is to collect a sample of 4×10^9 minimum bias events (10 times the current largest minimum bias sample at $\sqrt{s} = 7$ TeV) in pp collisions at $\sqrt{s} = 13$ TeV. This should be enough to perform the measurement down to $z < 0.3$ with a statistical uncertainty of 5%.

4. Summary

The current results on fragmentation function measurements in pp collisions performed with the ALICE detector were shown. The charged jet fragmentation function was measured in bins of $p_{\rm T}^{\rm jet,ch}$ in the region 20-80 GeV/c. Monte Carlo simulations do not reproduce the data in the region $\xi > 3.5 - 5$. The jet fragmentation in identified π, K, and p was also measured in the region $5 < p_{\rm T}^{\rm jet,ch} < 20$ GeV/c. Monte Carlo simulations fail in reproducing protons and pions at low-$p_{\rm T}^{\rm jet,ch}$ and low-$p_{\rm T}^{\rm track}$. The first studies for the measurement of the fragmentation function of D$^{*\pm}$ mesons were presented together with the outlook for Run2.

References

[1] D. d'Enterria, Landolt-Boernstein Vol. 1-23A, Springer Verlag (2010)
[2] Y. L. Dokshitzer and D. E. Kharzeev, Phys. Lett. B 519 (2001)
[3] M. Cacciari and G. P. Salam, Phys. Lett. B 641 1 (2006)
[4] T. Sjöstrand, S. Mrenna and P. Skands, JHEP 05 026 (2006)
[5] G. Corcella *et al.*, JHEP 01 010 (2001)
[6] S. Roesler, R. Engel and J. Ranft, SLAC-PUB-8740 (2000)
[7] P. Z. Skands, Phys. Rev. D 82 (2010) 074018
[8] B. Abelev (ALICE Collaboration), arXiv:1411.4969 [nucl-ex] (2014)
[9] X.-G. Lu, doctoral thesis CERN-THESIS-2013-179 (2013), X.-G. Lu, QM2014 proceedings, arXiv:1407.8385 [nucl-ex] (2014)
[10] B. Hess, LHCP 2014 proceedings, arXiv:1408.5723 [nucl-ex] (2014)
[11] R. Barate (ALEPH Collaboration), Eur. Phys. J. C 16 4 (2000)
[12] S. Chekanov (ZEUS Collaboration), Eur. Phys. J. C 44 351 (2005)
[13] G. Aad (ATLAS Collaboration), Phys. Rev. D 85 052005 (2012)
[14] K. Aamodt (ALICE Collaboration), JHEP 01 128 (2012)

Bottomonium measurements in pp, pPb and PbPb using the CMS detector

Nicolas Filipovic, on behalf of the CMS collaboration
Laboratoire Leprince-Ringuet, École polytechnique, 91128 PALAISEAU Cedex, France
E-mail: `filipovic@llr.in2p3.fr`

Abstract. Bottomonia are important probes of the quark-gluon plasma since they are produced at early times and propagate through the medium, mapping its evolution. The three Υ states (1S, 2S, 3S) were measured separately using the Compact Muon Solenoid (CMS) experimental apparatus and observed to disappear sequentially in PbPb collisions at 2.76 TeV. However, recent measurements in pp and pPb collisions, at 2.76 and 5.02 TeV respectively, show a surprising dependence of the excited state (2S or 3S) over the ground (1S) state ratio, as a function of event activity. The three Υ states are also observed to be individually more produced in events with more activity. We review the latest results from pp, pPb and PbPb collisions and highlight their possible interpretations.

1. Introduction

Quarkonium suppression is known to be one of the experimental signatures of the formation of a Quark-Gluon Plasma [1]. Additional forms of modification of a heavy $q\bar{q}$ pair may arise in nuclear interactions, such as gluon shadowing or absorption in spectator nuclear matter. In 2012, the CMS collaboration has published a centrality-dependent measure of Υ suppression at $\sqrt{s_{NN}} = 2.76$ TeV with 150 μb^{-1} of PbPb data [2]. However, the measurement of the nuclear modification R_{AA} in peripheral events was limited to a wide $50-100\%$ centrality range and no transition from cold to hot effects can be established. In a recent paper by CMS [3], the relative production of the excited states $\Upsilon(2S)$ and $\Upsilon(3S)$ over the ground $\Upsilon(1S)$ state is studied with 5.4 pb^{-1} of pp data and 34 nb^{-1} of pPb data, by the use of single yield ratios and double ratios, as well as self-normalised cross section ratios. These observables are evaluated for various event activities.

2. Sequential suppression of Υ in PbPb

Υ mesons are measured in CMS through their di-muon decay. A description of the CMS detector can be found elsewhere [4]. Muons are reconstructed in CMS using information from the muon chambers and the tracker. Events containing two muon tracks well matched to hits in the muon spectrometer are selected by the hardware level trigger and further filtered offline using the High Level Trigger of CMS [5]. The Υ yields are then extracted with a fit to the invariant mass spectrum of the di-muon event sample. A first observation of sequential Υ suppression was reported in [2], by computing the nuclear modification factor R_{AA}. With the use of PbPb and pp data at the same energy, the centrality-integrated R_{AA} values for each Υ state are: $R_{AA}(1S) = 0.56 \pm 0.08(\text{stat.}) \pm 0.07(\text{syst.})$, $R_{AA}(2S) =$

$0.12 \pm 0.04(\text{stat.}) \pm 0.02(\text{syst.})$, $R_{\mathrm{AA}}(3S) < 0.10$ (95%CL). The nuclear modification factor as a function of centrality (number of participating nucleons) is reported in Fig. 1. From these values, one can observe that the excited states are more suppressed than the ground state over all centrality ranges. The ground state yield is suppressed up to a level, which is compatible with the known feed-down fractions, suggesting that its suppression could come from the disappearance of higher energy states only. Comparing the R_{AA} of excited states to that of the ground state can be done by measuring the double ratio $\chi^{\mathrm{PbPb}}(2S/1S) = R_{\mathrm{AA}}(2S)/R_{\mathrm{AA}}(1S) = N(2S)/N(1S)|_{PbPb}/N(2S)/N(1S)|_{pp}$. In this ratio most of the detector effects cancel, as well as parton energy loss and gluon shadowing, which are expected to affect all $b\bar{b}$ states the same way. A clear suppression of the $\Upsilon(2S)$ and $\Upsilon(3S)$ states was measured over all centrality ranges with respect to the $\Upsilon(1S)$. The centrality-integrated values of the double ratios are $\chi^{\mathrm{PbPb}}(2S/1S) = 0.21 \pm 0.07(\text{stat.}) \pm 0.02(\text{syst.})$, $\chi^{\mathrm{PbPb}}(3S/1S) < 0.17$ (95%CL) as reported in Fig. 2.

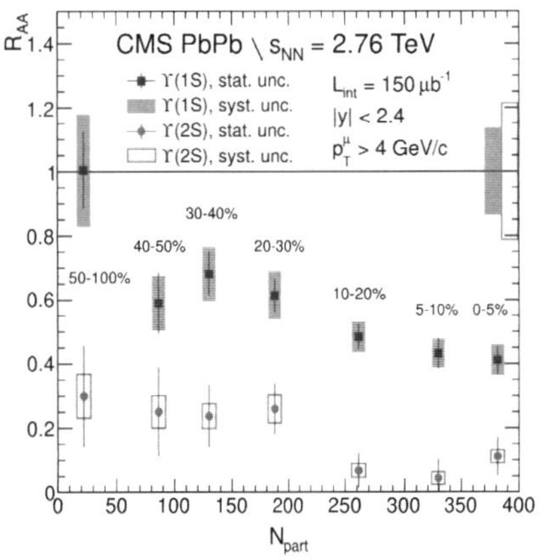

Figure 1. Nuclear modification factor R_{AA} for Υ states versus centrality.

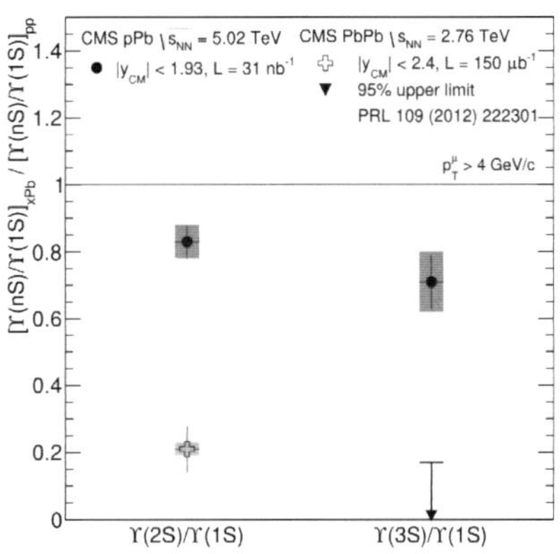

Figure 2. Double ratios $\chi(2S/1S)$, $\chi(3S/1S)$ for pPb and PbPb versus pp.

To extend this measurement of suppression to pPb collisions, double ratios are measured in pPb $\chi^{\mathrm{pPb}}(2S/1S)$, $\chi^{\mathrm{pPb}}(3S/1S)$ in [3] and reported in Fig. 2. The double ratio values are $\chi^{\mathrm{pPb}}(2S/1S) = 0.83 \pm 0.05 \pm 0.05$, $\chi^{\mathrm{pPb}}(3S/1S) = 0.71 \pm 0.08 \pm 0.09$ where the quoted uncertainties are statistical first and systematic second. This result indicates the excited states are more affected than the ground state. This suggests the presence of final state effects in pPb compared to pp, calling for a differential measurement as a function of event activity.

3. Event activity measurements in pPb

3.1. Transverse energy deposits in forward calorimeters: $E_T^{4<|\eta|<5.2}$

The CMS detector is equipped at forward pseudorapidities ($2.9 < |\eta| < 5.2$) with hadron forward ($\mathrm{HF}^{\pm}$) calorimeters, composed of steel absorbers embedded with quartz fibres emitting Čerenkov light, collected downstream by photomultipliers. The transverse energy deposit is used in PbPb for a measurement of the centrality (impact parameter) of the collision. In the context of pPb collisions, the total transverse energy deposit in the pseudorapidity range $4.0 < |\eta| < 5.2$ is used to separate the collision data in classes of forward event activity. This sampling at forward pseudo-rapidities provides an event activity measurement that should be independent of the measurement of the hard probe.

3.2. Inner tracker charged particle multiplicity: $N_{\rm Trk}^{|\eta|<2.4}$
The CMS tracking system is composed of an inner set of 66 million pixel detectors deployed in 3 layers around the beam pipe and silicon microstrips containing 9.9 million channels spread among a variable number of layers, within 4.8 pseudorapidity units around the primary vertex. This high precision device allows a very efficient tracking for charged particles with transverse momentum down to $p_T < 400$ MeV/c. The charged particle multiplicity associated with di-muon events is used to separate the data in classes of multiplicity.

4. Yield ratios

The single ratios $\Upsilon(2S)/\Upsilon(1S)$ and $\Upsilon(3S)/\Upsilon(1S)$ measured in pPb and pp are displayed in bins of charged particle multiplicity $N_{\rm Trk}^{|\eta|<2.4}$ in Fig. 3 and in bins of $E_T^{4<|\eta|<5.2}$ in Fig. 4.

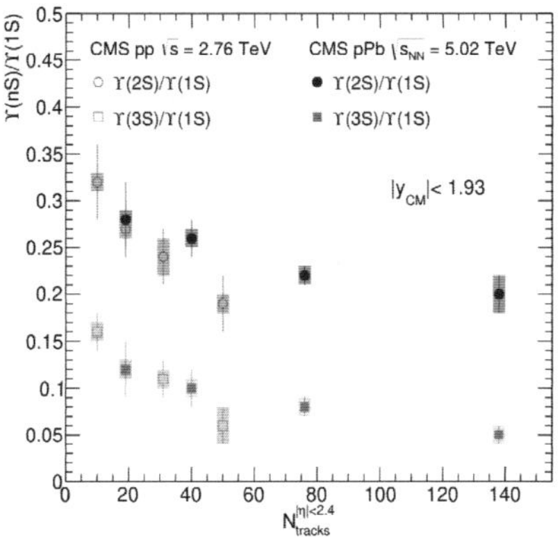

Figure 3. Single ratios $\Upsilon(2S)/\Upsilon(1S)$ and $\Upsilon(3S)/\Upsilon(1S)$ versus charged particles multiplicity in $|\eta_{\rm Trk}| < 2.4$.

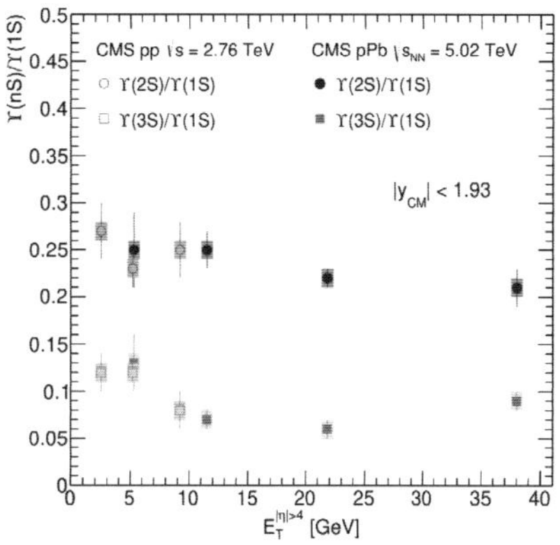

Figure 4. Single ratios $\Upsilon(2S)/\Upsilon(1S)$ and $\Upsilon(3S)/\Upsilon(1S)$ versus forward transverse energy.

The mean values of event activity observables are $\langle N_{\rm Trk}^{|\eta|<2.4}\rangle = (10, 41, 278)$, and $\langle E_T^{4<|\eta|<5.2}\rangle = (3.5, 15, 77)$ GeV for pp, pPb and PbPb, respectively. The ratios are fully corrected for detector inefficiencies regarding muon detection as well as event activity estimation. No clear dependence of the single ratios is observed when activity is measured in the forward region, while a variation is seen with increasing multiplicity at mid-rapidity. This dependence is unexpectedly steeper in pp than in pPb. Furthermore, the measured slope is compatible in both collision systems with the presence of two additional tracks on average in $\Upsilon(1S)$ events compared to $\Upsilon(2S, 3S)$. The number of additional charged particles does not vary when lowering the p_T threshold down to 200 MeV/c in the $N_{\rm Trk}^{|\eta|<2.4}$ computation, or when removing tracks located in a cone of radius $\Delta R = \sqrt{(\Delta\phi)^2 + (\Delta\eta)^2} = 0.3$ or 0.5 around the Υ momentum direction [3]. Moreover, the feed-down contribution from P-wave decays to Υ could not account for the full effect, given the low efficiency for electron pairs from low momentum photon conversion.

5. Self-normalised cross section ratios

To further investigate the Υ production rates, fully corrected cross sections were computed, normalised to their activity-integrated result. In Fig. 5, the horizontal axis displays multiplicity bins divided by the mean multiplicity. A positive correlation is observed with increasing charge-particle multiplicity. The self-normalised cross sections exhibit linear scaling with the forward transverse energy, as can be seen in Fig. 6.

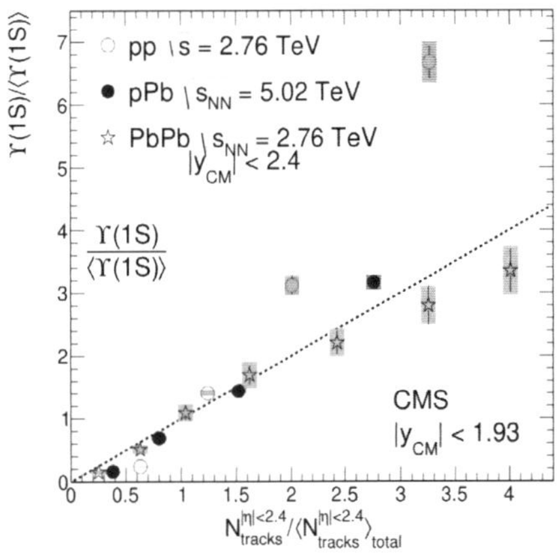

Figure 5. $\Upsilon(1S)$ yield, reconstructed with $|y^{CM}|$ <1.93 versus charged particle multiplicity in $|\eta|$ <2.4.

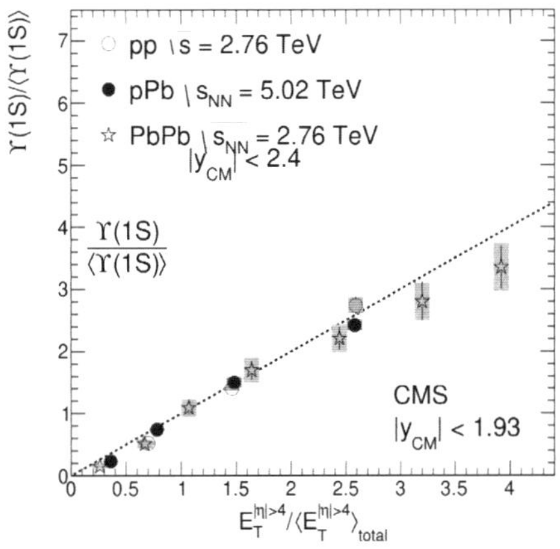

Figure 6. $\Upsilon(1S)$ yield, reconstructed with $|y^{CM}|$ <1.93 versus forward transverse energy.

6. Conclusion

We have presented the measurement of event-activity dependent production of Υ states by the CMS experiment in pp, pPb and PbPb collisions at the LHC. In the case of PbPb collisions, we observe a sequential pattern for suppression. Recent pPb data shows some suppression of the excited states with respect to the ground state, less pronounced than in PbPb collisions.

In the case of pp collisions, the normalised production rates exhibit correlations with associated particle multiplicity, a possible hint of quarkonium production being proportional to multi-parton interactions.

Acknowledgements

The speaker received funding from the European Research Committee, grant no. 259612.

References

[1] Matsui T and Satz H 1986 *Phys. Lett.* **B178** 416
[2] Chatrchyan S *et al.* (CMS Collaboration) 2012 *Phys. Rev. Lett.* **109** 222301 (*Preprint* 1208.2826)
[3] Chatrchyan S *et al.* (CMS Collaboration) 2014 *JHEP* **1404** 103 (*Preprint* 1312.6300)
[4] Adolphi R *et al.* (CMS) 2008 *JINST* **3** S08004
[5] Adam W *et al.* (CMS Trigger and Data Acquisition Group) 2006 *Eur. Phys. J.* **C46** 605–667 (*Preprint* hep-ex/0512077)

Measurement of Υ production in $p+p$ collisions at $\sqrt{s} = 500$ GeV in the STAR experiment

Leszek Kosarzewski
for the STAR Collaboration
Warsaw University of Technology
Koszykowa 75, 00-662 Warsaw, Poland
E-mail: `leszek.kosarzewski@gmail.com`

Abstract.
Quarkonium suppression was proposed as a signature of the quark-gluon plasma (QGP) formation. Different members of the quarkonium family are expected to melt at different temperatures providing an estimate of the temperature reached in the QGP. The Υ states offer a clean probe of the QGP, thanks to small secondary production and cold nuclear matter effects. Previous measurements at RHIC focused on colliding system and centrality dependence of Υ nuclear modification factor R_{AA}. The Υ R_{AA} as a function of p_T will provide additional information about Υ interaction with the QGP, which requires a p_T spectrum in $p+p$ collisions as a baseline. Measurement of the Υ p_T spectrum in $p+p$ collisions at $\sqrt{s} = 500$ GeV may provide such a baseline with high precision. It can be rescaled based on $pQCD$ calculations and used as a baseline for $Au+Au$ collisions at $\sqrt{s_{NN}} = 200$ GeV.

In this paper, the status of Υ measurements in $p+p$ collisions at $\sqrt{s} = 500$ GeV in the STAR experiment is presented. In addition, with the help of the newly installed Muon Telescope Detector (MTD) it will be possible to measure Υ 1S, 2S and 3S states separately.

1. Introduction
Suppression of J/ψ in $A+A$ collisions due to color screening was proposed as a signature of QGP formation [1]. This idea extends to other heavy quarkonia like Υ. Υ offers an advantage over J/ψ as a tool for QGP studies because it is less affected by the secondary production due to statistical recombination of heavy flavor quark-antiquark pairs in the QGP. In addition, cold nuclear matter effects (CNM) should be the same for all Υ states. The small differences in their masses makes the interpretation of CNM on Υ easier. The Υ studies in STAR in $p+p$ collisions at $\sqrt{s} = 500$ GeV offer an opportunity to obtain the high precision p_T spectrum. It may be rescaled and used as a baseline for $Au+Au$ collisions at $\sqrt{s_{NN}} = 200$ GeV to allow a study of dynamical aspects of Υ suppression.

2. Dataset
This analysis used data from $p+p$ collisions at $\sqrt{s} = 500$ GeV collected by the STAR experiment in 2011. The data were collected using a *High Tower* trigger, which selected events containing a high energy electron. This was done by implementing an online requirement (*L0* trigger) of at least one tower in the Barrel Electromagnetic Calorimeter (BEMC) [2] with $E > 4.3$ GeV (*L0 tower*) and a coincidence in Beam-Beam Counters (BBC). The tracks originating from

collisions were recorded by the Time Projection Chamber (TPC) [3]. After reconstruction the events with primary vertices within $|V_z| < 200\,cm$ of the center of the TPC were selected. The sample used in this analysis consists of 160 million events, which correspond to an integrated luminosity of $\mathcal{L} \approx 21.5\,pb^{-1}$.

3. Analysis

We reconstruct Υ in the e^+e^- channel ($BR = 2.38 \pm 0.11\%$) [4]. Electron candidates are selected using ionizing energy loss $\frac{dE}{dx}$ measured by the TPC. Figure 1 presents the distribution of $\frac{dE}{dx}$ and momentum p for all tracks. The $\frac{dE}{dx}$ is converted into a normalized $n\sigma_e$:

$$n\sigma_e = \log\left(\frac{\frac{dE}{dx}|_{Measured}}{\frac{dE}{dx}|_{Expected}}\right)/\sigma. \tag{1}$$

Here, σ is the $\frac{dE}{dx}$ resolution. Tracks with $-1.2 < n\sigma_e < 3$ are identified as electron candidates. Only tracks with $p > 1$ GeV/c are selected to reduce background.

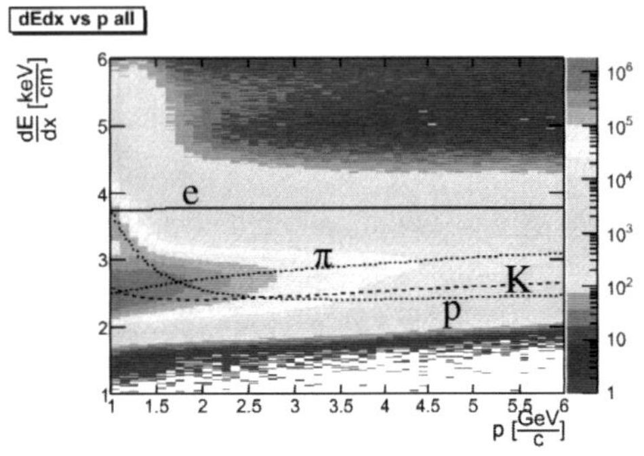

Figure 1. Energy loss $\frac{dE}{dx}$ as a function of momentum. The lines are theoretical values of $\frac{dE}{dx}$ from the Bichsel [5] functions for e, π, K and p.

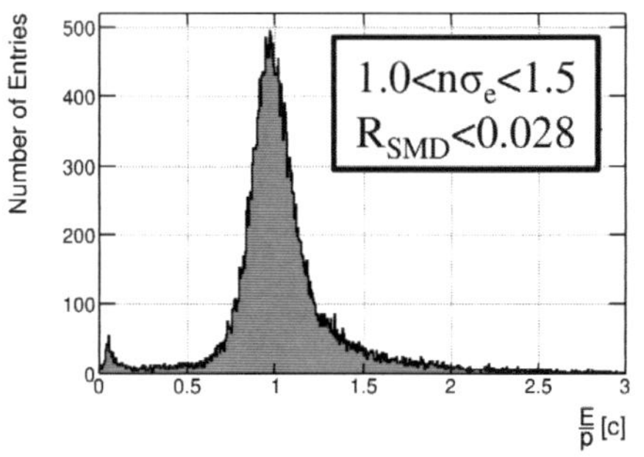

Figure 2. Distribution of $\frac{E_{CLU}}{p}$ for tracks with $1.0 < n\sigma_e < 1.5$ and $6 < p < 8$ GeV/c matched to BEMC towers and within $R_{SMD} < 0.028$ of a reconstructed cluster.

Information from the BEMC is used to further increase the purity of the electron sample. Each track is matched to the BEMC tower and energy in that tower is extracted as E_{TOW}.

Since electromagnetic showers could deposit energy in more than one tower, we reconstruct 3-tower clusters to recover full electron energy (E_{CLU}). Clusters are made by combining the tower pointed by a track and 2 neighbouring towers with the highest and second highest energies. The distance between the energy-weighted center of a cluster and a track projected to the Shower Maximum Detector layer of BEMC ($R_{SMD} = \sqrt{\Delta\eta^2 + \Delta\phi^2}$) is calculated.

Having reconstructed clusters, electron candidates are required to have $R_{SMD} < 0.028$. Electrons are expected to leave all their kinetic energies in the BEMC ($\frac{E}{pc} \sim 1$), so a cut $0.55 < \frac{E_{CLU}}{pc} < 1.45$ is applied. The distribution of $\frac{E_{CLU}}{p}$ for pure electron candidates with tight $\frac{dE}{dx}$ cut at $6 < p < 8$ GeV/c is presented in Fig. 2. Since the size of BEMC towers are close to the Moliére radius [2], most of the energy is expected to be deposited in a single tower. Thus it is required that $\frac{E_{TOW}}{E_{CLU}} > 0.5$.

The selected tracks are combined into pairs for Υ reconstruction and invariant mass is calculated. At least one track in a pair has to be matched to a *L0 tower*.

4. Signal

Υ signal is calculated by first obtaining the invariant mass distribution of e^+e^- pairs and a distribution of combinatorial background from both e^+e^+ and e^-e^- pairs. The e^+e^- distribution contains signal and both combinatorial and correlated background. The latter includes $b\bar{b}$ and Drell-Yan contributions. The background contributions are estimated by obtaining their functional parametrizations through fitting the combinatorial background in data or $b\bar{b}$ and Drell-Yan in Monte Carlo simulations. In addition, the shape for each of the Υ states is described by a Crystal Ball function [6], whose parameters are fixed based on MC simulation taking into account STAR detector responses. However, the relative contribution of each Crystal Ball function describing Υ 1S, 2S and 3S states is set free during fitting.

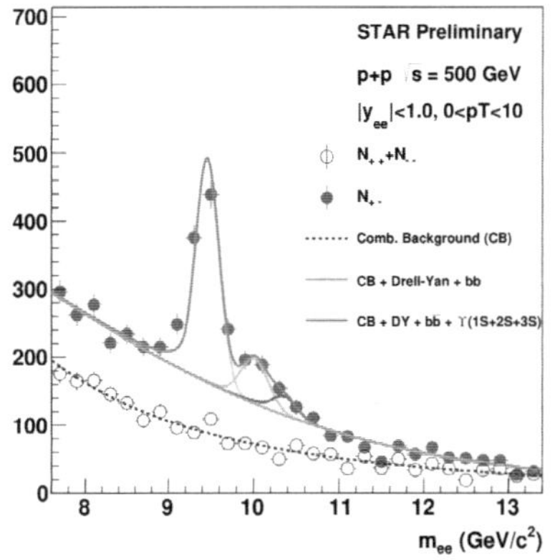

Figure 3. Invariant mass distribution of unlike-sign pairs (red closed circles) and like-sign pairs (blue open circles). The lines are various components of a simultaneous fit to both histograms. It includes combinatorial background (blue line), sum of: combinatorial+$b\bar{b}$+Drell-Yan (green line) and total Υ signal on top of background (red line). Also shown are the Crystal Ball functions for 1S (teal), 2S (orange) and 3S (violet) states.

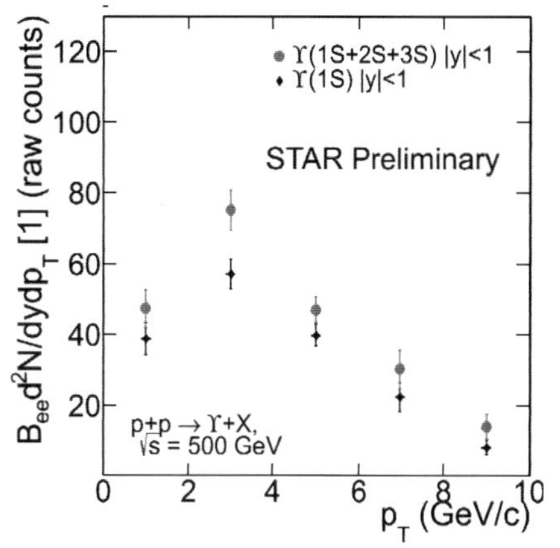

Figure 4. Uncorrected Υ p_T spectrum of combined 1S+2S+3S states (red closed circles) and 1S state (black diamonds) only.

The fit is done simultaneously to like-sign and unlike-sign pairs using RooFit [7] and its result is shown in Fig. 3. Finally, the Υ (1S+2S+3S) yield is calculated by integrating e^+e^- distribution and subtracting the integral of all background functions in the $8.8 < m_{ee} < 11\ \mathrm{GeV}/c^2$ mass range. The data allows separation of 1S state from the 2S and 3S, so the Υ 1S yield is calculated in the same way in the range of $8.8 < m_{ee} < 9.8\ \mathrm{GeV}/c^2$.

The total raw yield is divided into 5 p_T bins from $0 - 10\ \mathrm{GeV}/c$, which is shown in Fig. 4. No efficiency corrections are included. Results are summarized in Table 1 as well.

5. Outlook

STAR has undergone a few upgrades recently. The Muon Telescope Detector (MTD) [8] and Heavy Flavor Tracker (HFT) are the new detectors in STAR added for the 2014 run.

The MTD will allow precise measurements of quarkonia in the dimuon channel. Muons exhibit less bremsstrahlung than electrons and there is no background from γ conversion. The

Table 1. Summary of obtained raw Υ yield. The 1S+2S+3S yield is calculated in $8.8 < m_{ee} < 11$ GeV/c^2, while 1S in $8.8 < m_{ee} < 9.8$ GeV/c^2.

p_T [GeV/c]	1S+2S+3S Yield		1S Yield	
0-2	189 ± 22	(12%)	155 ± 18	(12%)
2-4	301 ± 23	(8%)	229 ± 18	(8%)
4-6	188 ± 15	(8%)	159 ± 13	(9%)
6-8	121 ± 22	(19%)	90 ± 16	(18%)
8-10	56 ± 15	(27%)	32.9 ± 8.4	(26%)

MTD will allow R_{AA} measurements separately for each of the Υ states. The statistical precision for Υ R_{AA} as a function of N_{part} are shown in Fig. 5.

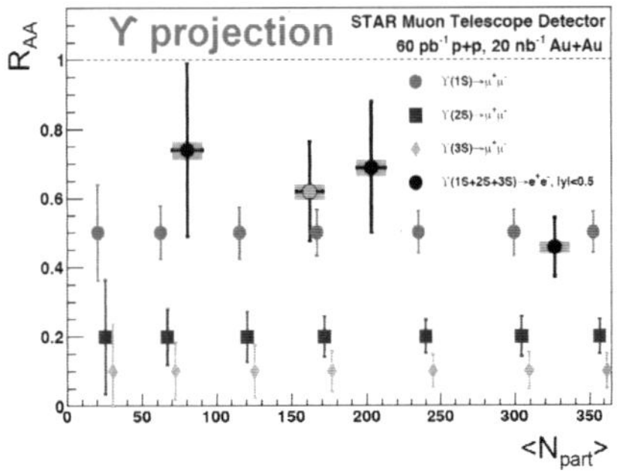

Figure 5. Expected precision for Υ R_{AA} as a function of N_{part} with MTD. The red circles are for 1S, the blue squares are for 2S and green diamonds are for 3S states. For comparison, recent STAR results are also shown as black closed and open circles [9].

The HFT is a silicon vertex detector, which will provide precision tracking for STAR. It will allow direct reconstruction of open charm and beauty decays. For Υ studies, the direct $b\bar{b}$ cross section measurement with the HFT will help understand the background.

6. Summary

The experimental method in STAR for Υ measurements in p+p at $\sqrt{s} = 500$ GeV was presented. The uncorrected yield of Υ 1S in $8.8 < m_{ee} < 9.8$ GeV/c^2 and combined Υ(1S+2S+3S) states in $8.8 < m_{ee} < 11$ GeV/c^2 were obtained. The results are summarized in Table 1. Based on that, the precision of the invariant yield is expected to be better than 20% for $p_T < 8$ GeV/c and even better than 10% for $2 < p_T < 6$ GeV/c. In addition prospects for Υ measurements in STAR were also reported.

References

[1] Matsui, T.; Satz, H., 1986, Phys. Lett. B, Volume 178, Issue 4, p. 416-422
[2] M. Beddo, et al. (STAR Collaboration), 2003, Nucl. Instr. and Meth. A499:721
[3] M. Anderson et al. (STAR Collaboration), 2003 Nucl.Instrum.Meth. A499:659-678
[4] K.A. Olive et al. (Particle Data Group), 2014, Chin. Phys. C, 38, 090001
[5] H. Bichsel, 2006, Nucl. Instr. Meth., A 562, 154
[6] T. Skwarnicki, 1986, DESY F31-86-02, Appendix E
[7] http://root.cern.ch/drupal/content/roofit
[8] L. Ruan et al., 2009, J. Phys. G 36, 95001
[9] L. Adamczyk et al. (STAR Collaboration), 2014 Phys. Lett. B 735, 127

Journal of Physics: Conference Series **612** (2015) 012023 doi:10.1088/1742-6596/612/1/012023

Five things you should remember about heavy flavor measurements

Daniel Kikoła

Faculty of Physics, Warsaw University of Technology, Warsaw, Poland

E-mail: `kikola@if.pw.edu.pl`

Abstract.
Heavy quarks are important tools to study the properties of the hot and dense nuclear matter created in heavy ion collisions. They are produced very early, in the initial interactions with large momentum transfer, and are expected to interact differently from light quarks with the nuclear medium. Measurements of heavy flavor production can provide insights into transport and thermodynamics properties of the created system. In this paper, I give as a brief introduction to the experimental studies of heavy flavor production in relativistic heavy ion collisions and discuss related challenges.

Introduction

Relativistic heavy ion collisions provide a unique opportunity for studying Quark-Gluon Plasma (QGP), a new state of nuclear matter with properties determined by quark and gluon degrees of freedom. In heavy ion collisions, we use an approach analogous to tomography for studying the QGP: We select an external probe, whose properties (production mechanism) are under experimental and theoretical control, which propagates through the medium. Then we can infer properties of the analyzed system from modification of the probe. Heavy quarks, charm and bottom, serve as such external-to-QGP probes. Because of their large masses, they are produced very early in the collision, in the initial interactions with large momentum transfer, before the QGP phase. Their production, both total and differential cross-sections, are well described by perturbative QCD (pQCD). Energy loss and elliptic flow of open heavy flavor are sensitive to the dynamics of the medium; such measurements could be used to determine the fundamental properties of the QGP, for instance transport coefficients (for details, see Ref. [1] and references therein). Elliptic flow of heavy quarks can shed new light on possible collective behavior and degree of thermalization of the created nuclear matter. Measurements of the production of various quarkonium states can provide insight into thermodynamic properties of the QGP ([2] and references therein).

In this paper, I give a brief introduction to the experimental studies of heavy flavor production in relativistic heavy ion collisions and challenges related to these measurements. I present experimental techniques and selected results and discuss their implications.

1. Quarkonium and open heavy flavors are different tools

Quarkonium and open heavy flavors are considered different tools for probing the QGP properties. Quarkonium, which is a bound state of $c\bar{c}$ or $b\bar{b}$ quarks, is a color-neutral object.

Its production can proceed via intermediate color-octet states [3, 4], but after a sufficiently long time a charmonium or bottomonium is formed. Quarkonium production is anticipated to be suppressed in the presence of the QGP, compared to yields in $p+p$ collisions scaled by the number of binary nucleon-nucleon collisions, due to the Debye screening of the quark-antiquark potential in the deconfined, partonic medium [5]. The Debye screening length depends on the temperature of the QGP. Different quarkonium states have different binding energies and thus different radii. Therefore, they are expected to disassociate at different temperatures [6, 7]. Thus measurements of production of various quarkonium states can provide insight into thermodynamic properties of the QGP i.e. its temperature. However, there are other effects that may affect the observed production, such as modifications of a parton distribution function in a nucleus, final state nuclear absorption or secondary production in the QGP, which I discuss in next sections.

Open heavy flavor hadrons are proxies for a single heavy quark, which carry a color charge and thus is expected to interact differently than quarkonium with the QGP. A fast parton traversing QGP will lose its energy due to interaction with the medium. The transport properties of the medium are extracted by comparing energy loss dE/dx observed in the data with models, where dE/dx is related to transport properties of the nuclear medium, for instance mean momentum transfer squared $\hat{q}$ [1].

2. We cannot measure heavy flavor quarks directly

Main complication in using heavy quarks for the QGP research is that we cannot study heavy quark production directly: we measure final state particles. Thus open heavy flavor measurement is always "contaminated" by light quarks. The best case scenario is a direct reconstruction of heavy flavor mesons which gives an access to the kinematics of the parent meson. Such measurement of D-mesons via their hadronic decay channels is feasible both at RHIC and LHC, but it is more difficult for bottom mesons and other means are used for that purpose (however, there are first measurements of exclusive B decays by the CMS collaboration in p+Pb collisions [8]). The $B \rightarrow J/\psi$ (so-called non-prompt J/ψ) is a popular approach, where J/ψ from B decays are separated from a direct J/ψ production using a pseudo-lifetime discriminating variable. Measurement of electrons from semi-leptonic decays of heavy flavor hadrons, e^{HF} (sometime called non-photonic electrons for historical reasons) is an alternative way to access heavy quarks' production. Such electrons are obtained by first selecting inclusive electron yields and then subtracting electrons from γ conversion (real photons conversion) or Dalitz decays of π^0 and η mesons (virtual photons conversion), called photonic background. Remaining electrons are dominated by open heavy flavor hadron decay, but contain other contributions as well. Hence a conservative name "non-photonic electrons" was invented for such sample, which has to be further corrected for contributions from weak kaon, Drell-Yan and quarkonium decays.

In the case of quarkonium, the main issue is a feed-down from excited states to ground state. Such decays amount to 40% of J/ψ (10% comes from ψ' and 30% from χ_c(1P)) and $\sim$ 40% for Υ(1S) mostly due to $\chi_{\rm b1}, \chi_{\rm b2}$ decays. This feed-down is extremely difficult to measure in heavy ion collisions because χ decays involve soft photons (energy $\sim$ 0.5 GeV) that are lost in a large background from the underlying event.

3. It is crucial to understand the reference

We use nuclear modification factors $R_{\rm pA}$ and $R_{\rm AA}$ to quantify modification of the production in p+A and A+A collisions (such as p + Pb, d + Au, Pb+Pb or Au+Au), respectively, compared to p+p interactions. They are defined as

$$R_{\rm pA}(R_{\rm AA})(p_T) = \frac{\sigma_{\rm inel}}{\langle N_{\rm coll}\rangle} \frac{{\rm d}^2 N_{\rm pA(AA)}/{\rm d}y{\rm d}p_T}{{\rm d}^2\sigma_{pp}/{\rm d}y{\rm d}p_T} \qquad (1)$$

where $\sigma_{\rm inel}$ is the inelastic cross section in $p+p$ collisions, $N_{\rm pA(AA)}$ is a yield in $p+A$ or $A+A$ collisions, respectively, σ_{pp} is the corresponding cross section in $p+p$ collisions, and $\langle N_{\rm coll}\rangle$ is an average number of binary collisions.

In the case of $p+A$ collisions, we generally do not expect a formation of a hot and dense matter. Thus p+Pb or d+Au measurements were introduced as a null experiments to quantify modification of the particle yields not related to the QGP formation, so called cold nuclear matter (CNM) effects. These include initial and final state effects. The former affect parton distributions in a nucleus (for instance shadowing) and thus the initial production yields. The latter modify yields or momentum distributions of final state particles. A Cronin effect (an enhancement of transverse momentum in $p+A$ with respect to $p+p$ collisions) is an example of such effect. These effects play an important role at SPS, RHIC and LHC. A significant quarkonium suppression was observed at midrapidity and forward rapidity at SPS [9], RHIC [10, 2] and at LHC [11, 12, 13, 14]. Overall, the suppression for J/ψ and Υ(nS) as a function of rapidity at LHC is consistent with model calculations based on a pure nuclear shadowing scenario [15, 16] and with models including a contribution from coherent partonic energy loss [17, 18]. An additional final breakup cross section is needed to explain the SPS [19] and RHIC data [10].

Experiments at RHIC and LHC recently reported collective behavior of light hadrons in high multiplicity d+Au [20] and p+Pb collisions [21], where these hadrons have significant elliptic flow v_2 with a characteristic mass-splitting pattern [22]. These observations triggered speculations that an enhancement of the e^{HF} production in central and minimum bias d+Au collisions at midrapidity at RHIC [23] could indicate a collective phenomena (radial flow) of heavy quarks in d+Au [24]. However, this phenomenon could also be owing to the CNM effects: The enhancement at low p_T is reproduced assuming Cronin effect for charm quarks [25]. No significant suppression/enhancement for open charmed and bottom hadrons is observed at LHC [26, 8, 11]: nuclear modification factor $R_{\rm pPb}$ for D-mesons measured by ALICE is shown in Fig. 4 (right panel).

For quarkonium, the most surprising CNM results are the strong suppression of ψ' production in d+Au collisions at $\sqrt{s_{\rm NN}}$ = 200 GeV [27] (Fig. 1) and p+Pb at $\sqrt{s_{NN}}$ = 5.02 TeV [28] and event activity dependence of Υ(nS) production in $p+p$ and p+Pb interactions [13]. The ψ' production is significantly more suppressed compared to J/ψ in p+Pb and d+Au collisions. One possible explanation is a breakup in the final state, either due to interaction with nuclear matter or with co-moving hadrons. Since this effect is significant in central p+Pb collisions at $\sqrt{s_{NN}}$ = 5.02 TeV, it has to be understood and taken into account in the interpretation of the quarkonium suppression pattern in heavy ion collisions. In the case of of Υ(nS) production in p+Pb (and also in $p+p$) interactions, the excited-to-ground-states cross section ratios Υ(nS)/Υ(1S) decrease with increasing charged-particle multiplicity measured at midrapidity (Fig. 2, left panel). The double ratios measured by the CMS collaborations, $[\Upsilon(\rm nS)/\Upsilon(1S)]_{\rm pPb}/[\Upsilon(\rm nS)/\Upsilon(1S)]_{p+p}$ points to final-state suppression effects in the p+Pb which affect more strongly the excited states than the ground state [13]. Another piece of the puzzle is Υ(1S) production in d+Au at 200 GeV collisions. A suppression of magnitude similar to that in central Au+Au collisions is observed there, which is difficult to explain by models.

Overall, the CNM effects are difficult to quantify because they depend on the kinematic range (longitudinal momentum fraction $x_{1,2}$ carried by the initial parton), energy and production mechanism (i.e. kinematics) [29]. The NA60 results for J/ψ production [9] provide an example how important it is to have a good handle on the reference measurement. Moreover, extrapolation of these effects from $p+A$ to $A+A$ is not straightforward [30].

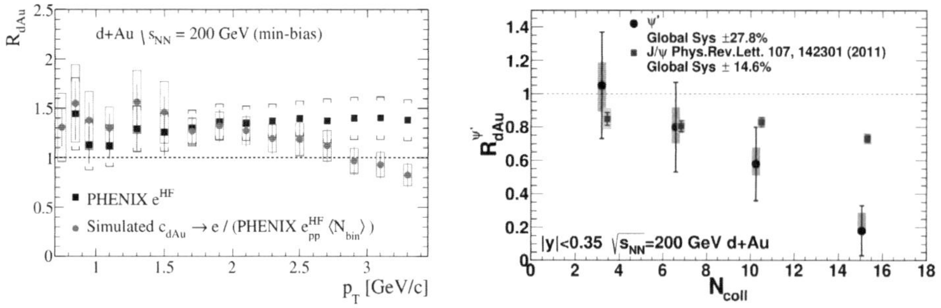

Figure 1. Cold nuclear matter effects at RHIC. Left panel: R_{dAu} for heavy flavor decay electrons compared to calculations which assume a Cronin effect for charm quarks [25]. Right panel: R_{dAu} for charmonium (ψ' and J/ψ) [27].

4. Color screening is a hypothesis

Quarkonium production is anticipated to be suppressed in the presence of the QGP and measurement of different quarkonium states in $A+A$ collisions gives a handle on the QGP temperature. This is a well established and popular idea, which served as a motivation for J/ψ measurements at SPS, RHIC and LHC and can be found in any conference talk on this topic. However, this hypothesis is still to be validated, because there are other possible scenarios. For the comprehensive review see [2, 31], here I list a few approaches successfully applied to describe the experimental data.

In the statistical hadronization model [32], the charm and bottom quarks thermalize in QGP and no quarkonium state is present in the deconfined state. All quarkonium states are produced during hadronization stage, together with all other hadrons. The comovers interaction model [33] does not assume thermal equilibrium and it does not use thermodynamical concepts. The quarkonium suppression is caused by the final state interactions of the $c\bar{c}$ pairs with the dense medium created in the collision. Quarkonium could be also created in the QGP, by (re)combination charm or bottom quarks [34, 35, 36]. In the transport approach [37, 38], the evolution of the quarkonium yield and spectrum is described by a transport relation, which includes terms for both quarkonium dissociation and formation reactions in the deconfined stage.

RHIC data for J/ψ production at $\sqrt{s_{NN}}$ = 200 GeV [39] were surprising because the level of suppression is similar as at SPS ($\sqrt{s_{NN}}$ = 17.3 GeV) and suppression at forward rapidity was larger than at midrapidity (Fig. 3, right panel). This observation was consistent with the hypothesis that there is a significant secondary production of J/ψ via coalescence of uncorrelated charm quarks, which is now an important ingredient of various models [2, 31]. Such a trend with rapidity can be also explained by Debye screening mechanism together with a larger suppression due to stronger shadowing (or other CNM effects) at forward rapidity compared to midrapidity.

The J/ψ suppression as a function of p_T has different trends at RHIC ($\sqrt{s_{\rm NN}}$ = 200 GeV) and LHC ($\sqrt{s_{\rm NN}}$ = 2.76 TeV). At RHIC, J/ψ is strongly suppressed at low p_T and then R_{AA} increases with transverse momentum, with a hint of suppression at high p_T in central collisions [40, 39]. At LHC, J/ψ is less suppressed at low p_T than at RHIC, and suppression increases with p_T [41, 42]. Such a trend is consistent with a significant contribution from J/ψ regeneration to the measured yields.

Upsilon states are better probes of the color screening hypothesis because they are less susceptible to CNM effects: Shadowing and final state absorption are expected to be smaller than in the case of J/ψ (the absorption due to interaction with co-moving hadrons was predicted to be minimal [43]). Moreover, the Υ(nS) regeneration is expected to be small and the initial-state effects are likely to cancel in a double ratio of excited-to-ground state in heavy ion and p+p collisions. Quarkonium suppression pattern in heavy ion collisions at RHIC and LHC is consistent with the color screening scenario: there is a larger suppression for loosely bound excited states compared to ground states J/ψ and Υ(1S) (Fig. 2(right panel)). However, there are observations which cannot be explained by color screening alone. Firstly, a ratio of excited-to-ground state as a function of event activity in p+p, p + Pb and Pb + Pb follows the same trend (Fig. 2(left panel)), which may suggest that there is a significant final state adsorption due to interactions with co-moving hadrons. Then, the nuclear modification factor Υ(1S) as a function of rapidity at LHC [44] (Fig. 3, left panel) exhibits a similar trend as J/ψ at RHIC: there is a larger suppression at forward rapidity than at midrapidity. These data present a challenge to models which successfully described J/ψ production at RHIC and LHC assuming a color screening and regeneration scenario [44] (Fig. 3) because secondary production of Υ(1S) via b-quarks coalescence is expected to be small. Such a trend could also originate from stronger CNM effects (shadowing) at forward rapidity compared to midrapidity (where even an anti-shadowing i.e. enhancement could be expected).

The jury is still out on the question of the nature of the quarkonium interactions with a hot nuclear matter.

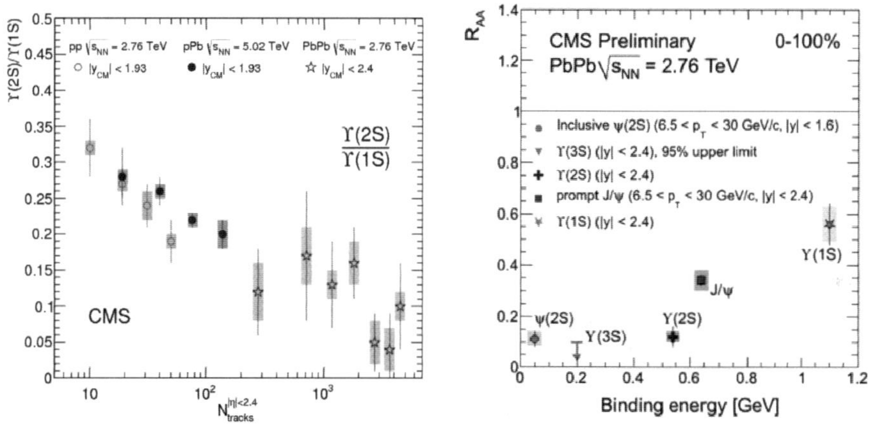

Figure 2. Left panel: Single cross section ratios Υ(2S)/Υ(1S) versus charged-particle multiplicity p+p collisions at $\sqrt{s}$ = 2.76 TeV, p+Pb collisions at $\sqrt{s_{NN}}$ = 5.02 TeV and Pb+Pb collisions at $\sqrt{s_{NN}}$ = 2.76 TeV. Right panel: Nuclear modification factor as a function of binding energy for various quarkonium states at LHC [45].

5. Open heavy flavor production at high p_T is suppressed at RHIC and LHC

Heavy flavor production at RHIC (D^0 and e^{HF}) in Au+Au collisions at $\sqrt{s_{NN}}$ = 200 GeV and at LHC (D-mesons and $B \rightarrow J/\psi$ in Pb+Pb collisions at $\sqrt{s_{NN}}$ = 2.7 TeV) is suppressed at high p_T (Fig. 4 and Fig. 5). This suppression is comparable to that for light hadrons. This was a surprising result. Models, which described the light flavor data well assuming gluon radiation (radiative energy loss, $dE/dx_{\rm rad}$) predicted that the heavy quark production will be less suppressed than light partons [46]. This lead to introduction of collisional energy loss $dE/dx_{\rm coll}$ due to binary interactions of partons with other objects in the QGP. Moreover, D-mesons at LHC and e^{HF} at RHIC have a significant positive elliptic flow [47, 48, 49]. These

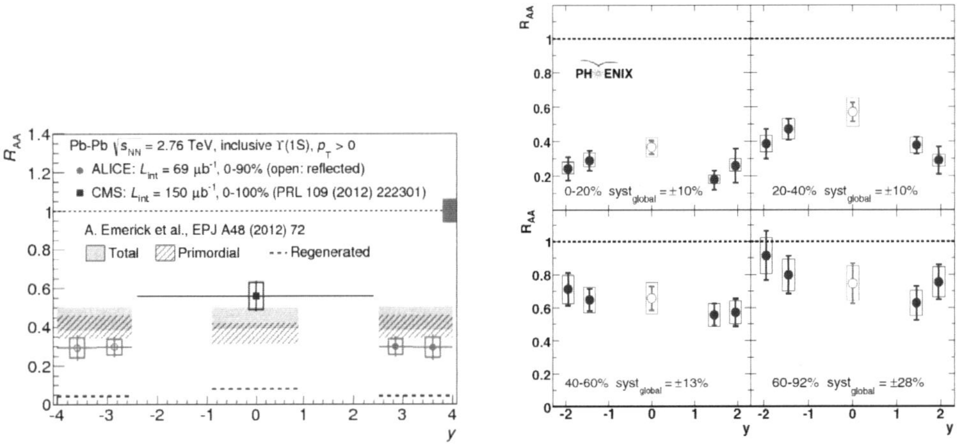

Figure 3. Nuclear modification factor as a function of rapidity for $\Upsilon(1S)$ at LHC [44] and J/ψ at RHIC [39].

observations suggest that charm quarks are strongly coupled with the medium: heavy quarks loose a large amount of energy and acquire a significant elliptic flow during interaction with matter created at top RHIC energy and at LHC.

At the moment it is challenging for models to describe the heavy flavor elliptic flow and high p_T suppression simultaneously. One difficulty is that the relative contributions of dE/dx_{coll} and dE/dx_{rad} is unknown. To overcome this problem, multiplication factors for dE/dx_{coll} and dE/dx_{rad} are adjusted based on the comparison with data. To understand better the interplay of these two processes, precise measurements of suppression of charm and bottom separately are necessary. New observables (for instance azimuthal correlations of charmed mesons [50]) could also help to address this issue.

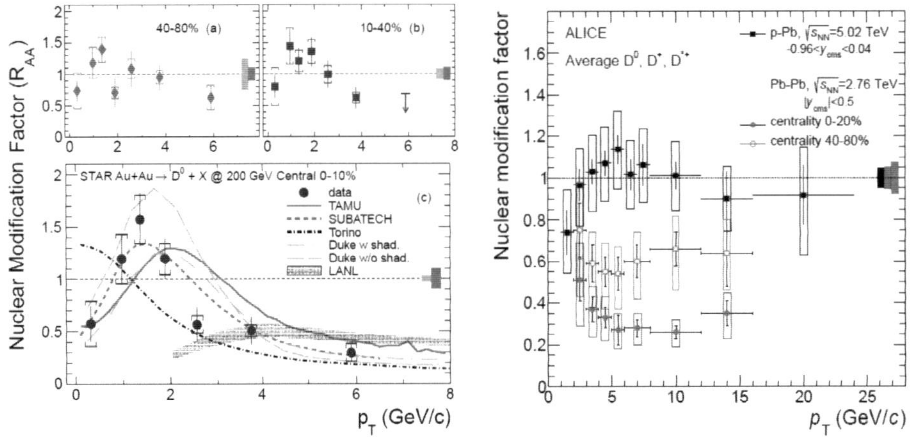

Figure 4. Charmed meson suppression at RHIC [51] and LHC [26].

The Beam Energy Scan (BES) program at RHIC was carried out recently to study the phase diagram of nuclear matter and search for a phase transition and a critical point. BES results show that elliptic flow of inclusive charged hadrons is approximately independent of beam energy (the difference is less than 10% for $0.5 < p_T < 3$ GeV/c) and light hadron production is suppressed at high p_T in the energy range of 39 – 200 GeV. In Au+Au collisions at $\sqrt{s_{NN}} = 62.4$ GeV, production of e^{HF} is not suppressed (Fig. 5, right panel). Moreover, contrary to results for light hadrons, where a positive v_2 is observed and the difference between 200 GeV and 39 GeV is small, STAR measurements in Au+Au collisions at $\sqrt{s_{NN}} = 62.4$ and 39 GeV [48] indicate that the v_2 of e^{HF} is consistent with zero. Furthermore, the difference between 200 GeV and results at those lower energies is statistically significant for $p_T < 1$ GeV/c.

These results indicate that there is a difference in the interactions of the heavy quarks with the surrounding nuclear matter at 200 GeV compared to the two lower energies and there is already a change of the nuclear matter properties in the energy range of 62.4 – 200 GeV.

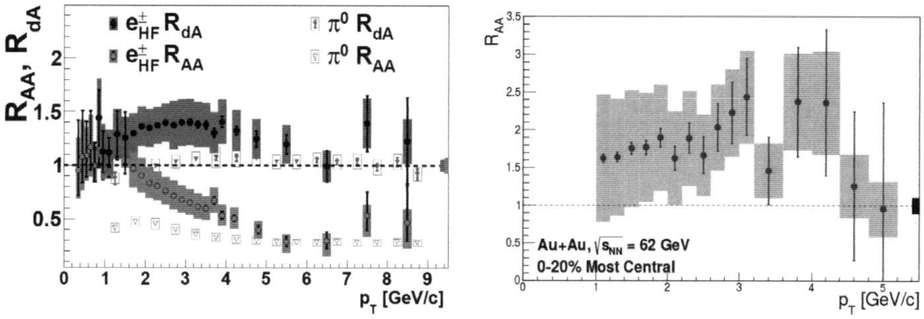

Figure 5. Nuclear modification factor for electrons from semi-leptonic heavy flavor hadron decays in Au+Au collisions at $\sqrt{s_{NN}} = 200$ GeV [23] and 62.4 GeV [52].

Summary

Heavy quarks are important tools to study the properties of the nuclear matter under extreme conditions but their measurements in heavy ion collisions are challenging and interpretation of the results is model dependent. Precision measurements of elliptic flow and production for charmed and bottom hadrons separately are necessary to further constrain models and to advance our understanding of the partonic medium properties. New vertex detectors at RHIC (Heavy Flavor Tracker at STAR and Silicon Vertex Tracker and Forward Silicon Vertex Detector at PHENIX) and LHC (new Inner Tracking System at ALICE) will provide data which will allow to address these issues.

Acknowledgements

This work was supported in part by the Foundation for Polish Science Grant HOMING PLUS/2013-7/8.

References

[1] Rapp R and van Hees H 2010 *Heavy Quarks in the Quark-Gluon Plasma in Quark-Gluon Plasma 4* (World Scientific) chap 3, p 111 (*Preprint* 0903.1096)
[2] Rapp R, Blaschke D and Crochet P 2010 *Prog. Part. Nucl. Phys.* **65** 209
[3] Lansberg J 2006 *Int. J. Mod. Phys. A* **21** 3857 (*Preprint* hep-ph/0602091)

[4] Conesa del Valle Z, Corcella G, Fleuret F, Ferreiro E, Kartvelishvili V *et al.* 2011 *Nucl. Phys. Proc. Suppl.* **214** 3 (*Preprint* 1105.4545)
[5] Matsui T and Satz H 1986 *Phys Lett. B* **178** 416
[6] Digal S, Petreczky P and Satz H 2001 *Phys. Lett. B* **514** 57 (*Preprint* hep-ph/0105234)
[7] Mocsy A 2009 *Eur. Phys. J. C* **61** 705 (*Preprint* 0811.0337)
[8] Kim H (CMS Collaboration) 2014 *Nucl. Phys. A* **931** 558
[9] Scomparin E (NA60 Collaboration) 2009 *Nucl. Phys. A* **830** 239C (*Preprint* 0907.3682)
[10] Adare A *et al.* (PHENIX Collaboration) 2011 *Phys. Rev. Lett.* **107** 142301 (*Preprint* 1010.1246)
[11] Yang Z 2014 *Nuclear Physics A* **931** 643
[12] Abelev B B *et al.* (ALICE Collaboration) 2014 *JHEP* **1402** 073 (*Preprint* 1308.6726)
[13] Chatrchyan S *et al.* (CMS Collaboration) 2014 *JHEP* **1404** 103 (*Preprint* 1312.6300)
[14] Abelev B B *et al.* (ALICE Collaboration) 2015 *Phys. Lett. B* **740** 105 (*Preprint* 1410.2234)
[15] Ferreiro E G, Fleuret F, Lansberg J P and Rakotozafindrabe A 2013 *Phys. Rev. C* **88** 047901
[16] Ferreiro E, Fleuret F, Lansberg J, Matagne N and Rakotozafindrabe A 2013 *Eur. Phys. J. C* **73**
[17] Arleo F and Peigné S 2012 *Phys. Rev. Lett.* **109** 122301
[18] Arleo F and Peign S 2013 *JHEP* **2013**
[19] Loureno C, Vogt R and Whri H K 2009 *JHEP* **2009** 014
[20] Adare A *et al.* 2013 *Phys. Rev. Lett.* **111** 212301
[21] Chatrchyan S *et al.* (CMS Collaboration) 2013 *Phys. Lett. B* **724** 213 (*Preprint* 1305.0609)
[22] Abelev B B *et al.* (ALICE Collaboration) 2013 *Phys. Lett. B* **726** 164 (*Preprint* 1307.3237)
[23] Adare A *et al.* (PHENIX Collaboration) 2012 *Phys. Rev. Lett.* **109** 242301 (*Preprint* 1208.1293)
[24] Sickles A M 2014 *Phys. Lett. B* **731** 51 (*Preprint* 1309.6924)
[25] Kikola D and Lipiec A 2014 (*Preprint* 1410.6503)
[26] Abelev B B *et al.* (ALICE Collaboration) 2014 *Phys. Rev. Lett.* **113** 232301 (*Preprint* 1405.3452)
[27] Adare A *et al.* (PHENIX Collaboration) 2013 *Phys. Rev. Lett.* **111** 202301 (*Preprint* 1305.5516)
[28] Abelev B B *et al.* (ALICE Collaboration) 2014 *JHEP* **1412** 073 (*Preprint* 1405.3796)
[29] Rakotozafindrabe A, Ferreiro E, Fleuret F and Lansberg J 2010 *J. Phys. G* **37** 094055 (*Preprint* 1002.2351)
[30] Kopeliovich B, Potashnikova I, Pirner H and Schmidt I 2011 *Phys. Rev. C* **83** 014912 (*Preprint* 1008.4272)
[31] Andronic A 2014 *Nucl. Phys. A* **931** 135 (*Preprint* 1409.5778)
[32] Andronic A, Braun-Munzinger P, Redlich K and Stachel J 2011 *J. Phys. G* **38** 124081 (*Preprint* 1106.6321)
[33] Capella A, Bravina L, Ferreiro E, Kaidalov A, Tywoniuk K *et al.* 2008 *Eur. Phys. J. C* **58** 437 (*Preprint* 0712.4331)
[34] Thews R L, Schroedter M and Rafelski J 2001 *Phys. Rev. C* **63** 054905
[35] Grandchamp L, Rapp R and Brown G E 2004 *Phys. Rev. Lett.* **92** 212301
[36] Gupta S and Sharma R 2014 *Phys. Rev. C* **89** 057901
[37] Liu Y, Qu Z, Xu N and Zhuang P 2009 *Physics Letters B* **678** 72
[38] Zhao X and Rapp R 2011 *Nuclear Physics A* **859** 114
[39] Adare A *et al.* (PHENIX Collaboration) 2007 *Phys. Rev. Lett.* **98** 232301 (*Preprint* nucl-ex/0611020)
[40] Adamczyk L *et al.* (STAR Collaboration) 2013 *Phys. Lett. B* **722** 55 (*Preprint* 1208.2736)
[41] Chatrchyan S *et al.* (CMS Collaboration) 2012 *JHEP* **1205** 063 (*Preprint* 1201.5069)
[42] Collaboration A 2014 *Physics Letters B* **734** 314
[43] Lin Z and Ko C 2001 *Physics Letters B* **503** 104
[44] Abelev B B *et al.* (ALICE Collaboration) 2014 *Phys. Lett. B* **738** 361 (*Preprint* 1405.4493)
[45] Caldern de la Barca Snchez M (CMS Collaboration) 2013 *J. Phys. Conf. Ser.* **458** 012011 (*Preprint* 1305.5560)
[46] Djordjevic M, Gyulassy M, Vogt R and Wicks S 2006 *Phys. Lett. B* **632** 81
[47] Abelev B *et al.* (ALICE Collaboration) 2013 *Phys. Rev. Lett.* **111** 102301 (*Preprint* 1305.2707)
[48] Adamczyk L *et al.* (STAR Collaboration) 2014 (*Preprint* 1405.6348)
[49] Adare A *et al.* (PHENIX Collaboration) 2011 *Phys. Rev. C* **84** 044905
[50] Nahrgang M, Aichelin J, Gossiaux P B and Werner K 2014 *Phys. Rev. C* **90**(2) 024907
[51] Adamczyk L *et al.* (STAR Collaboration) 2014 *Phys. Rev. Lett.* **113** 142301 (*Preprint* 1404.6185)
[52] Adare A *et al.* (PHENIX Collaboration) 2014 (*Preprint* 1405.3301)

Study of azimuthal correlations between D mesons and charged particles with the ALICE experiment

Fabio Colamaria, on behalf of the ALICE Collaboration
University of Bari "Aldo Moro" and INFN - Section of Bari
E-mail: fabio.colamaria@ba.infn.it

Abstract. A study of azimuthal correlations between D^0, D^+, and D^{*+} and charged particles in pp collisions at $\sqrt{s} = 7$ TeV and p–Pb collisions at $\sqrt{s_{NN}} = 5.02$ TeV is presented. D mesons were reconstructed from their hadronic decays at central rapidity ($|y| < 0.5$) in the transverse momentum range $3 \leq p_T^D \leq 16$ GeV/c and they were correlated with charged particles reconstructed in the pseudorapidity range $|\eta| < 0.8$. Perspectives for the measurement in Pb–Pb collisions at $\sqrt{s_{NN}} = 5.5$ TeV after the ALICE upgrade are presented as well.

1. Physics motivations

ALICE [1] has measured p_T-differential cross sections for D-meson production at central rapidity in pp [2, 3, 4], p–Pb [5] and Pb–Pb [6] collisions. A suppression of the D-meson yield by a factor 4–5 was observed in central Pb–Pb collisions for $p_T > 5$ GeV/c with respect to cross sections measured in pp collisions, scaled by the nuclear overlap function. This feature can be attributed to the interaction of heavy quarks with the Quark-Gluon plasma formed in such collisions. The analysis of angular correlations between D mesons and unidentified charged particles in Pb–Pb collisions can provide further insight into this topic, allowing us to address medium-induced modifications of charm fragmentation and hadronization and the dependence of the energy loss on the path length the parton travels through the medium.

Such a study is of great importance also in pp and p–Pb collisions. In pp collisions it allows us to characterize charm-quark jets and study their properties, providing at the same time a reference for the measurements in p–Pb and Pb–Pb collisions. In p–Pb collisions it enables the study of cold nuclear matter effects and the search, in the heavy-flavour sector, for long-range ridge-like structures ("double ridge") in near and away-side regions, already observed in hadron-hadron correlations [7, 8, 9].

2. Description of the analysis

The analysis was carried out for D^0, D^+ and D^{*+} mesons on minimum bias samples of $3.1 \cdot 10^8$ pp collisions at $\sqrt{s} = 7$ TeV and $1.0 \cdot 10^8$ p–Pb collisions at $\sqrt{s_{NN}} = 5.02$ TeV, for different ranges of the D-meson p_T and different thresholds on the charged particle p_T. D-meson candidates (*trigger* particles) were reconstructed from their hadronic decay channels ($D^0 \to K^-\pi^+$, $D^+ \to K^-\pi^+\pi^+$ and $D^{*+} \to D^0\pi^+ \to K^-\pi^+\pi^+$) and selected exploiting their displaced decay vertex topology, PID (particle identification) and reconstruction quality cuts on the daughter tracks. The values of S/B are in the range 0.25 to 2 (up to 5.6 for the D^{*+}), depending on the D-meson p_T. The selected D-meson candidates in the signal region ($\pm 2\sigma$ from the centre of the signal peak in the invariant mass distribution) were then correlated with charged particles (*associated tracks*) within $|\eta| < 0.8$ and satisfying a selection on their reconstruction quality. The difference in pseudorapidity $\Delta\eta$ and azimuthal angle $\Delta\varphi$ between the D-meson candidate and the charged-particle track was evaluated.

The contribution of background candidates was subtracted using the correlation distribution from the candidates in the sidebands of the invariant mass distribution (the regions $4-8\sigma$ away from the signal peak), normalized by the ratio of background candidates in signal and sideband regions. Several corrections were applied to the correlation distributions to account for: (i) *Associated track reconstruction efficiency*: each D–hadron pair was weighted by the inverse of the track reconstruction efficiency of the associated track, evaluated from simulations; (ii) *D-meson selection efficiency*: each correlation entry was weighted by the inverse of the D-meson reconstruction and selection efficiency, computed from simulations; (iii) *Limited detector acceptance and detector spatial inhomogeneities*: a correction factor was obtained exploiting the event mixing technique, i.e. correlating D-meson candidates from one event with tracks from other events with similar multiplicity and position of the primary vertex in the beam direction; (iv) *Beauty feed-down*: MC simulations based on PYTHIA were exploited to evaluate a template distribution of angular correlations between D mesons from B-meson decays and charged particles; this distribution was scaled to match the proper ratio of feed-down to prompt D mesons, calculated as described in [2], and subtracted from the inclusive D–hadron correlation distribution obtained from data; (v) *Secondary track contamination*, i.e. tracks from strange-hadron decays or produced in interactions of particles with the detector material. From a MC study, no visible angular dependence was found for this contamination. A global scale factor, corresponding to the fraction of primary particles in the sample of associated tracks, was thus applied to the correlation distributions.

Due to the limited statistics available, the fully corrected 2D correlations were projected onto the $\Delta\varphi$ axis, producing azimuthal correlation distributions, which were normalized to the number of trigger D mesons. A weighted average of the results for the three meson species, consistent among themselves within the uncertainties, was then performed. A fit function, composed of a constant term plus two Gaussian functions with a periodicity condition, was applied to the azimuthal correlation distributions. The two Gaussian functions, centered at $\Delta\varphi \sim 0$ and $\Delta\varphi \sim \pi$, model the near and away-side peaks in the correlation distribution. Physical observables (near-side yield, near-side peak width, height of the baseline of the distribution, defined as the average height of the points in the transverse region, far from the peaks) were extracted and compared between the two collision systems and with predictions from Monte Carlo simulations. Finally, a careful evaluation of all the sources of systematic uncertainties was performed.

3. Results in pp and p–Pb collision systems

The left panel of Fig. 1 shows a comparison of D–hadron correlations in pp and p–Pb collisions, after the subtraction of the baseline, for $5 < p_T^D < 8$ GeV/c and associated tracks with $p_T^{assoc} > 0.5$ GeV/c. In both systems, the near and away-side peaks are clearly visible, and the correlation pattern is similar for pp and p–Pb collisions. In the right panel of Fig. 1 the baseline-subtracted D–hadron correlation distribution from the pp analysis is compared with the outcome of PYTHIA simulations, at $\sqrt{s} = 7$ TeV, exploiting different tunes of the simulation parameters. The PYTHIA results with all three parameter tunes provide a fair description of the measured distribution within uncertainties. A more quantitative comparison between pp and p–Pb results for the near-side region can be performed by extracting the per-trigger yield in the near-side peaks from the fit function applied to the correlation distributions. The near-side associated yields obtained from pp and p–Pb correlation distributions are shown in Fig. 2 as a function of the D-meson p_T for associated tracks with $p_T^{assoc} > 0.3$ GeV/c (left) and $p_T^{assoc} > 1.0$

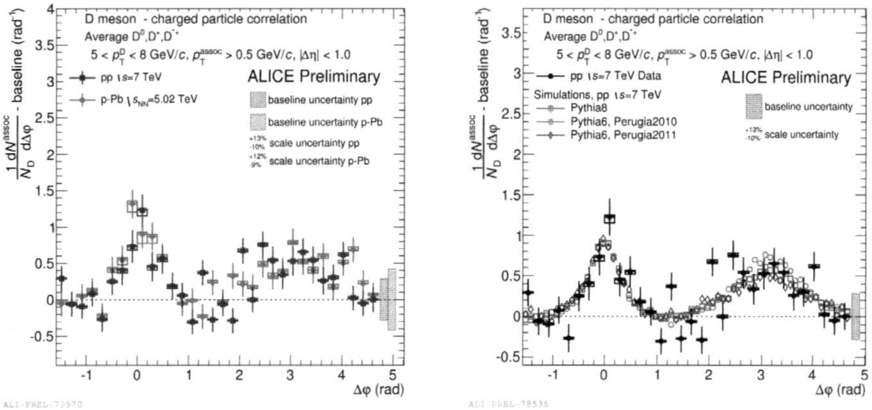

Figure 1. Left: comparison of D–hadron azimuthal correlation distributions in pp and p–Pb collisions, after baseline subtraction, from the weighted average of D^0, D^+ and D^{*+} measurements. Right: comparison of D–hadron azimuthal correlation distribution and predictions from different tunes of the PYTHIA generator, for pp collisions, after baseline subtraction. In both panels, statistical and uncorrelated systematic uncertainties are shown as error bars and boxes, respectively.

GeV/c (right). The pp and p–Pb yields are compatible within the total uncertainties. Given the size of the current uncertainties, no firm conclusions on possible modifications of D–hadron correlations due to cold nuclear matter effects can be drawn.

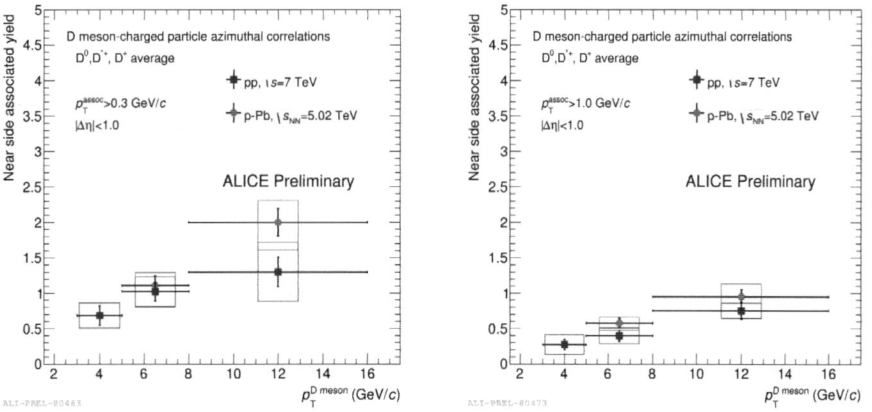

Figure 2. Comparison of near-side associated yields extracted in pp and p–Pb collisions as a function of the D-meson p_{T} for $p_{\mathrm{T}}^{\mathrm{assoc}} > 0.3$ (left panel) and 1 (right panel) GeV/c. Statistical and uncorrelated systematic uncertainties are shown as error bars and boxes, respectively.

4. Perspectives for Pb–Pb collisions

The D–hadron correlation measurement is not feasible on the currently available sample of Pb–Pb collisions, with the current performance of the ALICE detectors. The very low D-meson signal/background ratio and the large number of tracks from the underlying event, not correlated with D mesons, generate large uncertainties in the $\Delta\varphi$ distributions when the background is subtracted. This washes out any correlation structure possibly present.

After the ALICE upgrade (expected for 2018-2019) [10, 11], though, a dramatic improvement of tracking and vertexing performance is expected, which will lead to an increase of the S/B ratio by a factor up to 10 for the D-meson reconstruction. Combined with the increase of a factor $\approx$100 in the ALICE readout rate for minimum-bias Pb–Pb collisions, this will enable the study of D–hadron correlations also in this system. A simulation of the analysis performance on central (0-10%) Pb–Pb collisions was carried out using a template of correlation distributions from PYTHIA. As shown in Fig. 3, the results of this study indicate a very low statistical uncertainty on the evaluation of the near-side yield, below 1% for $p_{\mathrm{T}}^{\mathrm{D}^0} > 8$ GeV/c and about 10% for $3 < p_{\mathrm{T}}^{\mathrm{D}^0} < 5$ GeV/c, for an integrated luminosity of $L_{\mathrm{int}} = 10\ \mathrm{nb}^{-1}$. A substantial reduction of the uncertainties is expected also for pp and p–Pb collision systems. After the ALICE upgrade, thus, it should be possible to compare the results in the three collision systems and to quantify the effects of the charm quark energy loss in a QGP medium.

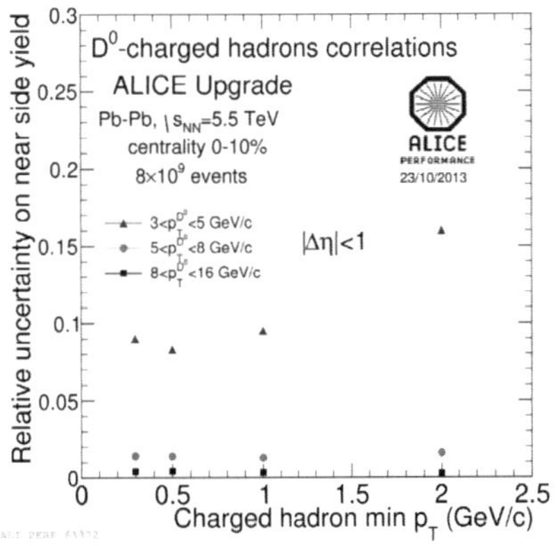

Figure 3. Estimates for the statistical uncertainty on the near-side yield from the analysis of the expected Pb–Pb statistics after the ALICE upgrade, for different kinematic ranges, obtained for an integrated luminosity of $L_{\mathrm{int}} = 10\ \mathrm{nb}^{-1}$.

5. Conclusions

ALICE measurements of D–hadron angular correlations in pp and p–Pb collisions are compatible, within uncertainties, in terms of the correlation shapes and the near-side associated yields. Consequently, no evident influence of cold nuclear matter on the charm fragmentation and hadronization can be claimed. The correlation distributions are well described by the outcome of PYTHIA simulations as well. While the analysis cannot be currently performed in Pb–Pb collisions, it will become feasible also in this collision system after the ALICE upgrade (2018-2019), when the tracking and vertexing performance of the experiment will be largely enhanced.

References

[1] ALICE Collaboration 2008, *JINST* **3** S08002
[2] ALICE Collaboration 2012, *JHEP* **01** 128
[3] ALICE Collaboration 2012, *Phys. Lett.* B **718** 279-94
[4] ALICE Collaboration 2012, *JHEP* **07** 191
[5] ALICE Collaboration 2014, *Phys. Rev. Lett.* **113** 232301
[6] ALICE Collaboration 2012, *JHEP* **09** 112
[7] ALICE Collaboration 2013, *Phys. Lett.* B **719** 29-41
[8] ATLAS Collaboration 2013, *Phys. Rev. Lett.* **110** 182302
[9] CMS Collaboration 2013, *Phys. Lett.* B **718** 795-814
[10] ALICE Collaboration 2012, CERN-LHCC-2012-012
[11] ALICE Collaboration 2013, CERN-LHCC-2013-024, ALICE-TDR-017

Low mass dimuon production in pp, p–Pb and Pb–Pb collisions with the ALICE muon spectrometer

Ester Anna Rita Casula
Università di Cagliari and INFN, Sezione di Cagliari, Italy
E-mail: ester.anna.rita.casula@cern.ch

Abstract. We present results on the low mass dimuon analysis in pp, p–Pb and Pb–Pb collisions.
In pp collisions at $\sqrt{s} = 2.76$ TeV the ϕ differential cross section as a function of the transverse momentum has been measured, while the ϕ yield and the nuclear modification factor R_{pPb} at forward and backward rapidity have been measured in p–Pb collisions at $\sqrt{s_{NN}} = 5.02$ TeV. The $\frac{BR\sigma_\phi}{BR\sigma_\rho + BR\sigma_\omega}$ ratio and the ϕ nuclear modification factor R_{AA} have been measured in Pb–Pb collisions at $\sqrt{s_{NN}} = 2.76$ TeV in the intermediate p_{T} region $2 < p_{\mathrm{T}} < 5$ GeV/c, as a function of the number of participating nucleons. Remarkable differences are observed in the comparison between these results and the ones measured in the same p_{T} range at midrapidity in the hadronic decay channel $\phi \to KK$.

1. Introduction

Low mass vector meson (ρ, ω, ϕ) production provides key information on the hot and dense state of strongly interacting matter produced in high-energy heavy-ion collisions. Among the different probes, strangeness enhancement can be accessed through the measurement of ϕ meson production [1], while the measurement of the ϕ nuclear modification factor provides a powerful tool to probe the production dynamics and hadronization process in relativistic heavy-ion collisions.
Vector meson production in pp collisions provides the reference for these studies and is interesting by itself, since it can be used to tune particle production models in the LHC energy range. The analysis in p–A collisions can give an insight on soft particle production in cold nuclear matter, being p–A a system where a hot, dense medium is not expected to be formed in the final state, while its initial state is similar to that of A-A collisions.

Vector mesons are reconstructed with the ALICE muon spectrometer [2] in the rapidity range $2.5 < y < 4$ through their decay into muon pairs.
The ALICE muon spectrometer is composed of a front hadron absorber, a set of cathode pad chambers (five stations, each one composed of two chambers) for the track reconstruction in a dipole field, an iron wall acting as a muon filter and two stations of two resistive plate chambers (RPC) for the muon trigger. The analyzed data were collected requiring the coincidence of an unlike sign dimuon trigger and a minimum bias trigger. The dimuon trigger requires two opposite sign tracklets in the muon trigger system. The minimum bias trigger, independent from the muon trigger, was based on a set of forward scintillators and on a silicon pixel detector placed

in the vertex region.

2. Analysis in pp collisions at $\sqrt{s}$ = 2.76 TeV

Data were collected in pp collisions in 2013 at $\sqrt{s}$ = 2.76 TeV. Muon tracks were selected requiring that the tracks reconstructed in the tracking stations matched the ones in the trigger chambers (single muon p_{T} trigger threshold ~ 0.5 GeV/c) and that their pseudorapidity was in the range $2.5 < \eta_\mu < 4$. Muon pairs were selected inside the dimuon rapidity interval $2.5 < y_{\mu\mu} < 4$.

The combinatorial background was evaluated through the event mixing technique. The invariant mass distribution after combinatorial background subtraction, shown in Fig. 1, left panel, for a dimuon $p_{\mathrm{T}} > 1$ GeV/c, is described as a superposition of light meson decays into muon pairs, with an additional contribution coming from charm and beauty semi-muonic decays. Low-mass resonances shapes come from a Monte Carlo simulation with a parametric generator [3], while open charm and beauty have been generated using a parametrization of PYTHIA [5].

Figure 1, right panel, shows the comparison of the ϕ differential cross section as a function of p_{T} with PHOJET [4] and several tunes of PYTHIA (Perugia0 [6], Perugia11 [7], ATLAS-CSC [8] and D6T [9]): the PYTHIA tunes Perugia0 and Perugia11 strongly underestimate the measured cross section; PYTHIA ATLAS-CSC also underestimates the data (even if to a lesser degree than Perugia0 and Perugia11), while PYTHIA D6T and PHOJET are in fair agreement with the measured values.

The integrated cross section σ_ϕ $(2.5 < y < 4, 2 < p_{\mathrm{T}} < 5$ GeV/$c) = 0.108 \pm 0.010$ (stat.) $\pm\, 0.007$ (syst.) mb has been used as baseline for the Pb–Pb analysis, while the value σ_ϕ $(2.5 < y < 4,$ $1 < p_{\mathrm{T}} < 5$ GeV/$c) = 0.542 \pm 0.052$ (stat.) $\pm\, 0.043$ (syst.) mb has been used as baseline for the interpolation of the ϕ cross section in the p–Pb analysis at 5.02 TeV.

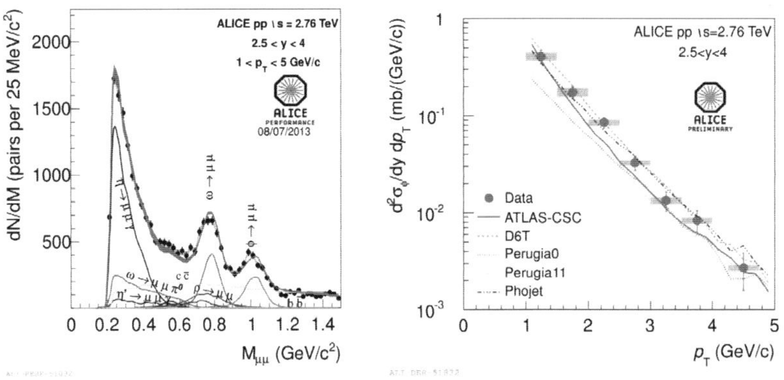

Figure 1. Left: Fit to the dimuon invariant mass spectrum for $p_{\mathrm{T}} > 1$ GeV/c in pp collisions at $\sqrt{s}$ = 2.76 TeV; blue band: systematic uncertainty from background subtraction; red band: uncertainty in the relative normalization of the sources. Right: ϕ differential cross section as a function of p_{T} in pp collisions compared with PHOJET and several tunes of PYTHIA.

3. Analysis in p–Pb collisions at $\sqrt{s_{NN}}$ = 5.02 TeV

The data for the p–Pb analysis were collected in 2013 at $\sqrt{s_{NN}}$ = 5.02 TeV, at forward rapidity, with the proton beam directed towards the muon spectrometer (p–Pb) and at backward rapidity, with the lead beam directed towards the muon spectrometer (Pb–p). Due to the different

energy of the proton (4 TeV) and of a nucleon in the lead nucleus (1.58 TeV), the rapidity of the nucleon-nucleon center of mass and the one of the laboratory do not coincide anymore: in p–Pb collisions, the rapidity of the center of mass is shifted by +0.465, and therefore the rapidity acceptance of the muon spectrometer is $2.03 < y_{cm} < 3.53$ (forward rapidity). In Pb–p collisions the rapidity of the center of mass is shifted by -0.465 and the rapidity acceptance of the muon spectrometer results to be $-4.46 < y_{cm} < -2.96$ (backward rapidity). A direct comparison between forward and backward is possible only in the rapidity window $2.96 < |y_{cm}| < 3.53$.
The criteria applied to the p–Pb analysis were the same as the ones applied in the pp analysis. The ϕ nuclear modification factor R_{pPb} has been calculated as $R_{pPb} = \frac{Y}{\sigma_{pp} < T_{pPb} >}$, where Y is the ϕ yield in p–Pb, σ_{pp} is the ϕ cross section at $\sqrt{s_{NN}} = 5.02$ TeV, calculated through an interpolation between the measurements at $\sqrt{s} = 2.76$ and 7 TeV [3], and $< T_{pPb} >$ is the nuclear overlap function in p–Pb collisions, calculated on the basis of the Glauber model [10].

The ϕ R_{pPb} at forward and backward rapidity as a function of p_T is shown in Fig. 2. The nuclear modification factor is compatible with unity for $p_T > 3$ GeV/c at forward rapidity and is larger than 1 with a Cronin-like peak at backward rapidity. This effect is due to the asymmetry in the particle production present between forward and backward rapidities in pA collisions.

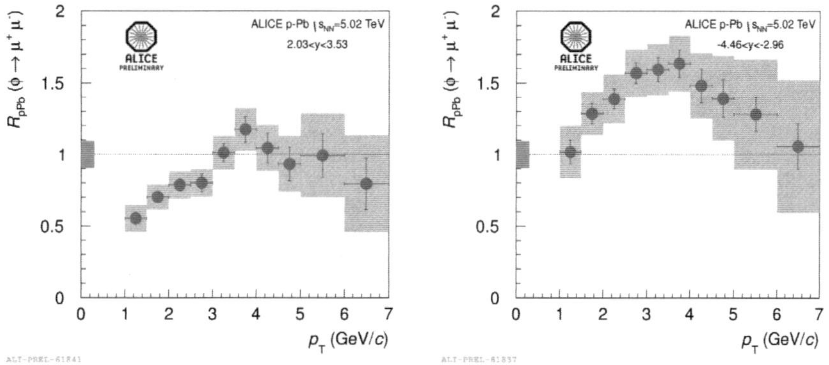

Figure 2. ϕ R_{pPb} at forward (left) and backward rapidity (right) as a function of p_T. Grey boxes: uncorrelated systematic uncertainties; lilac box at 1: correlated ones.

4. Analysis in Pb–Pb collisions at $\sqrt{s_{NN}} = 2.76$ TeV

Data in Pb–Pb collisions were collected in 2011 at $\sqrt{s_{NN}} = 2.76$ TeV. The selections applied were the same as in the pp and p–Pb analyses, with an additional cut on the single muon p_T at 0.85 GeV/c to reduce the background. The combinatorial background was also evaluated with the event mixing technique; a cut on the dimuon p_T at 2 GeV/c was applied, because for dimuon $p_T < 2$ GeV/c the acceptance for ϕ, ρ and ω is close to 0, since the single muon p_T trigger threshold in Pb–Pb collisions was set to 1 GeV/c.
In Fig. 3, left, the $\frac{BR\sigma_\phi}{BR\sigma_\rho + BR\sigma_\omega}$ ratio is shown as a function of the average number of participating nucleons $< N_{part} >$ in the intermediate p_T region $2 < p_T < 5$ GeV/c. The Pb–Pb values obtained in four different centrality classes are compared to the value obtained in the pp analysis at 2.76 TeV.
The $\frac{BR\sigma_\phi}{BR\sigma_\rho + BR\sigma_\omega}$ ratio in Pb–Pb increases with respect to the value in pp by a factor of about 2, and tends to saturate from peripheral towards central events, indicating an enhancement of the ϕ meson production with respect to ρ and ω mesons in central Pb–Pb collisions.

The ϕ R_{AA} as a function of $< N_{part} >$ in the intermediate $p_{\rm T}$ region is shown in Fig. 3, right. The R_{AA} is compatible with unity, within the uncertainties, in peripheral collisions and it is suppressed going towards central collisions.
The comparison with the results obtained at central rapidity in the KK decay channel shows a point-by-point agreement within the uncertainties. However, the most peripheral points of the R_{AA} at forward rapidity are higher than the ones at midrapidity, while the semicentral and the central points are lower, hinting thus two different behaviors: this issue is currently under investigation and may be due to a different hydrodynamic push that the particles are subjected to at forward and at midrapidity in the intermediate $p_{\rm T}$ region.

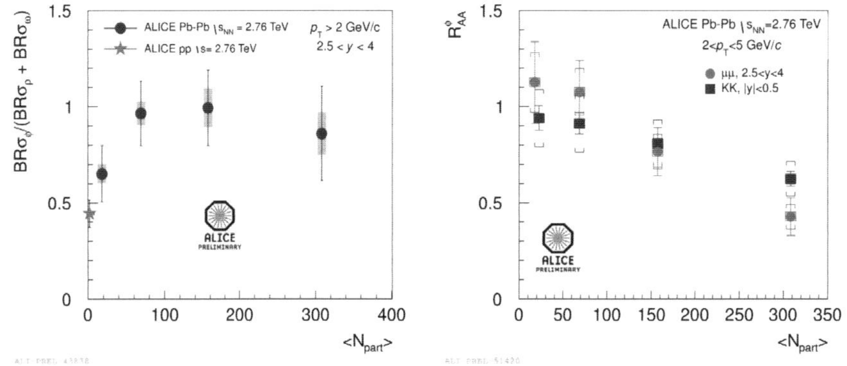

Figure 3. Left: $\frac{BR\sigma_\phi}{BR\sigma_\rho + BR\sigma_\omega}$ ratio as a function of the number of participating nucleons $< N_{part} >$. Right: Comparison of ϕ R_{AA} as a function of $< N_{part} >$ for $|y| < 0.5$ (KK channel) and for $2.5 < y < 4$ ($\mu\mu$ channel) in the intermediate $p_{\rm T}$ region $2 < p_{\rm T} < 5$ GeV/c.

In conclusion, we measured both integrated and $p_{\rm T}$-differential cross sections of ϕ in pp collisions at $\sqrt{s} = 2.76$ TeV. The integrated values were used as baseline for the Pb–Pb and for the p–Pb measurement. In the p–Pb analysis, we measured the ϕ R_{pPb} as a function of $p_{\rm T}$: R_{pPb} is compatible with unity for $p_{\rm T} > 3$ GeV/c at forward rapidity and larger than unity with a Cronin-like peak at backward rapidity. In the Pb–Pb analysis the $\frac{BR\sigma_\phi}{BR\sigma_\rho + BR\sigma_\omega}$ ratio and the ϕ nuclear modification factor R_{AA} as a function of $< N_{part} >$ have been measured in the intermediate $p_{\rm T}$ region. The $\frac{BR\sigma_\phi}{BR\sigma_\rho + BR\sigma_\omega}$ ratio increases from pp to Pb–Pb and tends to saturate towards central events. The comparison between the ϕ R_{AA} measured in the dimuon decay channel at forward rapidity and in the KK decay channel at midrapidity seems to indicate a different behavior, probably due to a different hydrodynamic push in the two rapidity domains in the intermediate $p_{\rm T}$ region $2 < p_{\rm T} < 5$ GeV/c.

References

[1] J. Rafelski, B. Müller *Phys. Rev. Lett.* *48* (1982) 1066.
[2] K. Aamodt et al. (ALICE Collaboration), *JINST* 3 (2008) S08002.
[3] B. Abelev et al. (ALICE Collaboration), *Phys. Lett. B* 710 (2012) 557.
[4] R. Engel, *Z. Phys. C* 66 (1995) 203; R. Engel, J. Ranft, *Phys. Rev. D* 54 (1996) 4244.
[5] T. Sjöstrand et al., *J. High Energy Phys.* 0605 (2006) 026.
[6] P.Z. Skands, *arXiv:0905.3418* (2009).
[7] P.Z. Skands, *Phys. Rev. D* 82 (2010) 074018.
[8] C. Buttar et al., *Acta Phys. Pol. B* 35 (2004) 433.
[9] R. Field, *Acta Phys. Pol. B* 39 (2008) 2611.
[10] K. Aamodt et al. (ALICE Collaboration), *Phys. Rev. Lett.* 106 (2011) 032301.

Effects of quark chemical equilibration on thermal photon elliptic flow

Akihiko Monnai
RIKEN BNL Research Center, Brookhaven National Laboratory, Upton, NY 11973, USA
E-mail: amonnai@riken.jp

Abstract. Large hadronic elliptic flow v_2 is considered as an evidence for the existence of a strongly-coupled QGP fluid in high-energy heavy-ion collisions. On the other hand, direct photon v_2 has recently been found to be much larger than that of hydrodynamic estimations, which is recognized as "photon v_2 puzzle". In this study, I discuss the implication of late production of quarks in an initially gluon-rich medium because photons are coupled to quarks. Numerical analyses imply that thermal photon v_2 can be visibly enhanced. This indicates that interplay of equilibration processes and collective expansion would be important.

1. Introduction

Azimuthal anisotropy of particle spectra is a unique observable in heavy-ion collisions. The second harmonics of Fourier expansion of hadronic spectra, elliptic flow v_2, is proposed as a useful quantity in non-central collisions because it would reflect the magnitude of interaction in the quark-gluon medium [1, 2]. Experimental data of Au-Au collisions at $\sqrt{s_{NN}} = 200$ GeV at RHIC show sizable hadronic v_2 which is in remarkable agreement with fluid dynamic description, leading to an understanding that the bulk medium is strongly-coupled in the vicinity of QCD crossover. This is reconfirmed in Pb-Pb collisions at $\sqrt{s_{NN}} = 2.76$ TeV at LHC, and higher-order harmonics v_n, which come from geometrical fluctuation, are also shown to be consistent with hydrodynamic description [3].

Given the success of hydrodynamic models so far, excess of direct photon v_2 compared with the model predictions was a surprising discovery [4, 5]. Here direct photons are defined as the sum of thermal photons, which are emitted softly from the medium, and prompt photons, which are created in the initial hard processes (Fig. 1). The QCD medium is considered to be transparent in terms of electromagnetic interaction. Direct photon v_2 was expected to be much smaller than the hadronic one because the momentum anisotropy of photons would be acquired only indirectly through that of emission sources, *i.e.*, quarks in the QGP phase and hadrons in the hadronic phase. There also is an experimental indication that photon v_3 is large. There are a number of possible solutions, which may be categorized as follows: (i) modification of thermal photon emission rates, (ii) modification of prompt photon emission rates, (iii) introduction of other photon emission sources, (iv) introduction of medium-photon interactions, (v) modification of the bulk medium evolution and (vi) improvement of the experimental data. Here I focus on (i) and (v) and discuss the effects of quark chemical equilibration in the bulk medium [6] on thermal photon emission rates in early stages of hydrodynamic evolution because the color glass theory indicates that the medium can be initially gluon-rich [7, 8].

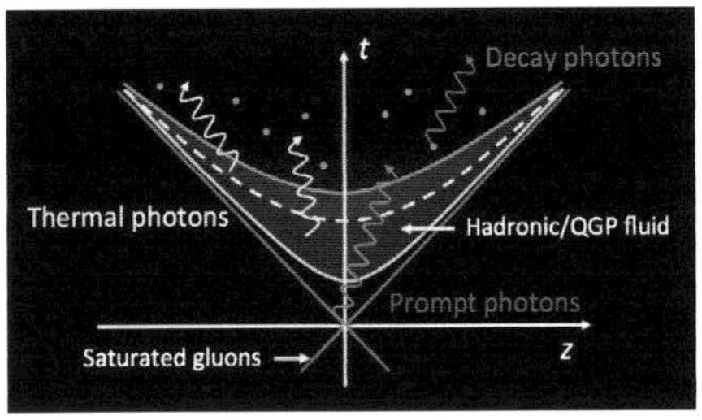

Figure 1. A schematic picture of the high-energy heavy-ion collision from the viewpoint of photons. Decay photons originate from hadronic decay.

2. Hydrodynamics in chemical non-equilibrium

Most modern hydrodynamic models assume that thermal and chemical equilibrations occur simultaneously even though the precise mechanism of the local equilibration is not well known. Since the colliding nuclei are considered to be described as saturated gluons, there are indications that chemical equilibration can be slower than thermalization [9], a requirement for defining the temperature in hydrodynamics. If the quark number is smaller than that at equilibrium at early times, photon emission is suppressed because photons are coupled to quarks but not to gluons. This could be an enhancement mechanism for thermal photon v_2 because the contribution of the photons with larger anisotropy is effectively enhanced [10, 11].

The chemically non-equilibrated hydrodynamic model is formulated as follows. The equation of motion for the system is energy-momentum conservation $\partial_\mu T^{\mu\nu} = 0$ and rate equations for the quark and the gluon number currents

$$\partial_\mu N_q^\mu = 2r_b n_g - 2r_b \frac{n_g^{\mathrm{eq}}}{(n_q^{\mathrm{eq}})^2} n_q^2, \tag{1}$$

$$\partial_\mu N_g^\mu = (r_a - r_b) n_g - r_a \frac{1}{n_g^{\mathrm{eq}}} n_g^2 + r_b \frac{n_g^{\mathrm{eq}}}{(n_q^{\mathrm{eq}})^2} n_q^2 + r_c n_q - r_c \frac{1}{n_g^{\mathrm{eq}}} n_q n_g, \tag{2}$$

where the processes $g \rightleftharpoons g + g$, $g \rightleftharpoons q + \bar{q}$ and $q(\bar{q}) \rightleftharpoons q(\bar{q}) + g$ (denoted by the subscripts a, b and c, respectively) are considered. Note that the four-gluon vertex process is suppressed by additional α_s. Here the net baryon density is assumed to be vanishing. An inviscid case is considered for simplicity, though introduction of viscosity to the formalism is straightforward. n_q and n_g are the quark and the gluon number densities defined in the tensor decomposition $N_q^\mu = n_q u^\mu$ and $N_g^\mu = n_g u^\mu$ where u^μ is the flow. r_a, r_b and r_c are the chemical reaction rates. Since the pair production/annihilation is the only quark number changing process considered here, $1/r_b$ roughly corresponds to the quark chemical relaxation time of the system. n_q^{eq} and n_g^{eq} are the equilibrium densities introduced for the parton numbers to be conserved in equilibrium. They are determined using the parton gas picture with $N_f = 2$. Thermalization is assumed to be quick enough to restore thermal equilibrium right after the splitting and merging processes occur. Note that the above rate equations are valid only in the QGP phase. The system is simply assumed to be equilibrated below the crossover temperature.

The QGP photon emission rate is modified by simply factoring the phase-space distributions by n_q/n_q^{eq} or n_g/n_g^{eq} [12]. The hadronic photon emission rate [13, 14] is unchanged because the system is assumed to be in chemical equilibrium in the hadronic phase. They are smoothly connected by a hyperbolic function at $T_c = 0.17$ GeV with $\Delta T_c = 0.017$ GeV [10]. The background medium is estimated with a boost-invariant (2+1)-dimensional hydrodynamic model. The hydrodynamic input for numerical estimations is summarized in Table 1.

3. Numerical results

Differential elliptic flow of the thermal photons $v_2(p_T)$ and time evolution of the quark number density $n_q(t)$ in and out of chemical equilibrium are shown in Fig. 2. $v_2(p_T)$ is found to become

Table 1. Input to chemically non-equilibrated hydrodynamic model

Fluid properties	Description
Hadronic EoS	Hadron resonance gas ($m_h \leq 2.5$ GeV)
QGP EoS	Parton gas ($N_f = 2$)
Chemical reaction rate	$r_i = c_i T$ ($i = a, b, c$)
Initial conditions	**Description**
Gluon energy density $e_g(\tau_0, x, y)$	Glauber model (Au-Au at $\sqrt{s_{NN}} = 200$ GeV, $b = 7$ fm) [15]
Quark energy density $e_q(\tau_0, x, y)$	0 GeV$\cdot$fm^{-3}
Gluon number density $n_g(\tau_0, x, y)$	$n_g^{\text{eq}} + n_q^{\text{eq}}/2$
Quark number density $n_q(\tau_0, x, y)$	0 fm^{-3}
Initial time τ_0	0.4 fm/c

visibly larger for slower quark production because the contribution of the hadronic photons becomes effectively large. The quark chemical reaction rate parameter is varied as $c_b = 0.2, 0.5$ and 2.0 where $c_a = c_c = 1.5$ are fixed. The equilibrium result is recovered in the large c_b limit. Estimated chemical relaxation times for the chosen parameters are $\tau_{\text{ch}} \sim 1/c_b T \sim 2.0, 0.5$ and 0.2 fm/c for the average temperature $T \sim 0.2$ GeV, which are consistent with the numerical results. It should be noted that n_q does not converge to n_q^{eq} from the parton gas picture near and below the crossover (indicated by thin lines in the figure) because the information of the hadronic phase is embedded in the EoS and thus is the hydrodynamic flow.

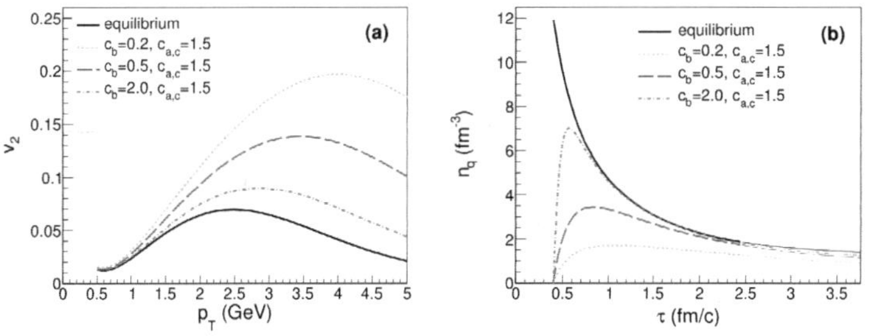

Figure 2. (a) Thermal photon $v_2(p_T)$ and (b) quark number density $n_q(t)$ for different quark chemical equilibration rates.

Figure 3 shows the numerical results for several different choices of the chemical equilibration rates for gluon splitting/merging and gluon emission from/absorption into quarks. The reaction rate parameters are varied as $c_a = c_c = 0.0, 1.5$ and 3.0 for the fixed $c_b = 0.5$. One can see that $v_2(p_T)$ becomes smaller for slower quark-number conserving chemical equilibration, though the dependences on c_a and c_c are weaker than that on c_b. This is because gluons are initially overpopulated and the recombination processes become dominant for gluons. The quark number density is indirectly suppressed by faster gluon recombination at early times.

A source of overestimation is the fact that the vanishing quark number density is assumed in the initial condition, while initial sea quarks and pre-thermal chemical equilibration will lead to a finite number of quarks at the beginning of the hydrodynamic stage. This is effectively taken into account by varying the reaction rates. Also dynamical effects on the EoS itself are important [16], as thermal photons are expected to be sensitive to them.

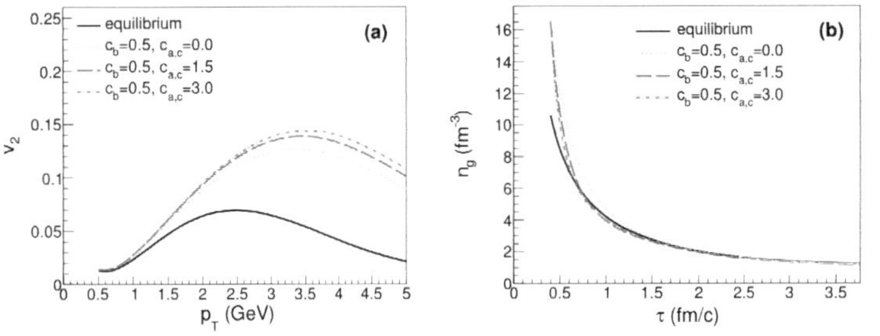

Figure 3. (a) Thermal photon $v_2(p_T)$ and (b) quark number density $n_g(t)$ for different gluon chemical equilibration rates.

4. Conclusion and outlook

I estimated the effects of quark chemical equilibration in the hydrodynamic stage on thermal photon v_2. The newly-developed (2+1)-dimensional ideal hydrodynamic model is coupled to the rate equations for the parton number densities in the QGP phase. Thermal photon v_2 is found to become visibly larger when the quark chemical equilibration is slow because of the initial suppression of the photons with small anisotropy. Here the medium can initially be gluon-rich due to the fact that colliding nuclei are described as the color glass condensate. This can partially explain the recently observed excessive photon v_2. Late gluon chemical equilibration slightly reduces v_2 but the dominant process for the modification is the quark pair production and annihilation. The results imply the importance of considering chemical imbalance of a QCD medium and subsequent equilibration processes in heavy-ion physics.

Future prospects include the estimation of prompt photons, which might wash some of the enhancement effects out of direct photon v_2. Also the chemical equilibration rates, like the equation of state and viscous coefficients, should be constrained by microscopic theories. Modification of the transport coefficients due to off-equilibrium chemical composition of the bulk medium would be worth-investigating. An additional v_2 enhancement mechanism [17] may also be required to avoid too much suppression of the total yield of photons.

Acknowledgments

The work is inspired by fruitful discussion with Müller B. The author would like to thank for valuable comments by Akiba Y and McLerran L on the paper. The work is supported by RIKEN Special Postdoctoral Researcher program.

References

[1] Ollitrault J -Y 1992 *Phys. Rev. D* **46** 229
[2] Poskanzer A M and Voloshin S A 1998 *Phys. Rev. C* **58** 1671
[3] Schenke B, Jeon S and Gale C 2011 *Phys. Rev. Lett.* **106** 042301
[4] Adare A *et al.* [PHENIX Collaboration] 2012 *Phys. Rev. Lett.* **109** 122302
[5] Lohner D [ALICE Collaboration] 2013 *J. Phys. Conf. Ser.* **446** 012028
[6] Gelis F, Kajantie K and Lappi T 2006 *Phys. Rev. Lett.* **96** 032304
[7] McLerran L D and Venugopalan R 1994 *Phys. Rev. D* **49** 2233
[8] McLerran L D and Venugopalan R 1994 *Phys. Rev. D* **49** 3352
[9] Monnai A and Müller B 2014 *arXiv:1403.7310* [hep-ph]
[10] Monnai A 2014 *Phys. Rev. C* **90** 021901
[11] Monnai A 2014 *arXiv:1410.8621* [nucl-th]
[12] Traxler C T and Thoma M H 1996 *Phys. Rev. C* **53** 1348
[13] Turbide S, Rapp R and Gale C 2004 *Phys. Rev. C* **69** 014903
[14] Arleo F *et al.* 2003 *arXiv:hep-ph/0311131*
[15] Kolb P F, Sollfrank J and Heinz U W 2000 *Phys. Rev. C* **62** 054909
[16] Gelis F, Niemi H, Ruuskanen P V and Rasanen S S 2004 *J. Phys. G* **30** S1031
[17] Monnai A 2014 *arXiv:1408.1410* [nucl-th]

Hot Quarks 2014 IOP Publishing
Journal of Physics: Conference Series **612** (2015) 012027 doi:10.1088/1742-6596/612/1/012027

J/ψ production in the statistical hadronization model - implications and limitations

P B Bahavar, J Uphoff and C Greiner

Institute for Theoretical Physics, Johann Wolfgang Goethe-Universität, Max-von-Laue-Str. 1, 60438 Frankfurt am Main

E-mail: bahavar@th.physik.uni-frankfurt.de

Abstract. The Statistical Hadronization Model can be extended to include hadrons which are produced far away from chemical equilibrium or in an earlier phase of the evolution of a heavy ion collision. The effects of J/ψ production above the critical temperature are explored, with special focus on the mass shift for the thermal dressed mass of the charm quarks in the quark-gluon plasma.

1. Introduction

The statistical hadronization model (SHM) has proved to describe well the production of hadrons in experiments in which a quark-gluon plasma (QGP) is formed [1, 2, 3, 4, 5], for both hadrons containing only light quarks [6, 7], and in an extension of the original idea also for hadrons containing charm quarks [8, 9]. This enables the calculation of the J/ψ multiplicity, an important observable in heavy ion collicions, since its suppression is regarded as a signature of the formation of the QGP [10]. One of the fundamental assumptions of statistical hadronization, the instantaneous production of all particle species at one point in the phase diagram of strongly interacting matter, is called into question in the case of the J/ψ by results from lattice calculations suggesting survival of the bound state up to 1.9 times the critical temperature T_c [11, 12], though more recent works contest this interpretation[13].

Following the deliberations of Grandchamp *et al.* [14], we propose a method for treating J/ψ production in the earlier phase of a heavy ion collision by assuming equilibration of the charm and anti-charm quarks and the J/ψ under detailed balance constraints, and compare the resulting yields to both those measured in experiments and obtained from traditional statistical hadronization model calculations [15].

2. Model overview

In the Statistical Hadronization Model, hadronic states and resonances are treated as states of grand-canonical ensemble which are occupied according to the equilibrium distribution governed by the chemical freeze-out temperature T and the baryochemical potential μ_B, which is derived from the grand-canonical partition function Z^{GC} [7, 16], where

$$\ln Z_i^{\mathrm{GC}} = \pm \frac{V g_i}{2\pi^2} \int_0^\infty \mathrm{d}p\, p^2 \ln\left(1 \pm e^{-\frac{E_i-\mu_i}{T}}\right) \tag{1}$$

$$= \frac{V T g_i}{2\pi^2} \sum_{n=1}^{\infty} \frac{(\mp 1)^{n+1}}{n^2} \lambda_i^n m_i^2 K_2\left(\frac{n m_i}{T}\right). \tag{2}$$

The upper sign applies to fermions and the lower sign to bosons, while considering only $n = 1$ in eq. (2) corresponds to Boltzmann statistics and is sufficient for all species i with $m_i \gg T$. From $\ln Z^{\mathrm{GC}}$, the thermodynamic properties of the system are calculated.

$$N_i^{\mathrm{th}}(T,\mu) = T \frac{\partial \ln Z_i^{\mathrm{GC}}}{\partial \mu} = \frac{V T g_i}{2\pi^2} \sum_{n=1}^{\infty} (\mp 1)^{n+1} \frac{m_i^2}{n} \cdot \lambda_i^n(T,\mu) \cdot K_2\left(\frac{n m_i}{T}\right), \tag{3}$$

$$S_i(T,\mu) = \frac{\partial}{\partial T}\left(T \ln Z_i^{\mathrm{GC}}\right) = \frac{g_i V}{2} \pi^2 m_i^2 \sum_{n=1}^{\infty} (\mp 1)^{n+1} \lambda_i^n \frac{\mathcal{G}_{i,n}(T,\mu)}{n^2}, \tag{4}$$

where

$$\mathcal{G}_{i,n}(T,\mu) = (2T - \mu n) K_2\left(\frac{n m_i}{T}\right) + \frac{n m_i}{2}\left[K_1\left(\frac{n m_i}{T}\right) + K_3\left(\frac{n m_i}{T}\right)\right]. \tag{5}$$

Fitting eq. (3) to different ratios $N_i^{\mathrm{th}}(T,\mu)/N_j^{\mathrm{th}}(T,\mu)$, the model parameters T and μ_B can be determined. The hadronization source volume V_{SHM} is then adjusted to the measured production of charged particles in a heavy ion collision experiment such that the thermal densities n_{ch} match the observed multiplicities N_{ch}. A thermodynamically consistent correction for finite hadron volumes is applied according to the method provided in [17, 18].

To calculate J/ψ production in the SHM, the fact that charm quarks are not in chemical equilibrium in the pre-hadronic phase of a heavy ion collision is accounted for by enforcing exact charm conservation and taking the number of $c\bar{c}$ pairs produced in the initial hard collisions $N_{c\bar{c}}^{\mathrm{dir}}$ as an additional input [9, 16, 19] , which leads to the charm balance equation for open (oc) and hidden ($c\bar{c}$) charm,

$$N_{c\bar{c}}^{\mathrm{dir}} = \frac{1}{2} g_c N_{oc}^{\mathrm{th}} \frac{I_1\left(g_c N_{oc}^{\mathrm{th}}\right)}{I_0\left(g_c N_{oc}^{\mathrm{th}}\right)} + g_c^2 N_{c\bar{c}}^{\mathrm{th}}, \tag{6}$$

where g_c is the charm fugacity factor to ensure that the thermal multiplicities are scaled according to the constraint of exact charm conservation. Consequently, the J/ψ multiplicity is

$$N_{J/\psi}^{\mathrm{SHM}} = g_c^2 \cdot n_{J/\psi}^{th} \cdot V_{\mathrm{SHM}}. \tag{7}$$

Note that due to the high uncertainties for the charm production cross section in heavy ion collisions and the quadratic contribution of g_c to $N_{J/\psi}^{\mathrm{SHM}}$, this is subject to equivalent uncertainties.

3. J/ψ in the QGP

Results from lattice QCD suggest survival of heavy quarkonia in the QGP at temperatures up to 1.9 times the critical temperature [11, 12]. Even if more recent analyses call this interpretation into question [13], the implications regarding statistical J/ψ production before the hadronization of the thermal medium can be studied under the assumption of entropy conservation at the chemical freeze-out temperature, justified by the assumed instantness of the hadronization process in the SHM, and by approximating the plasma phase in the Stefan-Boltzmann limit. Analogous to the hadronic case, a source volume is defined by $V_{\rm QGP} = S_{\rm SHM}/s_{\rm QGP}$ using eq. (4). Instead of quantifying the deviation of open and hidden charm hadrons from chemical equilibrium with g_c, the deviation of the charm and anti-charm quarks from equilibrium is quantified by the plasma charm fugacity λ_c, which is calculated from

$$N_{c\bar{c}}^{\rm dir} = \lambda_c \frac{2V_{\rm QGP}}{2\pi^2} (m_c^* T)^{\frac{3}{2}} \exp\left(-m_c^*/T\right), \tag{8}$$

again under the constraint of exact charm conservation. Note that the result is strongly dependent on the charm quark mass in the medium m_c^*, which might differ from the bare quark mass of 1275 MeV [20]. It follows that

$$N_{J/\psi}^{\rm QGP} = \lambda_c^2 \cdot n_{J/\psi}^{th} \cdot V_{\rm QGP}. \tag{9}$$

4. Results

The J/ψ multiplicities obtained from evaluating eqs. (7) and (9) should be the same, as the J/ψ color singlet states should 1) equilibrate with the medium under the constraint of detailed balance ensured by λ_c and 2) be mostly unaffected by any processes occurring at the hadronization of the medium itself. By comparing both results in the range for T suggested by fits to experimental data, see fig. 1, the value of m_c^* needed for compatible results in both pictures can be obtained. The model parameters were chosen as to represent central Au-Au collisions at top RHIC energy [9]. Generally, a low dependence of the J/ψ yield on the temperature as a primary model parameter can be observed, in contrast, the dependence on m_c^* is very high in the partonic description. For a freeze-out temperature of $T \simeq 160\,\text{MeV}$, a dressed charm quark mass of $m_c^* \simeq 1550\,\text{MeV}$ is needed for the multiplicities to be compatible with the standard SHM result.

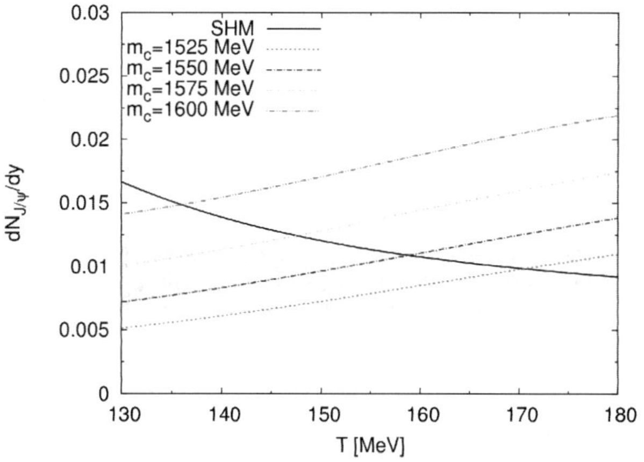

Figure 1. Comparision of the results from eqs. (7) and (9) for m_c^* = 1525-1600 MeV at RHIC energy. The shaded band represents the experimental results [21].

5. Conclusions and outlook

Assuming conservation of entropy across the hadronization process and the exact conservation of charm and anti-charm quarks over the evolution of the fireball, the treatment of J/ψ

mesons above the chemical freeze-out temperature is possible. By comparing to the standard treatment of J/ψ yields, the dressed mass of the charm quarks is fixed to values in the vicinity $m_c^* \simeq 1550\,\mathrm{MeV}$, much higher than the bare quark mass of 1275 MeV, but also much lower than the value of $m_c^* = m_D = 1870\,\mathrm{MeV}$ used in transport codes which describe the experimental results for the J/ψ very well [22, 23, 24]. We will check the influence of a more realistic entropy parametrization as for example provided by [25, 26], where we expect m_c^* to be higher and therefore compatible with the result of $m_c^* = 1600$–$1700\,\mathrm{MeV}$ found in [14]. We will further investigate the equilibration process between the J/ψ and the charm quarks in the partonic transport model BAMPS (*Boltzmann Approach of MultiParton Scatterings*) [27, 28], which simulates the 3+1 space-time evolution of the QGP.

Acknowledgments

This work was supported by the Bundesministerium für Bildung und Forschung (BMBF) and by the Helmholtz International Center for FAIR within the framework of the LOEWE program launched by the State of Hesse. PBB acknowledges financial support by the Stiftung Polytechnische Gesellschaft Frankfurt am Main.

References

[1] Adams J *et al.* (STAR Collaboration) 2005 *Nucl. Phys. A* **757** 102–183 (*Preprint* nucl-ex/0501009)
[2] Adcox K *et al.* (PHENIX Collaboration) 2005 *Nucl. Phys. A* **757** 184–283 (*Preprint* nucl-ex/0410003)
[3] Arsene I *et al.* (BRAHMS Collaboration) 2005 *Nucl. Phys. A* **757** 1–27 (*Preprint* nucl-ex/0410020)
[4] Back B, Baker M, Ballintijn M, Barton D, Becker B *et al.* 2005 *Nucl. Phys. A* **757** 28–101 (*Preprint* nucl-ex/0410022)
[5] Muller B, Schukraft J and Wyslouch B 2012 *Ann. Rev. Nucl. Part. Sci.* **62** 361–386 (*Preprint* 1202.3233)
[6] Braun-Munzinger P, Magestro D, Redlich K and Stachel J 2001 *Phys. Lett. B* **518** 41–46 (*Preprint* hep-ph/0105229)
[7] Andronic A, Braun-Munzinger P and Stachel J 2006 *Nucl. Phys. A* **772** 167–199 (*Preprint* nucl-th/0511071)
[8] Braun-Munzinger P and Stachel J 2009 (*Preprint* 0901.2500)
[9] Andronic A, Braun-Munzinger P, Redlich K and Stachel J 2003 *Phys. Lett. B* **571** 36–44 (*Preprint* nucl-th/0303036)
[10] Matsui T and Satz H 1986 *Phys. Lett. B* **178** 416
[11] Asakawa M and Hatsuda T 2004 *Phys. Rev. Lett.* **92** 012001 (*Preprint* hep-lat/0308034)
[12] Karsch F, Datta S, Laermann E, Petreczky P, Stickan S *et al.* 2003 *Nucl. Phys. A* **715** 701–704 (*Preprint* hep-ph/0209028)
[13] Mocsy A and Petreczky P 2007 *Phys. Rev. Lett.* **99** 211602 (*Preprint* 0706.2183)
[14] Grandchamp L, Rapp R and Brown G E 2004 *Phys. Rev. Lett.* **92** 212301 (*Preprint* hep-ph/0306077)
[15] Bahavar P B, Uphoff J and Greiner C 2014 (*Preprint* 1404.4517)
[16] Braun-Munzinger P, Redlich K and Stachel J 2003 (*Preprint* nucl-th/0304013)
[17] Rischke D H, Gorenstein M I, Stoecker H and Greiner W 1991 *Z. Phys. C* **51** 485–490
[18] Yen G D, Gorenstein M I, Greiner W and Yang S N 1997 *Phys. Rev. C* **56** 2210–2218 (*Preprint* nucl-th/9711062)
[19] Andronic A, Braun-Munzinger P, Redlich K and Stachel J 2007 *Nucl. Phys. A* **789** 334–356 (*Preprint* nucl-th/0611023)
[20] Beringer J *et al.* (Particle Data Group) 2012 *Phys. Rev. D* **86** 010001
[21] Adare A *et al.* (PHENIX Collaboration) 2007 *Phys. Rev. Lett.* **98** 232301 (*Preprint* nucl-ex/0611020)
[22] Liu Y p, Qu Z, Xu N and Zhuang P f 2009 *Phys. Lett. B* **678** 72–76 (*Preprint* 0901.2757)
[23] Zhuang P, Yan L and Xu N 2007 *J. Phys. G* **34** S487–S494
[24] Zhou K, Xu N, Xu Z and Zhuang P 2014 (*Preprint* 1401.5845)
[25] Bazavov A, Bhattacharya T, Cheng M, Christ N, DeTar C *et al.* 2009 *Phys. Rev. D* **80** 014504 (*Preprint* 0903.4379)
[26] Borsanyi S, Fodor Z, Hoelbling C, Katz S D, Krieg S *et al.* 2014 *Phys. Lett. B* **730** 99–104 (*Preprint* 1309.5258)
[27] Xu Z and Greiner C 2005 *Phys. Rev. C* **71** 064901 (*Preprint* hep-ph/0406278)
[28] Xu Z and Greiner C 2007 *Phys. Rev. C* **76** 024911 (*Preprint* hep-ph/0703233)

Low-Mass Dielectron Production in pp, p–Pb and Pb–Pb Collisions with ALICE

Patrick Reichelt (for the ALICE Collaboration)
Institut für Kernphysik, Goethe-Universität Frankfurt am Main, Germany
E-mail: preichelt@ikf.uni-frankfurt.de

Abstract. The ALICE Collaboration measures the production of low-mass dielectrons in pp, p–Pb and Pb–Pb collisions at the LHC. The main detectors used in the analyses are the Inner Tracking System, Time Projection Chamber and Time-Of-Flight detector, all located around mid-rapidity. The production of virtual photons relative to the inclusive yield in pp collisions is determined by analyzing the dielectron excess with respect to the expected hadronic sources. The direct photon cross section is then calculated and found to be in agreement with NLO pQCD calculations. Results from the invariant mass analysis in p–Pb collisions show an overall agreement between data and hadronic cocktail. In Pb–Pb collisions, uncorrected background-subtracted yields have been extracted in two centrality classes. A feasibility study for LHC run 3 after the ALICE upgrade indicates the possibility for a future measurement of the early effective temperature.

1. Introduction

The measurement of electron-positron pairs (dielectrons) in the low invariant mass region allows studying the vacuum and in-medium properties of light vector mesons. Additionally, low-mass dielectrons are produced by internal conversion of virtual direct photons. They are excellent probes to study all collision stages, since they pass through the created medium almost unaffected. To quantify modifications of the dielectron production in heavy-ion collisions, measurements in pp collisions serve as a reference, while the analysis of p-A collisions allows disentangling cold from hot nuclear matter effects. In ALICE [1] at the LHC, dielectron measurements are performed using the central barrel detectors around mid-rapidity. Electrons can be identified via their specific energy loss in the Inner Tracking System (ITS) and the Time Projection Chamber (TPC), combined with time-of-flight information from the TOF detector [2]. In this proceedings we present a virtual direct photon measurement in pp collisions and the invariant mass analyses in p–Pb and Pb–Pb collisions. Prospects of a future measurement in Pb–Pb after the ALICE upgrade for LHC run 3 are also discussed.

2. Virtual direct photon production in pp collisions at $\sqrt{s} = 7$ TeV

The analysis presented here is based on $3 \cdot 10^8$ minimum bias pp collisions at $\sqrt{s} = 7$ TeV, recorded in 2010. Fiducial cuts on transverse momentum ($p_{\mathrm{T}} > 0.2$ GeV/c) and pseudorapidity ($|\eta| < 0.8$) are applied to electron candidates. A clean electron sample is achieved by cuts on the time-of-flight and on the $\mathrm{d}E/\mathrm{d}x$ in the TPC. Dielectron spectra are created for unlike-sign and like-sign combinations of these particles. The unlike-sign distribution contains a superposition

of signal and combinatorial background, while the like-sign is used to provide an estimate of such a background. The acceptance may be different for unlike- and like-sign pairs due to detector effects. Therefore, the signal yield is obtained as $N_{\rm ULS}^{\rm same} - N_{\rm LS}^{\rm same} \cdot R$, where R is built via a mixed-event technique as $R = N_{\rm ULS}^{\rm mix}/N_{\rm LS}^{\rm mix}$.
Figure 1 shows the differential dielectron cross-section in a kinematic region $p_{\rm T}^{\rm ee} \gg m_{\rm ee}$ which is useful to study direct photon production. Also shown in Fig. 1 are a hadronic cocktail and its components as well as the expected mass distribution of dielectrons coming from virtual direct photons, after having applied the single-electron fiducial cuts on $p_{\rm T}$ and η. A fit to the data is performed to determine the virtual photon fraction r by using the function $f_{\rm combined} = (1-r) \cdot f_{\rm cocktail} + r \cdot f_{\rm photon}$ in the range $0.1 < m_{\rm ee} < 0.4$ GeV/c^2. This fit is done in four pair-$p_{\rm T}$ regions and the extracted virtual photon fraction is then multiplied by the inclusive photon cross section, also measured by ALICE via the photon conversion method (PCM) [3]. In Figure 2 the resulting direct photon cross section is compared to NLO pQCD calculations. Reasonable agreement is found, within uncertainties, between data and model. Details on this analysis can be found in [4].

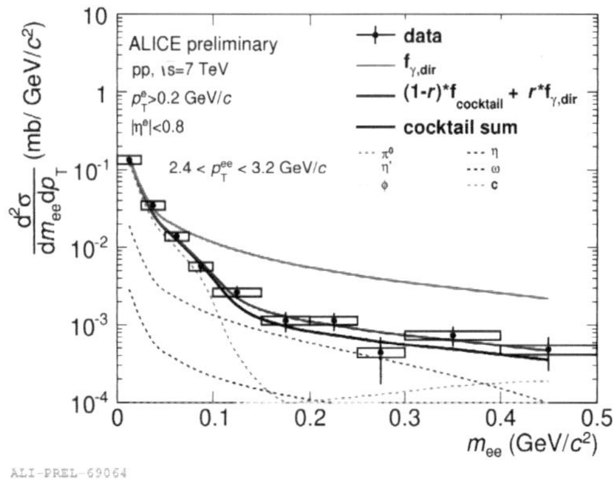

Figure 1. Low-mass region of the dielectron cross section for $2.4 < p_{\rm T}^{\rm ee} < 3.2$ GeV/c compared to the hadronic cocktail, as well as to a fit to extract the virtual direct photon fraction r.

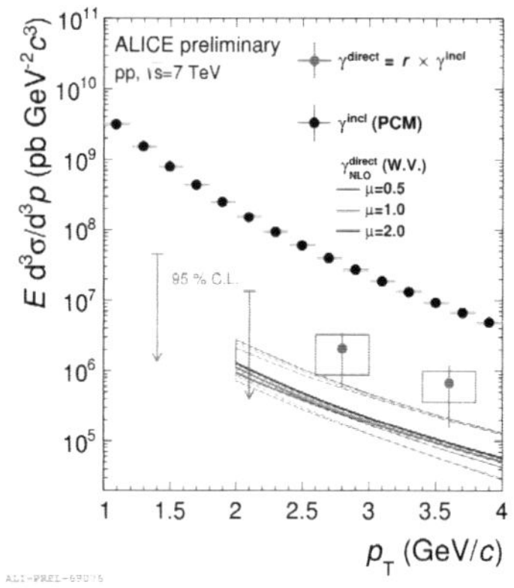

Figure 2. Direct photon cross section compared to NLO pQCD calculations, and the inclusive photon cross section extracted from PCM.

3. Low-mass dielectron production in p–Pb collisions at $\sqrt{s_{\rm NN}} = 5.02$ TeV

The analyzed data sample consists of 10^8 minimum bias p–Pb collisions. The analysis is carried out in $|\eta| < 0.8$ and two different track selections $p_{\rm T} > 0.2$ GeV/c or $p_{\rm T} > 0.4$ GeV/c are performed, to enable comparisons with both the pp and Pb–Pb data. The PID scheme was modified to increase the electron efficiency: a TOF signal is not required, yet used if available to purify the electron sample. Kaons and protons are instead rejected by an electron inclusion cut on the dE/dx measured by the Silicon Drift and Silicon Pixel Detectors of the ITS. In addition, electrons from photon conversions are identified and rejected either by their displaced vertex or by the orientation of the pair plane relative to the magnetic field. More in detail, since a conversion pair does not have an intrinsic opening angle, an apparent opening after reconstruction may only appear perpendicular to the magnetic field vector, while true finite-mass signal pairs have a uniform azimuthal opening angle. Therefore, an exclusion cut on the pair plane normal vector $\phi_{\rm V} > 2.5$ is applied for $m_{\rm ee} < 0.05$ GeV/c^2.

The signal extraction is performed as described in section 2. The signal-to-background ratio is smaller than in pp data by a factor two to four, and range from 50 in the π^0 mass region to 0.02 at $m_{ee} \approx 0.5$ GeV/c^2 and to ≈ 4 in the J/ψ region. The corrected dielectron yield for $p_T > 0.2$ GeV/c and integrated over pair-p_T, is shown in Fig. 3. The data points and systematic uncertainties are extracted from the mean values and the spread of results obtained with 22 different combinations of analysis cut settings (inspired by [5]). The data are compared to the hadronic cocktail, which uses the charged pion measurement [6] as π^0 input and m_T-scaling for the other light mesons. Open heavy-flavour and J/ψ contributions are calculated from PYTHIA, tuned to independent ALICE measurements in pp and p–Pb. Uncertainties on the input of the cocktail sources are propagated to the shapes shown in Fig. 3. Data and cocktail are in reasonable agreement within their uncertainties over the full mass range.

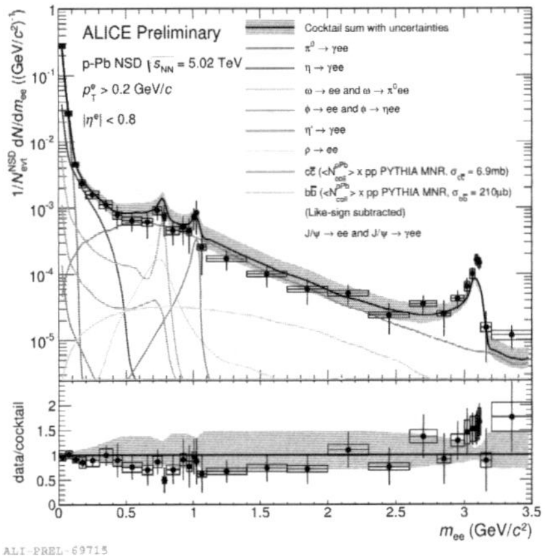

Figure 3. Dielectron mass spectrum from p–Pb collisions in comparison to the hadronic cocktail.

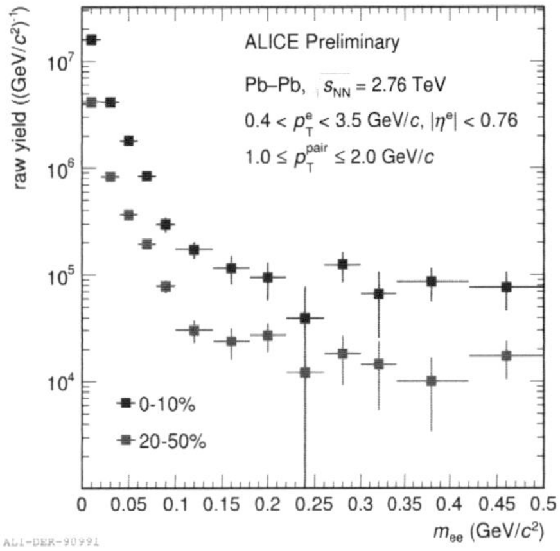

Figure 4. Dielectron raw subtracted yield in central and semi-central Pb–Pb collisions after background subtraction.

4. Low-mass dielectron continuum in Pb–Pb collisions at $\sqrt{s_{NN}} = 2.76$ TeV

From the 2011 Pb–Pb run, $1.7 \cdot 10^7$ and $1.2 \cdot 10^7$ events in central (0-10%) and semi-central (20-50%) Pb–Pb collisions, respectively, are selected for this analysis. The PID scheme is as described for the p–Pb analysis. A track selection cut $0.4 < p_T < 3.5$ GeV/c ensures exclusion of charged pions and reduces the contribution from soft conversion and π^0-Dalitz electrons to the combinatorial background. A residual hadron contamination at the $\approx 10\%$ level also contributes to the background pairs but Monte-Carlo studies show that it does not distort the signal. A conversion rejection cut $\phi_V > 2.36$ is applied for $m_{ee} < 0.1$ GeV/c^2. The signal-to-background ratio for $0.2 < m_{ee} < 0.5$ GeV/c^2 decreases from 0.01 for 20-50% to 0.003 for 0-10% centrality. The combinatorial-background-subtracted uncorrected yields (*raw subtracted yield*) are shown in Fig. 4 in the low-mass region for $1 < p_T^{ee} < 2$ GeV/c for both centralities. Efficiency correction and cocktail comparison are in progress.

5. Upgrade Study for future dielectron measurements in Pb–Pb collisions

A fine-binning differential low-mass dielectron measurement in Pb–Pb collisions is a major physics case for the ALICE upgrade for LHC run 3 [7]. We evaluated to which precision a possible thermal excess yield [8] could be reconstructed in central and peripheral Pb–Pb collisions at full LHC energy ($\sqrt{s_{\rm NN}} = 5.5$ TeV). Electron tracking and PID efficiencies from a Monte-Carlo simulation using the current ALICE setup, operated at reduced solenoid magnetic field, are used and modified according to the expected performance after the upgrade. A statistical significance is calculated from realistically constructed signal and background spectra, and is then used for Poisson sampling of data points around the model input. Figures 5 and 6 show the expected spectra after subtraction of hadronic sources, for $2.5 \cdot 10^7$ and $2.5 \cdot 10^9$ events using the current and upgraded setup, respectively. While the measurement is dominated by uncertainties in the first case, the upgrade is expected to allow for a temperature measurement in the intermediate mass region (IMR) within 10% statistical and 20% systematic uncertainty [9].

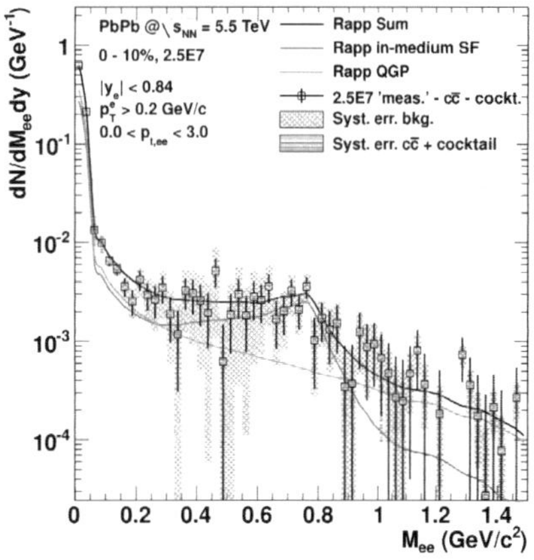

Figure 5. Expected dielectron excess yield for $\sqrt{s_{\rm NN}} = 5.5$ TeV after hadronic cocktail subtraction, using the current experimental setup of ALICE.

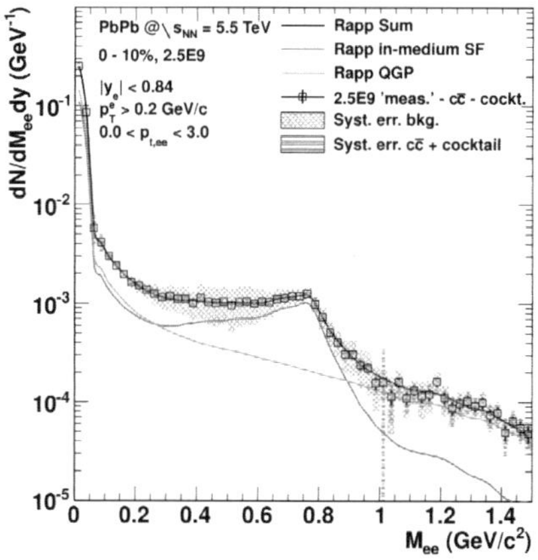

Figure 6. Same as Fig. 5, but with upgraded ITS and TPC. Reduced statistical and systematic uncertainties make a temperature measurement in the IMR feasible.

6. Conclusions

Virtual direct photons are observed in pp collisions at $p_{\rm T}^{\rm ee} \gg m_{\rm ee}$. The extracted direct photon cross section agrees with NLO pQCD calculations within uncertainties. The pair-$p_{\rm T}$-integrated invariant mass spectrum measured in p–Pb collisions is in agreement with the expected hadronic sources. From Pb–Pb collisions, raw subtracted yields have been extracted in two centrality classes and a further analysis is ongoing. A feasibility study indicates that the ALICE upgrade for LHC run 3 will facilitate a measurement of the early effective temperature.

References

[1] K. Aamodt et al. (ALICE collaboration), *JINST* **3** (2008) S08002.
[2] B. Abelev et al. (ALICE Collaboration), *Int. J. Mod. Phys.* A **29** (2014) 1430044.
[3] M. Wilde (for the ALICE Collaboration), *arXiv*:1210.5958 (2012).
[4] M. K. Köhler, Doctoral Thesis, TU Darmstadt (2014).
[5] R. Barlow, *arXiv*:hep-ex/0207026 (2002).
[6] B. Abelev et al. (ALICE Collaboration), *Phys. Lett.* B **728** (2014) 25.
[7] B. Abelev et al. (ALICE Collaboration), CERN-LHCC-2012-012 / LHCC-I-022 (2012).
[8] R. Rapp and J. Wambach, *Eur. Phys. J.* A **6** (1999) 415 and private communication.
[9] B. Abelev et al. (ALICE Collaboration), *J. Phys.* G **41** (2014) 087002.

Hot Quarks 2014 IOP Publishing
Journal of Physics: Conference Series **612** (2015) 012029 doi:10.1088/1742-6596/612/1/012029

Study of Charmonia Production vs. Charged Track Multiplicity in $p + p$ Collisions at PHENIX

Haiwang Yu (for PHENIX Collaboration)

School of Physics, Peking University, Beijing, 100871, China
Department of Physics, New Mexico State University, Las Cruces NM 88003, US
P-25 Group, Bldg. 1, TA53, Los Alamos National Lab. Los Alamos, NM 87544, US

E-mail: yuhw.pku@gmail.com

Abstract. Charmonia production is a unique probe to explore the dynamics of particle production in hot and cold nuclear matter. Traditionally, charmonia suppression has been studied in heavy-ion collisions to quantify various nuclear effects. Since the "long-range, near-side angular correlation in $p + p$ collisions" discovered at the LHC, collective QCD phenomena in smaller systems have been a hot research topic. A recent PHENIX $d + Au$ results also show intriguing hints that the comover effect may play an important role along with other competing effects. At PHENIX the newly upgraded silicon detector FVTX provides improved dimuon mass resolution at forward rapidity as well as the capability of measuring charged track multiplicity directly. In these proceedings, we present the latest result on $\psi(2S)$ production in d+Au collisions at midrapidity, the first $\psi(2S)$ measurement at forward rapidity at RHIC and the expected precision for the charmonia production vs. charged track multiplicity in $p + p$ collisions at PHENIX.

1. Introduction

Charmonia production in relativistic nuclear collisions has been the subject of intense theoretical and experimental investigation for decades [1]. Charmonia measurements serve as a unique probe of the dynamics of particle production in hot and cold nuclear matter. The PHENIX experiment was designed to measure charmonia production through the di-electron channel at midrapidity and di-muon channel at forward rapidity [2]. In these proceedings, we present final PHENIX results on $\psi(2S)$ production in d+Au collisions at midrapidity [3], and the first $\psi(2S)$ measurement at forward rapidity at RHIC. We also show the expected precision for the charmonia production as a function of charged track multiplicity in $p + p$ collisions at PHENIX.

2. Study Charmonia Suppression in $d + Au$ Collisions

In nuclear collisions charmonia production is sensitive to a variety of competing initial and final state effects: the modifications of the parton distribution function in the nucleus, Cold Nuclear Matter (CNM) initial and final state effects, QGP color screening effects, regenerations, etc. Net charmonia suppression observed in central heavy ion collisions is the result of multiple effects on the total cross-section. Recently after the observation of two particle angular correlations at LHC in $p + p$ collisions [9], the study of collective QCD phenomena in small systems such as $p + p$ began to arouse a major interest in both theoretical and experimental areas. In the context of the charmonia suppression study, the effect of nucleus modification of the parton distribution does not exist in $p + p$ colliding system. In heavy ion collisions at RHIC, the estimated J/ψ formation time, ~0.15 fm/c [4], is much larger than the proper time the $c\bar{c}$ pair spends in the nucleus, ~ 0.05 fm/c [3]. Since the charm quarks exist only as a precursor state while in the nucleus, no breakup mechanism that acts within that time scale can affect J/ψ and $\psi(2S)$ production. However, effects which occur after hadronization may very well be more significant for the $\psi(2S)$, as its small binding energy of ~50 MeV may cause it to be more easily dissolved. With a smaller system size in $p + p$ collisions, the breakup effect in the precursor state is also minimized. Therefore the breakup mechanism after haronization can be better isolated and studied.

Fig. 1 shows the di-electron invariant mass spectrum in central and peripheral $d + Au$ collisions [3]. In conjunction with results measured in $p + p$ collisions [6] Fig. 2 shows that $\psi(2S)$ production in central $d + Au$ collisions is clearly more suppressed than in peripheral collisions compared with the J/ψ. Even with the relatively large uncertainty for the $\psi(2S)$ data points on the plot, we obtain evidence that the $\psi(2S)$ suppression as a function of centrality has very different trend with regarding to the J/ψ.

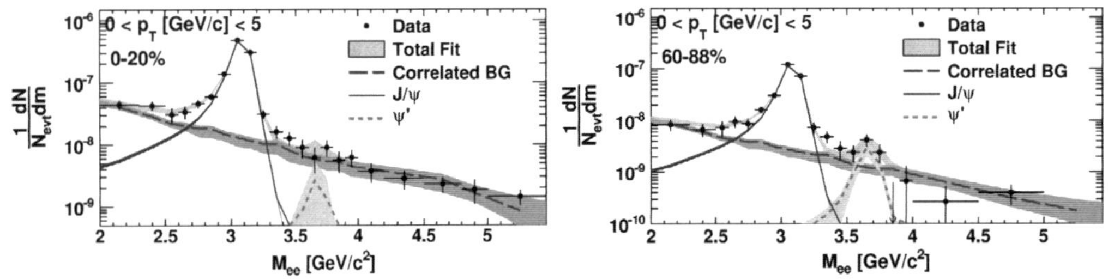

Figure 1. (color online) The e^+e^- mass distribution in $d + Au$ collisions, for $(0 - 20)\%$ (left) and (right) $(60 - 88)\%$ [3]

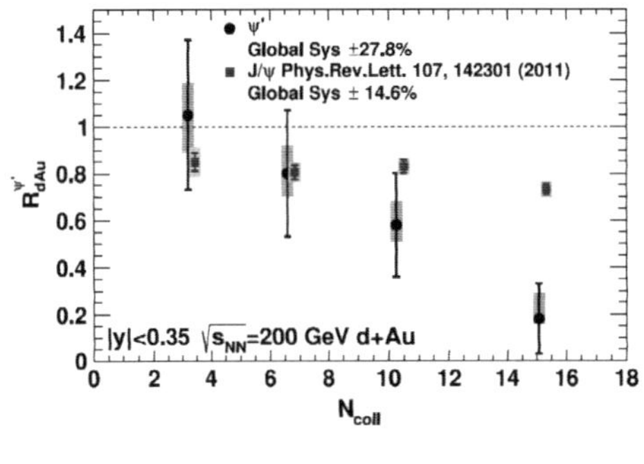

Figure 2. (color online) The $\psi(2S)$ and J/ψ R_{dAu} as a function of N_{coll} at midrapidity in $d + Au$ collisions at $\sqrt{s_{NN}} = 200$ GeV [3].

3. $\psi(2S)$ Measurement at Forward Rapidity

In 2012, PHENIX added new Forward Silicon Vertex detector (FVTX) to the muon arms. It provides us the capability to separate the J/ψ and $\psi(2S)$ peak at forward rapidity and to measure charged track multiplicity directly. Therefore we could extend the charmonia suppression measurement to the forward rapidity. The FVTX provide precise tracking measurements of charged particles in front of the hadron absorbers, before these particles undergo any multiple scattering in the absorber material. By matching the FVTX tracks and the existing muon spectrometer tracks, the muon pair opening angle is more precisely determined than using the muon spectrometer alone, which greatly improves the dimuon mass resolution.

Fig. 3 shows the dimuon mass distribution. A fit function consisting of a Crystal Ball plus Gaussian is used to extract the J/ψ and $\psi(2S)$ yields, with the background represented by the mixed-event combinatorial contribution plus an exponential to describe the open heavy flavor combinations. During fitting, the difference between the peak centers is set to the PDG value of 0.589 MeV$/c^2$. The ratio of the yields of these two states is corrected for the difference in detector effciency, and shown in Fig. 4, along with world data from other experiments (see [6]–[8] and references therein). The error bar on the ratio is the quadrature sum of the statistical uncertainty and the systematic uncertainty due to uncertainties in the relative widths of the Gaussian component of the fit, dimuon trigger efficiency, background contributions, and detector efficiency. The new PHENIX measurement is consistent with the world data and its uncertainty is comparable to other precise measurements.

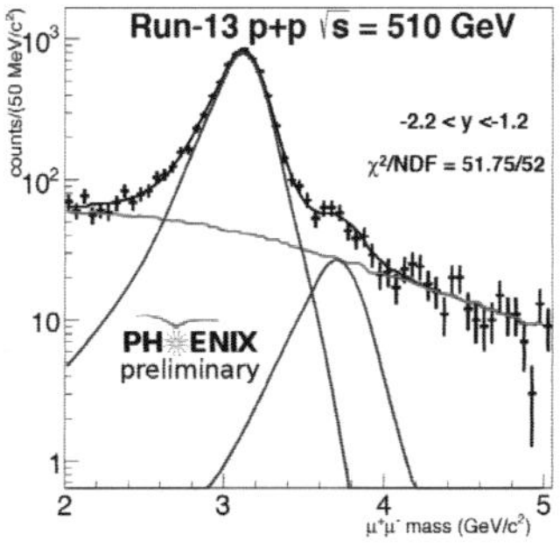

Figure 3. (color online) Fit to dimuon mass spectra used to extract the J/ψ and $\psi(2S)$ peaks based on PHENIX 2013 $\sqrt{s} = 510$ GeV $p + p$ data.

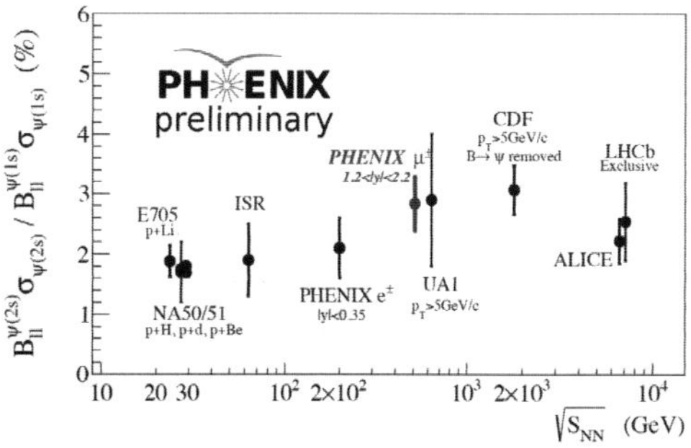

Figure 4. (color online) World data on the ratio of $\psi(2S)/J/\psi$ production as a function of collision energy.

4. Progress of the Study on Charged Track Multiplicity Using FVTX

In addition to improving the dimuon mass resolution at forward rapidity and providing multiple vertexing capability for PHENIX, the FVTX can also measure charged track multiplicity directly. Fig. 5 shows that the charged track multiplicity distribution measured by FVTX using PHENIX 2013 $p + p$ data. However in this run, due to high luminosity, we encountered multiple collision issue which we are still working on at this time. In Fig. 6, we provide the statistical uncertainty projection we expect for the normalized J/ψ yield as a function of normalized charged multiplicity once these cross-checks are completed.

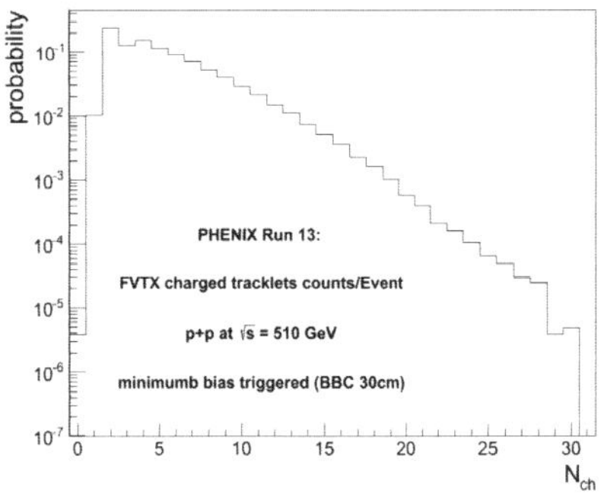

Figure 5. (color online) FVTX charged track multiplicity distribution per bunch crossing for PHENIX 2013 $p + p$ run with $\sqrt{s} = 510$ GeV.

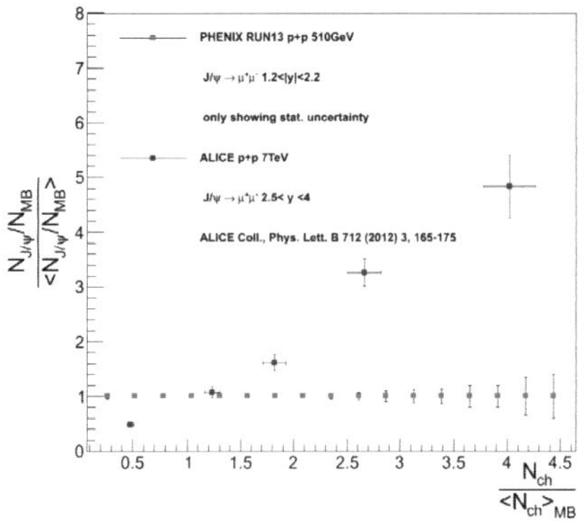

Figure 6. (color online) Normalized J/ψ yield vs. normalized charged track multiplicity, only showing the statistical uncertainty projection for PHENIX 2013 $p+p$ run with $\sqrt{s} = 510$ GeV. As a reference, the ALICE result ([11]) is shown as well.

5. Conclusions and Outlook

The midrapidity data on $\psi(2S)$ production in $d + Au$ collisions at PHENIX indicates that there are additional mechanisms that suppress $\psi(2S)$ more than the J/ψ in $d + Au$ collisions. The short crossing time at RHIC suggests that this suppression occurs outside the nucleus, possibly through interactions with comoving particles. Comparing results from different colliding systems also provide extra evidence to support the comover models. In order to further investigate this at forward rapidity at PHENIX, we utilize the new FVTX detector to measure $\psi(2S)$ to J/ψ ratio and charged track multiplicity. We already get the preliminary $\psi(2S)$ result at forward rapidity for the first time at RHIC. A study of the charmonia production versus the charged track multiplicity is currently under investigation.

References

[1] N. Brambilla, et al. 2011 *Eur. Phys. J.* C **71** 1534 1–178
[2] K. Adcox, et al. 2003 *Nucl. Instrum. Meth.* A **499** 469–479
[3] A. Adare, et al. (PHENIX Collaboration) 2013 *Phys. Rev. Lett.* **111** 202301
[4] F. Arleo, P.-B. Gossiaux, T. Gousset and J. Aichelin 2000 *Phys. Rev.* C **61** 054906
[5] C. Aidala, et al. 2014 *Nucl. Instrum. Meth.* A **755** 44–61
[6] A. Adare, et al. (PHENIX Collaboration) 2012 *Phys. Rev.* D **85** 092004
[7] B. Abelev, et al. (ALICE Collaboration) 2014 *Eur. Phys. J.* C **74** 2974 (*Preprint* 1403.3648 [nucl-ex])
[8] R. Aaij, et al. (LHCb Collaboration) 2013 *J. Phys.* G **40** 045001
[9] V. Khachatryan, et al. (CMS Collaboration) 2010 *J. High Energy Phys.* 09 **2010** 091
[10] A. Capella, A. Kaidalov, A. Kouider Akil, and C. Gerschel 1997 *Phys. Lett. B* **393** 431
[11] B. Abelev, et al. (ALICE Collaboration) 2012 *Phys. Lett. B* **712** 3, 165–75

Hot Quarks 2014 IOP Publishing
Journal of Physics: Conference Series **612** (2015) 012030 doi:10.1088/1742-6596/612/1/012030

Inclusive J/ψ production in pp, p-Pb and Pb-Pb collisions at forward rapidity with ALICE at the LHC

Javier Martín Blanco (for the ALICE Collaboration)
SUBATECH, Ecole des mines de Nantes, Université de Nantes, CNRS-IN2P3, Nantes, FRANCE

E-mail: javier.martin.blanco@cern.ch

Abstract. The ALICE Collaboration has measured the inclusive J/ψ production at forward rapidity in pp, p-Pb and Pb-Pb collisions at the LHC. The pp measurements are crucial for a deeper understanding of the physics involving hadroproduction processes and provide the baseline for p-Pb and Pb-Pb measurements. The comparison of the results on the nuclear modification factor in p-Pb collisions as a function of rapidity or transverse momentum with theoretical predictions shows a fair agreement with a shadowing based model and energy loss models with or without a shadowing contribution. The nuclear modification factor in Pb-Pb as a function of collision centrality or transverse momentum shows a weaker suppression than at lower energies and can be described by statistical hadronization or transport models, suggesting a contribution to the J/ψ production due to the (re)combination of charm quarks, especially at low transverse momentum. The extrapolation to Pb-Pb collisions of the cold nuclear matter effects evaluated in p-Pb collisions supports the (re)combination mechanism interpretation.

1. Introduction

ALICE is the LHC experiment dedicated to the study of heavy-ion collisions. The peculiar properties of some of the charmonium states, like their small size (< 1 fm), strong binding energy (several hundred MeV) and large mass of the charm quarks, make them ideal probes of the strongly interacting Quark-Gluon Plasma (QGP) produced in high-energy heavy-ion collisions. It was predicted that for a high temperature medium, the color-screening of the heavy-quark potential would lead to a sequential suppression of the production of quarkonium states with increasing temperature [1] . However, the experimental results in Pb-Pb collisions at LHC energies have revealed that other effects like quarkonium regeneration, in the QGP or at the boundary with the hadronic phase, may contribute to the charmonium production. In order to measure the effects related to cold nuclear matter (CNM), p-Pb collisions have also been studied at the LHC. The study of these effects is important to disentangle hot (QGP related) and cold nuclear matter effects in Pb-Pb collisions.

In this contribution, results on the inclusive J/ψ production at forward rapidity, such as the nuclear modification factor as a function of transverse momentum (p_{T}) in p-Pb collisions at a centre-of-mass energy per nucleon pair $\sqrt{s_{\mathrm{NN}}} = 5.02$ TeV, and the nuclear modification factor in Pb-Pb collisions at $\sqrt{s_{\mathrm{NN}}} = 2.76$ TeV as a function of centrality or p_{T} are presented and compared to theoretical models. Finally, an estimation of the CNM effects in Pb-Pb, extrapolated from p-Pb measurements, is discussed. The results on the J/ψ production in pp collisions are not presented here but are discussed in [2, 3, 4].

2. ALICE detector and analysed data sample

The main ALICE detectors used for the measurements of muon pairs from the J/ψ decay are the Muon Spectrometer, covering the forward rapidity region ($-4 < \eta < -2.5$); the two VZERO hodoscopes ($2.8 < \eta < 5.1$ and $-3.7 < \eta < -1.7$), used for triggering and centrality determination in Pb-Pb collisions, and the Silicon Pixel Detector (SPD) used for vertex reconstruction.

The energy asymmetry of the LHC beams in p-Pb collisions leads to an asymmetric rapidity coverage for the two different beam configurations: $-4.46 < y_{\rm cms} < -2.96$ at backward rapidity (Muon Spectrometer in the Pb-going direction) and $2.03 < y_{\rm cms} < 3.53$ at forward rapidity (Muon Spectrometer in the p-going direction). The integrated luminosity is 5.8 $\rm nb^{-1}$ (5.0 $\rm nb^{-1}$) for the p-Pb data sample at backward (forward) rapidity and 69 μb^{-1} for the Pb-Pb one.

3. Results

3.1. Nuclear modification factor in p-Pb collisions

The nuclear modification factor in heavy-ion collisions, $R_{\rm AA}$, is an observable that quantifies the nuclear effects and is defined as: $R_{\rm AA} = Y^{J/\psi}/(\langle T_{\rm AA}\rangle \cdot \sigma_{\rm pp}^{J/\psi})$, where $Y^{J/\psi}$ is the J/ψ yield per minimum bias event, $\langle T_{\rm AA}\rangle$ is the nuclear overlap function estimated through the Glauber model and $\sigma_{\rm pp}^{J/\psi}$ is the J/ψ cross section in pp collisions at the same energy[1]. For p-Pb collisions, an equivalent relation is used to define $R_{\rm pPb}$.

The J/ψ production in p-Pb collisions was found to be suppressed at forward rapidity while at backward rapidity $R_{\rm pPb}$ is consistent with binary scaling [6]. At forward rapidity (Fig. 1, right) the $R_{\rm pPb}$ increases with $p_{\rm T}$, reaching unity at $p_{\rm T} \geq 7$ GeV/c. At backward rapidity (Fig. 1, left) the increase is less pronounced and $R_{\rm pPb}$ becomes larger than 1 at $p_{\rm T} \geq 7$ GeV/c, although the systematic uncertainty remains large. Model calculations based on the EPS09 nuclear parton distribution functions parameterization (shadowing model) [7] are in fair agreement with data at high $p_{\rm T}$ for both rapidity ranges, but they seem to underestimate the suppression in the lowest $p_{\rm T}$ region covered by the model ($p_{\rm T} \sim 3$ GeV/c) at forward rapidity. Coherent parton energy loss model calculations [8], with or without a shadowing contribution, describe the data at backward rapidity within the uncertainties. At forward rapidity, the calculation with only energy loss is not able to reproduce the $p_{\rm T}$ dependence, but the one including also a shadowing contribution describes the data for most of the $p_{\rm T}$ range, except at low $p_{\rm T}$. Finally, the CGC-based calculation [9] overestimates the suppression at forward rapidity.

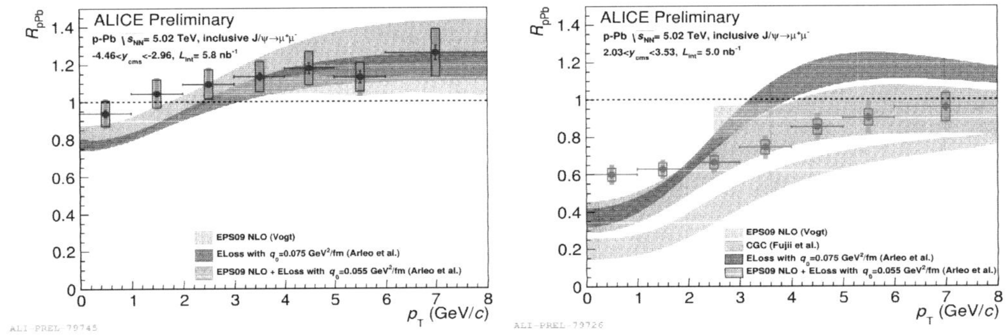

Figure 1. (Color online) J/ψ nuclear modification factor in p-Pb collisions at $\sqrt{s_{\rm NN}} = 5.02$ TeV as a function of $p_{\rm T}$, at backward (left) and forward (right) rapidity, compared to theoretical models.

[1] Extrapolation/interpolation of pp results to $\sqrt{s_{\rm NN}} = 5.02$ TeV is discussed in [5].

Journal of Physics: Conference Series **612** (2015) 012030 doi:10.1088/1742-6596/612/1/012030

3.2. Nuclear modification factor in Pb-Pb collisions

The inclusive J/ψ nuclear modification factor $R_{\rm AA}$ in Pb-Pb collisions as a function of the number of participant nucleons in the collision ($N_{\rm part}$), compared with the PHENIX measurement in Au-Au collisions at $\sqrt{s_{\rm NN}} = 200$ GeV, was presented in [10]. The comparison showed less suppression of the J/ψ at $\sqrt{s_{\rm NN}} = 2.76$ TeV, than that at $\sqrt{s_{\rm NN}} = 200$ GeV for large $N_{\rm part}$, suggesting a (re)generation component in the J/ψ production at LHC energies. The ALICE result does not exhibit a significant centrality dependence for $N_{\rm part} \geq 70$. In Fig. 2 (left), it is compared with theoretical models that include a (re)generation component from deconfined charm quarks in the medium. In the Statistical Hadronization model [11], the charmonium production is fully ascribed to the statistical recombination of charm quarks at the phase boundary, since this model assumes the deconfinement and thermal equilibration of the bulk of the charm quarks. The transport models [12, 13], propose a dynamical competition between the suppression in the QGP and the regeneration mechanisms. The latter model includes shadowing, charmonium dissociation by interaction with the co-moving matter and recombination [14]. As can be observed in Fig. 2 (left), models with a full or partial J/ψ production from deconfined charm quarks can reproduce the centrality dependence observed in the experimental data. A precise determination of the amount of shadowing in p-Pb collisions as well as a measurement of the open charm cross section in Pb-Pb collisions are needed in order to reduce the uncertainties on the model calculations.

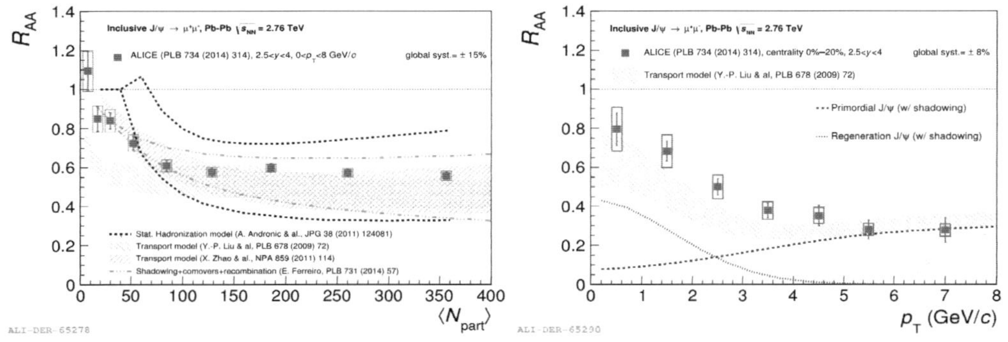

Figure 2. (Color online) Left: J/ψ nuclear modification factor in Pb-Pb collisions at $\sqrt{s_{\rm NN}} = 2.76$ TeV as a function of $N_{\rm part}$, compared with theoretical models. Right: J/ψ $R_{\rm AA}$ in Pb-Pb collisions as a function of $p_{\rm T}$ compared with a transport model calculation [12].

In [15], the $p_{\rm T}$ dependence of the Pb-Pb $R_{\rm AA}$ was compared with the Au-Au PHENIX measurement. It was observed that the ALICE $R_{\rm AA}$ decreases at high $p_{\rm T}$, indicating a stronger suppression in that kinematic range. Furthermore, the ALICE $R_{\rm AA}$ increases at low $p_{\rm T}$, where it becomes up to a factor four larger than at PHENIX. Since the bulk of the charm quarks is produced at low $p_{\rm T}$, regeneration effects should indeed lead to an increase of $R_{\rm AA}$ at low $p_{\rm T}$, as observed by ALICE. The transport model calculations are in agreement with the ALICE result (Fig. 2, right), and confirm that the recombination component dominates at low $p_{\rm T}$.

The p-Pb results presented in section 3.1 can be used to extrapolate the CNM effects to Pb-Pb measurements under the following assumptions: (i) the J/ψ production mechanism is $g+g \to J/\psi$, which can be shown to lead to a comparable gluon longitudinal momentum fraction in the nucleus in p-Pb and Pb-Pb collisions, despite the different energies and rapidity domains [6], (ii) shadowing is the dominant effect, which allows one to consider CNM effects factorizable in p-Pb and Pb-Pb collisions. Under these assumptions we can evaluate the CNM effects in $R_{\rm AA}$

at forward rapidity as the product $R_{\rm pPb}(2.03 < y_{\rm cms} < 3.53) \times R_{\rm Pbp}(-4.46 < y_{\rm cms} < -2.96)$, (denoted $R_{pPb}^{forw} \cdot R_{pPb}^{backwd}$). In Fig. 3 (left), we present the comparison of the $R_{\rm AA}$ with the extrapolated CNM effects. For J/ψ $p_{\rm T}$ above 4 GeV/c CNM effects are small, while below they lead to a suppression of the J/ψ production. The CNM effects can be disentangled from the Pb-Pb nuclear modification factor by computing the ratio $S_{J/\psi} = R_{PbPb}/(R_{pPb}^{forw} \cdot R_{pPb}^{backw})$. $S_{J/\psi}$ (Fig. 3 (right)) increases at low $p_{\rm T}$, further confirming the (re)generation scenario. It then decreases with $p_{\rm T}$ and, for $p_{\rm T} > 4$ GeV/c, the J/ψ suppression reaches a factor three.

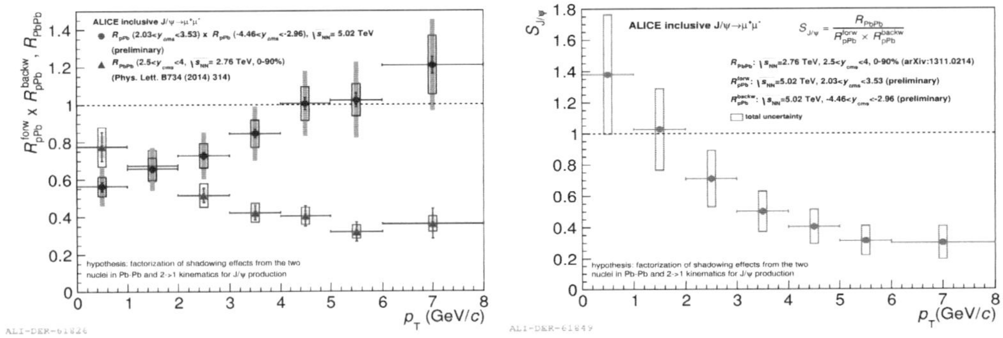

Figure 3. (Color online) Left: J/ψ nuclear modification factor in Pb-Pb collisions as a function of $p_{\rm T}$ and its comparison with the CNM effects extrapolated from the p-Pb measurements. Right: J/ψ nuclear modification factor in Pb-Pb collisions corrected for CNM effects.

4. Conclusions

The inclusive J/ψ production measurements at forward rapidity in p-Pb collisions at $\sqrt{s_{\rm NN}} = 5.02$ TeV and Pb-Pb collisions at $\sqrt{s_{\rm NN}} = 2.76$ TeV have been presented. The nuclear modification factor in p-Pb as a function of $p_{\rm T}$ shows a reasonable agreement with theoretical models based on nuclear shadowing and with models based on coherent parton energy loss. In Pb-Pb collisions the nuclear modification factor as a function of centrality shows less suppression than in lower energy measurements. Theoretical models with a full or partial J/ψ production from deconfined charm quarks describe the data. Less suppression than the one measured at lower energies has been observed at low $p_{\rm T}$. These observations support the (re)combination scenario in Pb-Pb collisions at LHC energies, which appears to be dominant at low $p_{\rm T}$.

References

[1] T. Matsui, H. Satz, Phys. Lett. B178 (1986) 416.
[2] B. Abelev et al. (ALICE Collaboration), Phys.Lett. B718 (2012) 295-306.
[3] B. Abelev et al. (ALICE Collaboration), Eur.Phys.J. C74 (2014) 8, 2974.
[4] B. Abelev et al. (ALICE Collaboration), Phys.Rev.Lett. 108 (2012) 082001.
[5] ALICE and LHCb Collaborations, ALICE-PUBLIC-2013-002.
[6] B. Abelev et al. (ALICE Collaboration), JHEP 1402 (2014) 073.
[7] R. Vogt, Int. J. Mod. Phys. E22 (2013) 1330007 and priv. comm.
[8] F. Arleo and S. Peigné , JHEP 1303 (2013) 122.
[9] H. Fujii and K. Watanabe, Nucl. Phys. A915 (2013) 1.
[10] B. Abelev et al. (ALICE Collaboration), Phys. Rev. Lett. 109 (2012) 072301.
[11] A. Andronic et al., J.Phys. G38 (2011) 124081.
[12] Y.-P. Liu et al., Phys.Lett. B678 (2009) 72-76.
[13] X. Zhao et al., Nucl.Phys. A859 (2011) 114-125.
[14] E. G. Ferreiro, Phys.Lett. B731 (2014) 57-63.
[15] B. Abelev et al. (ALICE Collaboration), Phys.Lett. B734 (2014) 314-327.

Measurement of electrons from charm and beauty-hadron decays in p-Pb collisions at $\sqrt{s_{\mathrm{NN}}} = 5.02$ TeV with ALICE at the LHC

Jan Wagner

Research Division and ExtreMe Matter Institute, GSI Helmholtzzentrum für Schwerionenforschung GmbH, Darmstadt, Germany

E-mail: j.wagner@cern.ch

Abstract. The p_{T}-differential production cross section of electrons from heavy-flavour hadron decays in the rapidity range $-1.06 < y_{\mathrm{cms}} < 0.14$ in p-Pb collisions at $\sqrt{s_{\mathrm{NN}}} = 5.02$ TeV has been measured with ALICE. The cross section of electrons from beauty-hadron decays, isolated based on their larger average displacement from the interaction vertex, is presented as well as the nuclear modification factor R_{pPb} of heavy-flavour and beauty-hadron decay electrons. Theoretical predictions including the effects due to the nuclear modification of the parton distribution functions are discussed with the results.

1. Introduction

The characterization of the Quark-Gluon Plasma (QGP), the deconfined state of strongly-interacting matter produced in high-energy heavy-ion collisions, is one of the main purposes of ALICE at the LHC. The medium can be probed by charm and beauty quarks since they are produced in initial hard partonic interactions and experience its complete evolution. The transverse momentum (p_{T}) distribution of heavy-flavour hadrons and of their decay leptons is an observable sensitive to the energy loss of heavy quarks in the hot and dense medium [1]. Furthermore, a quark-mass dependence of the in-medium energy loss is predicted by theory [2]. By separating leptons from charm-hadron and beauty-hadron decays it is possible to study their in-medium modification and to compare to theoretical predictions.

A significant suppression of electrons from heavy-flavour hadron decays has been observed in heavy-ion collisions at $\sqrt{s_{\mathrm{NN}}} = 2.76$ TeV with ALICE [3]. In addition to the QGP, the presence of cold nuclear matter in the initial state may affect the production of heavy-flavour hadrons through e.g. shadowing/saturation effects and could contribute to the suppression observed in Pb-Pb collisions. This can be investigated by studying the nuclear modification factor R_{pPb} measured in p-Pb collisions, where an extended hot medium is not expected to form:

$$R_{\mathrm{pPb}} = \frac{1}{< T_{\mathrm{pPb}} >} \frac{dN_{\mathrm{pPb}}/dp_{\mathrm{T}}}{d\sigma_{\mathrm{pp}}/dp_{\mathrm{T}}}$$

where $dN_{\mathrm{pPb}}/dp_{\mathrm{T}}$ is the invariant yield in p-Pb collisions, $d\sigma_{\mathrm{pp}}/dp_{\mathrm{T}}$ is the reference cross section measured from pp collisions and $< T_{\mathrm{pPb}} >$ is the nuclear overlap function [4]. Using the tracking and electron identification capabilities of ALICE, it is possible to measure electrons from heavy-flavour hadron decays over a wide momentum range. By taking advantage of the resolution of the silicon vertex detectors, electrons from charm and beauty-hadron decays can be discriminated based on their displacement from the interaction vertex.

2. Analysis

In 2013 ALICE recorded p-Pb collisions at $\sqrt{s_{\rm NN}} = 5.02$ TeV. In this analysis about 10^8 events corresponding to an integrated luminosity $L_{\rm int} = (48.6 \pm 1.6)\mu{\rm b}^{-1}$ were selected using a minimum bias (MB) trigger condition and requiring an event vertex within ± 10 cm from the detector center. The MB trigger for the p-Pb data taking was set to require at least one hit in both of the forward scintillator arrays V0A and V0C [5]. The central barrel of ALICE in the rapidity range $-0.6 < y_{\rm lab} < 0.6$ was used to measure electrons coming from charm and beauty-hadron decays.

For tracking and particle identification (PID) the Time Projection Chamber (TPC) was used. Exploiting the particle specific energy loss $\mathrm{d}E/\mathrm{d}x$ in the TPC, electron identification is possible over a large momentum range. The $\mathrm{d}E/\mathrm{d}x$ is similar for electrons and hadrons at specific momenta, thus other detectors are also required. At low momentum ($p_{\rm T} < 2.5\,{\rm GeV}/c$) electrons are required to be within 3σ of the electron hypothesis in the Time Of Flight detector (TOF). At high momentum ($p_{\rm T} > 6\,{\rm GeV}/c$) the energy deposit E in the electromagnetic calorimeter (EMCal) relative to the momentum p is required to be $0.8 < E/p < 1.2$. The remaining hadron contamination is estimated and subtracted by fitting the TPC $\mathrm{d}E/\mathrm{d}x$ distributions to a superposition of the expected shapes for the various hadrons and then subtracting them.

Two methods are used to distinguish electrons from heavy-flavour hadron decays and electrons from other sources.

The first method is a cocktail-based approach where the different contributions of electrons from background sources are added up and then subtracted from the electron raw yield. The main ingredient is a pion spectrum measured with ALICE, which is used to calculate a spectrum of electrons from pion decays. Contributions from other light mesons are added by applying $m_{\rm T}$ scaling. Electrons from photon conversions are simulated knowing the precise material budget of the beam pipe and the detectors. Other contributions such as weak kaon decays and di-electron charmonium decays are taken into account as well using simulations and measured spectra. A detailed description of the method can be found in published results for pp collisions at $\sqrt{s} = 7$ TeV measured with ALICE [6].

The second method is based on the measurement of the dominant background contribution from photon conversions and Dalitz decays. The contribution of this kind of background electrons, denoted 'photonic' electrons, can be quantified by reconstructing electron-positron pairs with low invariant mass. Each electron (or positron) forming with an opposite-charge partner an e^+e^- pair with an invariant mass smaller than the pion mass is tagged as background. The uncorrelated contribution to this background estimation is removed by subtracting the same-charge combinations passing the same invariant mass cut. The resulting yield of photonic electrons is corrected for the tagging efficiency for finding the partner. This "tagging" efficiency is calculated with Monte Carlo simulations by measuring the fraction of pairs which pass all selection cuts. Because the light meson kinematics in the MC generator are not necessarily the same as in nature, the meson $p_{\rm T}$ distributions were re-weighted with the pion spectrum measured with ALICE and $m_{\rm T}$ scaled spectra for the other meson species. The remaining background electron contributions (weak kaon decays and quarkonium di-electron decays) are estimated in the same way as described in the cocktail-based approach above.

To separate the contribution of electrons from beauty-hadron decays one takes advantage of the relatively large mean proper decay length of beauty hadrons ($c\tau \cong 500\mu{\rm m}$). Since these decays involve a neutrino, the decay vertex of the beauty hadron can not be determined. Instead electron candidates with a p_T dependent minimum impact parameter (IP) to the primary vertex were selected, to enhance the signal of electrons from beauty-hadron decays. This IP cut also suppressed the electron background from conversions and Dalitz decays, since the origin of these contributions is the primary vertex. The remaining background is calculated as a cocktail consisting of electrons from conversions, light meson decays and charm hadron decays. In the cocktail simulation, the measured pion spectrum is used for the non-charm background, the measured D-meson p_T-differential cross-sections [7] are used for the charm background and the impact parameter selection is applied. A detailed description of the method can be found in published results for pp collisions at $\sqrt{s} = 7$ TeV measured with ALICE [8].

3. Results

After subtracting the electrons from background sources the spectra are corrected for the detector geometrical acceptance and efficiency to obtain a cross section. Figure 1 shows the p_T-differential cross section of electrons from heavy-flavour hadron decays (black squared markers) and the cross section of electrons from beauty-hadron decays (red circled markers). For $p_T \sim 6\,\mathrm{GeV}/c$ electrons from beauty hadron decays become the dominant contribution to the inclusive spectrum of heavy-flavour electrons. One of the main contributions to the systematic uncertainty is in the determination of the electron background due to the uncertainties of the measured light meson and charm meson spectra. To calculate the nuclear modification factor R_{pPb} a reference spectrum in pp is needed. For this analysis reference spectra for heavy-flavour decay electrons (Fig. 2) and beauty decay electrons measured at $\sqrt{s} = 7$ TeV [6, 8] are extrapolated to $\sqrt{s_{\mathrm{NN}}} = 5.02$ TeV based on the $\sqrt{s}$ dependence from Fixed Order Next-to-Leading Log (FONLL) pQCD calculations [9].

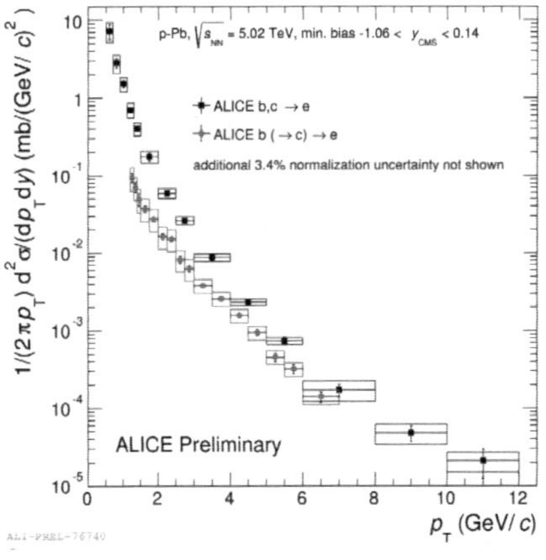

Figure 1. p_T differential cross section of electrons from heavy-flavour hadron decays (black squares) and from beauty-hadron decays only (red circles).

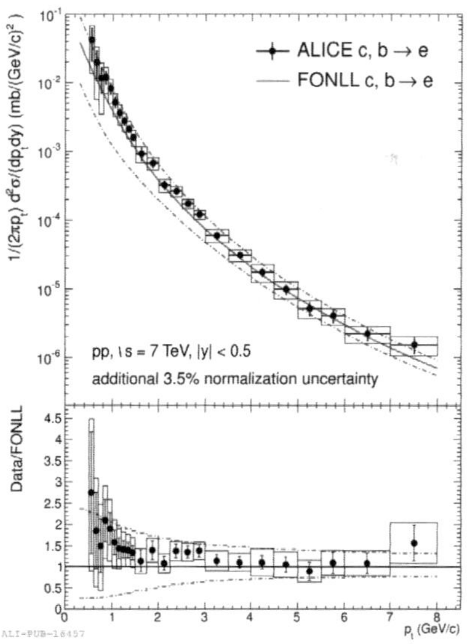

Figure 2. pp reference at $\sqrt{s} = 7$ TeV used for extrapolation to $\sqrt{s} = 5.02$ TeV and compared to FONLL calculations.

For $p_T > 8\,\mathrm{GeV}/c$ an extrapolation with FONLL calculations fitted to the spectrum was used because of the lack of a measured reference at these momenta. In Fig. 3 the measured R_{pPb} is shown as a function of p_T. It is found to be compatible with unity. Also shown in Fig. 3

is a prediction for $R_{\rm pPb}$ calculated using FONLL $p_{\rm T}$-differential cross sections of heavy-flavour decay electrons and shadowing effects based on the EPS09 parametrization [10]. Uncertainties on the FONLL predictions as well as on the shadowing effects are propagated. Within the large uncertainties the measured $R_{\rm pPb}$ is in agreement with the shadowing calculations. The results are also compatible with measurements by the PHENIX collaboration in d-Au collisions at $\sqrt{s_{\rm NN}} = 0.2$ TeV [11].

In Fig. 4, the nuclear modification factor of electrons from beauty-hadron decays measured from p-Pb collisions is shown to be in agreement with the inclusive (charm+beauty) $R_{\rm pPb}$ within the uncertainties.

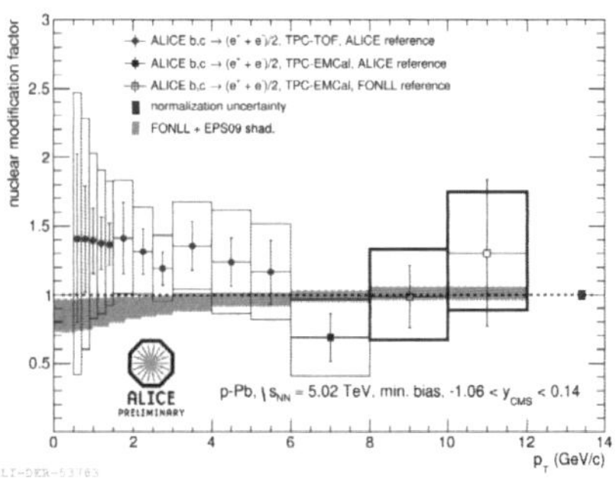

Figure 3. $R_{\rm pPb}$ of heavy-flavour hadron decay electrons together with EPS09 nuclear shadowing predictions [10].

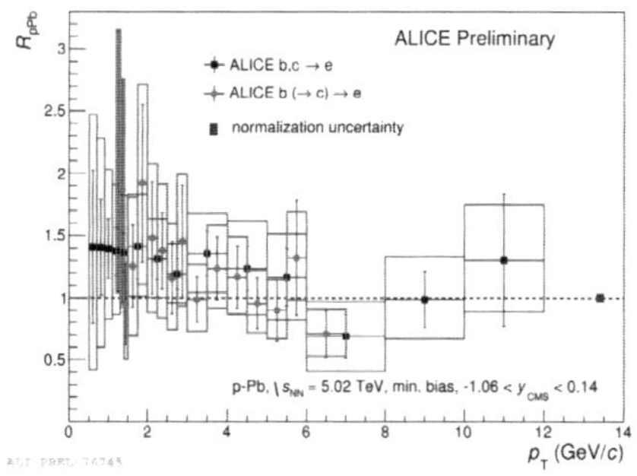

Figure 4. Inclusive (charm+beauty) $R_{\rm pPb}$ (black squares) together with $R_{\rm pPb}$ for electrons from beauty-hadron decays (red circles).

4. Summary

ALICE has measured the $p_{\rm T}$ differential cross section for electrons from heavy-flavour hadron decays as well as the cross section for electrons from beauty-hadron decays in p-Pb collisions. The nuclear modification factor $R_{\rm pPb}$ was measured for both cases.

As shown in Figures 3 and 4 the results are compatible with the absence of nuclear effects in the measured $p_{\rm T}$ ranges, although the experimental uncertainties are substantial. In contrast to the nuclear modification factor measured in central Pb-Pb collisions no hint for a suppression of the electron yields from heavy-flavour decays is observed at high $p_{\rm T}$ in p-Pb collisions, suggesting that in-medium energy loss of partons is responsible for the suppression in Pb-Pb collisions. A quantification of cold nuclear matter effects requires more precise data from p-Pb collisions, though.

References

[1] Baier R, Dokshitzer Y L, Mueller A H, Peigne S and Schiff D 1997 *Nucl.Phys.* **B483** 291–320
[2] Dokshitzer Y L and Kharzeev D 2001 *Phys.Lett.* **B519** 199–206 (*Preprint* hep-ph/0106202)
[3] Sakai S (ALICE Collaboration) 2013 *Nucl.Phys.* **A904-905** 661c–664c
[4] Abelev B *et al.* (ALICE Collaboration) 2013 *Phys.Rev.Lett.* **110** 082302 (*Preprint* 1210.4520)
[5] Abelev B B *et al.* (ALICE Collaboration) 2014 *Int.J.Mod.Phys.* **A29** 1430044 (*Preprint* 1402.4476)
[6] Abelev B *et al.* (ALICE Collaboration) 2012 *Phys.Rev.* **D86** 112007 (*Preprint* 1205.5423)
[7] Abelev B B *et al.* (ALICE Collaboration) 2014 (*Preprint* 1405.3452)
[8] Abelev B *et al.* (ALICE Collaboration) 2013 *Phys.Lett.* **B721** 13–23 (*Preprint* 1208.1902)
[9] Averbeck R, Bastid N, del Valle Z C, Crochet P, Dainese A *et al.* 2011 (*Preprint* 1107.3243)
[10] Eskola K, Paukkunen H and Salgado C 2009 *JHEP* **0904** 065 (*Preprint* 0902.4154)
[11] Adare A *et al.* (PHENIX Collaboration) 2012 *Phys.Rev.Lett.* **109** 242301 (*Preprint* 1208.1293)

Search for muonic atoms at RHIC

Kefeng Xin
Physics & Astronomy Department, Rice University, Houston, Texas 77005
E-mail: kefeng.xin@rice.edu

Abstract. We report the search results of muonic atoms from the STAR experiment at RHIC. The invariant mass distributions of hadron-muon pairs show peaks at the expected atom masses. Two particle correlation functions indicate the hadrons and muons that are close in phase space are emitted at the same space time point, which is a possible signature of muonic atom ionization at the beam pipe. The pion-muon correlation functions are further decomposed into two contributions, and the fraction of direct muons is obtained.

1. Introduction

Muonic atoms are exotic atoms made of hadrons and muons bounded with the Coulomb force. In such atoms, the electrons in ordinary atoms are replaced by muons. The simplest muonic atom, made of a proton and a muon, also known as muonic hydrogen, has been widely used in precise measurement of the proton structure [1] . The π-μ muonic atom has been produced from K_L^0 decay from very intense beams [2, 3]. However, muonic atoms with more exotic cores, such as anti-matter muonic hydrogens and muonic atoms with strangeness have never been observed. The high-energy heavy-ion experiments provide a high multiplicity environment of hadrons and leptons. It is predicted that right after the collisions, the produced hadrons and muons can be bound together by the Coulomb force and form muonic atoms [4].

In heavy-ion experiments, leptons are considered a good probe of the quark-gluon plasma (QGP) by providing measurement of the electromagnetic emission from the medium without strong interaction in the later stages. However, these early produced leptons are mixed with leptons from hadronic weak decays. The later, undergoing complex processes, carry indirect information of the hot and dense matter and thus complicate the study of the lepton probe. Using muonic atoms addresses this issue as hadronic weak decay muons mostly are produced relatively late in the evolution of the heavy-ion collision without much opportunity to bind with hadrons into atomic systems [4, 5]. Therefore the muonic atoms are believed to be an ideal tool to explore the QGP properties.

2. Data Analysis and Results

This study used the dataset collected by the STAR experiment at RHIC at $\sqrt{s_{NN}} = 200$ GeV from Au+Au collisions in year 2010. About 231 million central triggered events passed the event selections and were used in this analysis. Two main detectors were used for particle identification: the Time-Projection Chamber (TPC) and the Time-of-Flight detector (TOF). The energy loss measured from the TPC and timing information measured from the TOF are combined to select a very pure muon sample between 0.15-0.25 GeV/c. The corresponding kaons and protons which fall into the momentum ranges 0.70-1.17 GeV/c and 1.33-2.22 GeV/c can also be cleanly identified.

2.1. *Invariant Mass*

With the proper particle identification, the atom invariant mass distributions can be studied. Three combinatorial methods were used, the unlike-sign, the like-sign, and the mixed-event. The like-sign pair mass was corrected from the acceptance difference between positive and negative charged particles:

$$LS_{+-}(corrected) = \sqrt{LS_{++}LS_{--}}\frac{ME_{+-}}{\sqrt{ME_{++}ME_{--}}}, \tag{1}$$

where LS stands for like-sign, ME for mixed-event, and the indices of each term stand for the electric charges for hadrons and leptons. The origin of the correction is the different bending direction of particles with different charges in the magnetic field inside the detector. The consequence is that pairs with certain opening angles are lost at the TPC sector boundaries or dead TPC readout units. The opening angles are different between same-charge and opposite-charge pairs, and lead to lost pairs in different mass regions. More discussion of this correction can be found in [8].

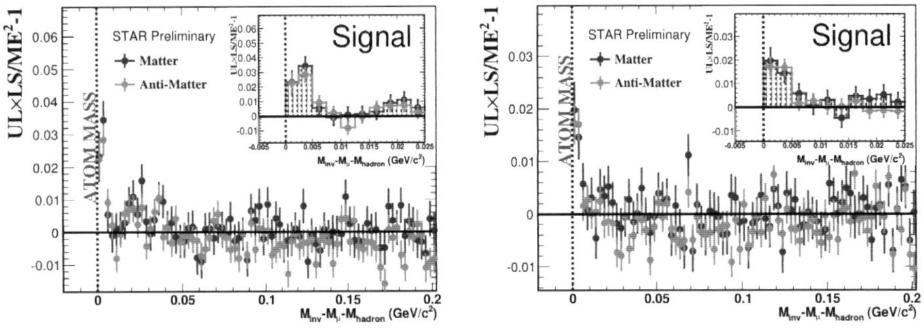

(a) p-μ^- and $\bar{p}$-μ^+ pairs. (b) K^--μ^+ and K^+-μ^- pairs.

Figure 1: The pair invariant mass distributions of $UL \times LS/ME^2 - 1$ show peaks at the atom masses.

The unlike-sign method, which pairs a hadron and a muon with opposite electric charges, contains the atom signal, a combinatorial background, and the attractive Coulomb effect. The like-sign method, which pairs a hadron and a muon with the same change, contains a combinatorial background and repulsive Coulomb effect. The mixed-event method, which pairs a hadron and a muon with opposite charges from different events, contains only a combinatorial background. Note that the attractive Coulomb force in the unlike-sign method will enhance the distributions; and the repulsive Coulomb force in the like-sign method will suppress the distributions. The following variable was used to cancel the Coulomb effect in like-sign and unlike-sign and reveals the correct signal:

$$UL \times LS/ME^2 - 1, \tag{2}$$

where $UL \times LS$ stands for unlike-sign $\times$ like-sign, which cancels the Coulomb effect, and ME stands for mixed-event for normalization. Figure 1 shows sharp peaks at zero net mass in $UL \times LS/ME^2 - 1$ distributions after the Coulomb effect was removed. Here the net mass is the atom mass subtracting from hadron and muon masses.

2.2. Femtoscopic Correlations

Two particle correlations have been used to study the interactions between the two daughter particles. The correlation function $C(k^*)$, where k^* is the magnitude of momentum in the pair rest frame, describes the strength of the correlation. A typical two-particle correlation function shows large deviations from unity if there are only Coulomb interactions [9]. For non-identical particles, a particular particle species can be chosen as a leading particle, and two scenarios can be discussed. The first scenario is that the leading particle is produced on average closer to the center of the collision. If it travels faster than the other particle, the leading particle tends to catch up with the other particle. As their distance gets smaller, the correlation, denoted as C_+, gets stronger because of more interactions. If the leading particle travels slower than the other particle, it tends to move relatively away from the other particle. Large distance then leads to weaker interactions and correlations, noted as C_-. The double ratios C_+/C_- in this scenario deviate from unity, and fall on the same side as the correlation function with respect to unity. In the second scenario, the leading particle is produced further away from the center of the collision than the other particle. In this scenario, following the similar logic, C_+/C_- also deviates from unity. The double ratio was used as an indication of the space time asymmetry of the production of the two particles.

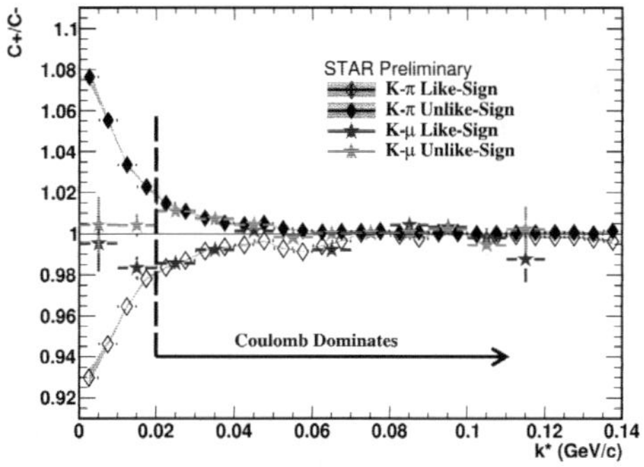

Figure 2: The double ratio of the K-π and K-μ systems show significant difference at low k^*. The convergence to unity of K-μ suggests atom ionization at the beam pipe, which is a beryllium foil and was used to disassociate the hadron and the muon in an atom.

The K-π system has been thoroughly studied [9] and we calculated the correlation functions as a reference. The pion momenta are selected to match the muons. We observed the double ratio C_+/C_- in K-π system deviates from unity as shown in Fig. 2. The smaller the k^* is, the more deviation from unity. This monotonic behavior comes from the Coulomb effect, and is an indication of the space-time asymmetry of the production of the pions and the kaons. Compared with the reference system, the K-μ system can be divided into two regions indicated in Fig. 2 by the dashed line. The Coulomb effect dominates at high k^*, and we observed consistency between K-π and K-μ systems. At very low k^*, where we expect the signal from atoms to appear, we observe significant differences between the two systems. The double ratio C_+/C_- in K-μ system converges to unity instead of diverging. This is an indication that the two particles are emitted at the same space-time point, which is consistent with the dissociation of the hadron and muon at the detector beam pipe. The correlation study provides a signature of muonic atom ionization when the atoms hit the detector beam pipe.

2.3. π-μ Correlations

The π-μ correlations were used to extract the fraction of direct muons. The measured correlation functions, denoted as B, have two major contributions, the correlation from directly produced muons and pions, and the correlation from weak decay muons and pions. The later one inherits the correlations from two pions. The two pion correlation, besides the Coulomb effect, has large quantum interference effect from identical particles, which appears to be an attractive force and enhances the correlation [10]. We estimated the first contribution from π-π correlations, considering the pion mass is relatively close to muon mass. And we used the reversed unlike-sign π-π correlation to avoid the quantum effect, noted as $1/C$. The weak decay muon and pion correlation function was obtained with simulation technique. One of the two pions taken from the data is manually set to decay to a muon and a neutrino based on energy momentum conservation. Then the correlation function between this artificial muon and the other pion is calculated, denoted as A. A linear relationship is established: $B = \alpha \times 1/C + \beta \times A$, where the coefficient α stands for the fraction of direct muons. A χ^2 minimization is then performed to extract the parameters. Note that when two pions have very similar momentum, the helices that they produce in the TPC will overlap and cannot be separated. The consequence is that one pion is missing in data and cannot be saved. The lost pions caused lost muons at very low k^*. So in the fitting, the low k^* region was discarded, and the fitting range is indicated on Fig. 3. The fitting results show $22.0 \pm 0.4\%$ of the muons out of the inclusive muons are direct muons.

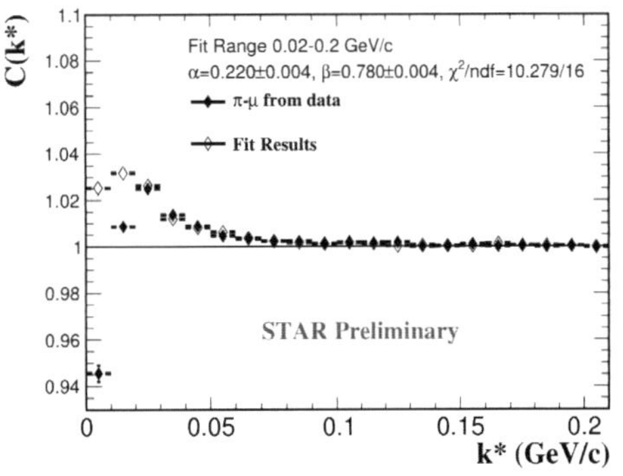

Figure 3: Measured π-μ correlation function, fitted by π-π correlation function and simulated π-μ_{decay}. The coefficient α represents the weight of the contribution from direct muon and pion correlation.

3. Summary

We presented the first search results on antimatter muonic atoms and K-μ atoms from Au+Au collisions at $\sqrt{s_{NN}} = 200$ GeV at the STAR experiment. The invariant mass distributions show sharp peaks at the expected mass positions. The peaks are consistent across different combinations of hadrons and muons. The correlation functions show a signature of muonic atom ionization at the beam pipe, indicating the two constituent particles are disassociated at the same space and time. The two components of the π-μ correlation function were extracted and the fraction of direct muons out of all produced muons is found to be $22.0 \pm 0.4\%$.

References

[1] A. Antognin, *et al.*, Science **339** 417 (2013); Y. Tanaka and B. M. Steffen, Phys. Rev. Lett. **51** 18 (1983).
[2] R. Coombes, *et al.*, Phys. Rev. Lett. **37** 5 (1976).
[3] S. H. Aronson, *et al.*, Phys. Rev. Lett. **48** 16 (1982).
[4] G. Baym, G. Freidman, R. J. Hughes, and B. Jack, Phys. Rev. **D48** 9 (1993).
[5] J. Kapusta and A. Mocsy, Phys. Rev. C **59** 5 (1999).
[6] M. Anderson, *et al.*, Nucl. Instrum. Methods Phys. Res., Sect. A, Accel. Spectrom. Detect. Assoc. Equip., 499 (2003), p. 659.
[7] J. Adams, *et al.*, Phys. Rev. Lett. 94 (2005) 062301; J. Adams, *et al.*, Phys. Lett. B 616 (2005) 8.
[8] A. Adare, *et al.*, Phys. Rev. C **81** 5 (2010); L. Adamczyk, *et al.*, Phys. Rev. C **86** 5 (2012).
[9] J. Adams, *et al.*, Phys. Rev. Lett. **91** 26 (2003).
[10] J. Adams, *et al.*, Phys. Rev. C **71** 044906 (2005).

Forward rapidity $\psi(2S)$ meson production in pp, p-Pb and Pb-Pb collisions with ALICE at the LHC

Marco Leoncino for the ALICE Collaboration
INFN and University, Via Pietro Giuria 1, Torino, Italy
E-mail: marco.leoncino@cern.ch

Abstract. The ALICE Collaboration has studied the inclusive $\psi(2S)$ meson production in pp, p-Pb and Pb-Pb collisions at the CERN LHC. The $\psi(2S)$ is detected through its decay to a muon pair, using the forward Muon Spectrometer, which covers the pseudo-rapidity range $-4 < \eta < -2.5$. The $\psi(2S)$ production cross sections in pp collisions are presented as a function of rapidity (y) and transverse momentum (p_T). In p-Pb collisions, $\psi(2S)$ results are compared to the J/ψ ones by the ratio of their production cross sections as a function of rapidity, transverse momentum and event activity. The $\psi(2S)$ nuclear modification factor, R_{pA}, is also discussed. The results show a $\psi(2S)$ suppression compared to the one observed for the J/ψ meson and are not described by theoretical models including cold nuclear matter effects as nuclear shadowing and energy loss. Finally, the preliminary results of $\psi(2S)$ meson production in Pb-Pb collisions are shown in two p_T ranges as a function of the collision centrality.

1. Introduction
The study of charmonia (bound states of c and $\bar{c}$ quarks), in different collision systems, is the object of intense theoretical and experimental investigations [1]. Proton-proton (pp) collisions are fundamental to evaluate the production cross section and to test production models. In proton-nucleus (p-A) collisions, several initial and final state effects, related to the presence of cold nuclear matter (shadowing, energy loss and nuclear absorption) can influence the observed charmonium yields [2, 3]. Finally, in nucleus-nucleus (A-A) collisions, a deconfined phase of quarks and gluons (QGP) is expected to play an important role on the charmonium production [4]. Among the charmonium the $\psi(2S)$ meson is receiving a lot of attention since it is more weakly-bound than the J/ψ and intriguing results have been already obtained at lower collision energies [5]. ALICE data can improve the understanding of $\psi(2S)$ production in hadronic collisions.

2. ALICE detector and data samples
The ALICE detector consists of a central barrel dedicated to particle tracking and identification (in the pseudo-rapidity range of $|\eta| < 0.9$) and a forward spectrometer for the detection of muons (in the interval of $-4 < \eta < -2.5$). More details about the experimental setup can be found in [6]. Charmonium states are detected in the dimuon decay channel using the Muon Spectrometer. The pp analysis is performed in the rapidity interval of $-4 < y_{lab} < -2.5$ using a data sample obtained at the center of mass energy of $\sqrt{s}$=7 TeV and corresponding to an integrated luminosity of $L_{int}^{pp} = 1.35 \pm 0.07$ pb^{-1} . In p-Pb collisions data have been collected

at $\sqrt{s_{NN}}$=5.02 TeV in two configurations with inverted beam directions, with the following rapidity coverages: $-4.46 < y_{\rm cms} < -2.96$ ($L_{\rm int}^{\rm Pbp} = 5.81 \pm 0.18$ nb^{-1}, Pb-going direction) at backward rapidity and $2.03 < y_{\rm cms} < 3.53$ ($L_{\rm int}^{\rm pPb} = 5.01 \pm 0.19$ nb^{-1}, p-going direction) at forward rapidity. Finally, the ψ(2S) production in Pb-Pb collisions is studied at $\sqrt{s_{NN}}$=2.76 TeV ($L_{\rm int}^{\rm PbPb} = 68.8 \pm 0.9\,\mu$b^{-1}) in the rapidity region of $-4 < y_{\rm lab} < -2.5$.

3. Results

The ψ(2S) cross section is obtained as: $\sigma^{\psi(2S)} = N^{\psi(2S)}/(L_{\rm int} \cdot {\rm BR}_{\mu^+\mu^-} \cdot \langle A\epsilon \rangle)$, where $N^{\psi(2S)}$, the number of reconstructed ψ(2S), is divided by the branching ratio $BR_{\mu^+\mu^-}$, the detector mean acceptance times efficiency $\langle A\epsilon\rangle$ and finally normalized to the integrated luminosity $L_{\rm int}$.

3.1. pp collisions

The results in pp collisions [7] are shown in Fig.1: the $p_{\rm T}$-differential cross section is compared to LHCb results [8]. A good agreement is observed between the two experiments (small differences are visible at low $p_{\rm T}$, but the comparison is not trivial because of the different rapidity coverage of the two detectors).

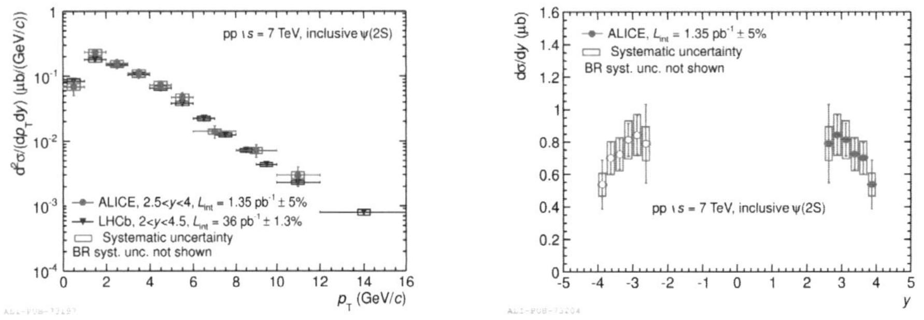

Figure 1. Differential ψ(2S) production cross section as a function of $p_{\rm T}$ (left) and y (right). The $p_{\rm T}$ differential results are compared to LHCb measurements [8].

3.2. p-Pb collisions

The cross section ratio between the tightly bound J/ψ and the loosely bound ψ(2S) charmonium states, ${\rm B.R.}_{\psi(2S)\to\mu^+\mu^-}\sigma_{\psi(2S)}/{\rm B.R.}_{J/\psi\to\mu^+\mu^-}\sigma_{J/\psi}$ is shown in the left panel of Fig.2. These ratios are significantly lower than the ones in pp, both at forward and backward rapidity, pointing to a bigger ψ(2S) suppression (compared to the J/ψ) in p-Pb collisions than in pp.
The double ratios together with that of PHENIX, $[\sigma_{\psi(2S)}/\sigma_{J/\psi}]_{\rm pPb}/[\sigma_{\psi(2S)}/\sigma_{J/\psi}]_{\rm pp}$ is shown in the right panel of Fig.2. These results indicate thate the ψ(2S) suppression is more than the J/ψ to a level of 2.1σ at forward-rapidity and 3.5σ at backward-rapidity. At midrapidity, PHENIX results [9], from $\sqrt{s_{\rm NN}}$ = 200 GeV d-Au collisions, are in qualitative agreement with ALICE data [10].
The nuclear modification factor $R_{\rm pA}$, i.e. the ratio of the ψ(2S) production yield in p-A to the one in pp scaled by the number of binary collisions, is another useful quantity to study the effects of nuclear matter on the ψ(2S) production. The $R_{\rm pA}$ of ψ(2S) and J/ψ, are shown in Fig.3, left, in the two rapidity intervals, indicating a stronger ψ(2S) suppression than that of the J/ψ, both at backward and forward rapidity.

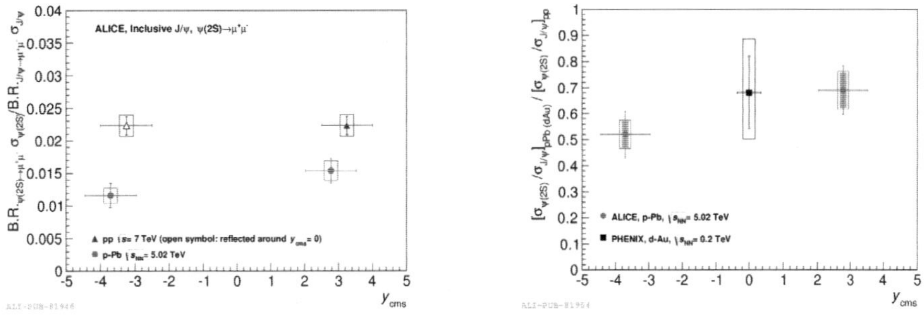

Figure 2. Left: the cross section ratios compared with the corresponding pp results at $\sqrt{s} = 7$ TeV. Right: the double ratios compared to the corresponding PHENIX result [9].

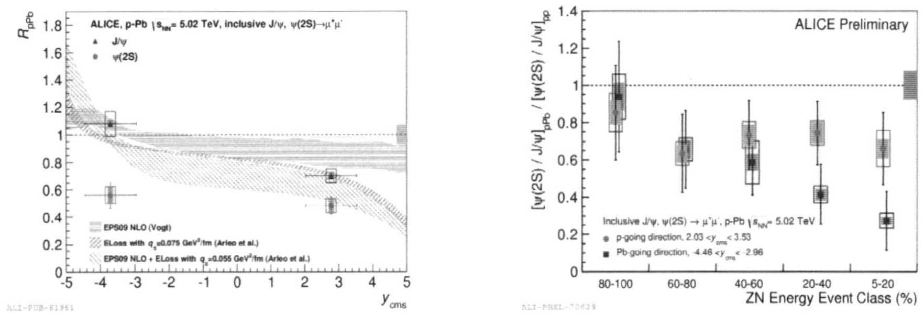

Figure 3. Left: the nuclear modification factor for ψ(2S) compared to the corresponding J/ψ one. Model calculations tuned on J/ψ and including nuclear shadowing and coherent energy loss are also shown. Right: double ratios as a function of the event activity in p-Pb and Pb-p collisions.

ALICE results are compared with theoretical predictions including shadowing only [11] or coherent energy loss, with or without a shadowing contribution [12]. These calculations correspond to the ones performed for the J/ψ: shadowing effects are expected to be similar (within 2-3%), because of the similar gluon distributions that produce the $c\bar{c}$ state, while no dependence on the final state is expected for coherent energy loss. The predictions are in disagreement with the ψ(2S) data and indicate that other final state effects should be considered to explain the observed ψ(2S) suppression. The break-up of the resonance in the nuclear medium depends on the binding energy of the charmonium states and could be considered a cause of the larger ψ(2S) suppression. However, the break-up is relevant only if the charmonium formation time τ_f is smaller than the time τ_c spent by the $c\bar{c}$ pair in the nucleus. Estimates for τ_f [13] are in the range 0.05-0.15 fm/c, while $\tau_c = \langle L \rangle / (\beta_z \gamma)$ [14] (where $\langle L \rangle$ is the average length of nuclear matter crossed by the pair, $\beta_z = \tanh y^{rest}_{c\bar{c}}$ and $\gamma = E_{c\bar{c}}/m_{c\bar{c}}$) is about 10^{-4}fm/c at forward rapidity and about $7 \cdot 10^{-2}$fm/c at backward rapidity. In this situation, the strong ψ(2S) suppression cannot be explained in terms of the $c\bar{c}$ pair break-up (expecially at backward rapidity where the difference between the J/ψ and ψ(2S) R_{pA} is bigger). Finally, the double ratio $[\sigma_{\psi(2S)}/\sigma_{J/\psi}]_{\mathrm{pPb}}/[\sigma_{\psi(2S)}/\sigma_{J/\psi}]_{\mathrm{pp}}$ is presented as a function of the event activity (i.e. the event multiplicity based on a measurement from the Zero Degree Calorimeters) in the two rapidity intervals (see Fig.3, right panel). When compared to the J/ψ, the ψ(2S) is more suppressed with increasing event activity, in particular at backward rapidity. This could be another hint of final state effects that can affect the ψ(2S) production, in particular at backward rapidity.

3.3. Pb-Pb collisions
The double ratio $[\sigma_{\psi(2S)}/\sigma_{J/\psi}]_{PbPb}/[\sigma_{\psi(2S)}/\sigma_{J/\psi}]_{pp}$ has been studied by ALICE as a function of the collision centrality in two p_T intervals (see Fig.4). In the interval $0 < p_T < 3$ GeV/c, the $\psi(2S)$ signal can be extracted in three centrality classes, while, in the interval $3 < p_T < 8$ GeV/c the upper limit at 95% confidence level is shown for the most central collisions. ALICE results are compared with the CMS double ratios presented in two p_T intervals corresponding to two different rapidity ranges. However, the large statistical and systematic uncertainties of the ALICE results prevent a firm conclusion on the $\psi(2S)$ behaviour in Pb-Pb and the comparison with the CMS values [15] is not straightforward, given also the different kinematic coverage.

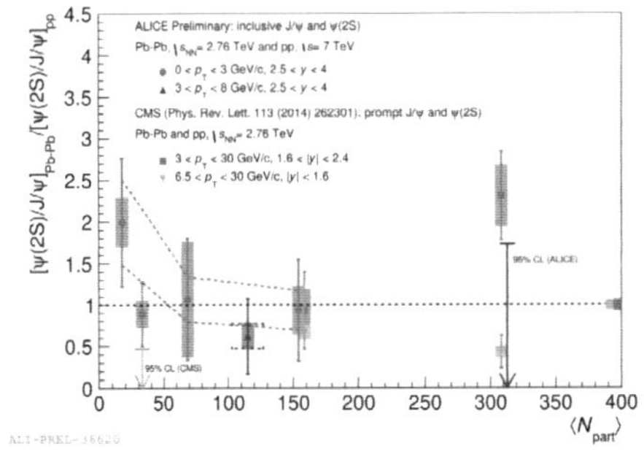

Figure 4. Double ratios $[\sigma_{\psi(2S)}/\sigma_{J/\psi}]_{PbPb}/[\sigma_{\psi(2S)}/\sigma_{J/\psi}]_{pp}$ as a function of the event centrality, in two p_T intervals. CMS measurements [15], in two p_T intervals corresponding to two different rapidity coverages, are also shown.

4. Conclusions
In summary, ALICE collaboration has studied the $\psi(2S)$ production in pp, p-Pb and Pb-Pb collisions. In pp collisions the $\psi(2S)$ production cross sections have been obtained as a function of p_T and y, and are in good agreement with the LHCb measurements. In p-Pb collisions the $\psi(2S)$ is more suppressed than the J/ψ at both forward and backward rapidity. Theoretical models based on shadowing and/or energy loss are in disagreement with data and the break-up of the $c\bar{c}$ pair can hardly explain the strong $\psi(2S)$ suppression, indicating that other final state effects are required. Finally, preliminary results in Pb-Pb collisions have been shown: large uncertainities prevent to make definitive conclusions.

5. References
[1] Brambilla N. et al., *Eur. Phys. J.* C **71** (2011) 1534
[2] de Florian D. et Sassot R., *Phys. Rev.* D **69** (2004) 074028
[3] Arleo F. et Peigne S., *Phys. Rev. Lett.* **109** (2012) 122301
[4] Matsui T. et Satz H., *Phys. Lett.* B **178** (1986) 416
[5] Leitch M. J. et al., (E866 Collaboration) *Phys. Rev. Lett.* **84** (2000) 3256
[6] Aamodt K. et al., (ALICE Collaboration) *JINST* **3** (2008) S08002
[7] B. Abelev et al. (ALICE Collaboration), *Eur. Phys. J.* C **74** (2014) 2974
[8] Aaij R. et al., (LHCb Collaboration) *Eur. Phys. J.* C **72** (2012) 2100
[9] Adare A. et al., (PHENIX Collaboration) *Phys. Rev. Lett.* **111** (2013) 202301
[10] B. Abelev et al. (ALICE Collaboration), arXiv:1405.3796, accepted by JHEP
[11] Albacete J. et al., *Int. J. Mod. Phys.* E **22** (2013) 1330007
[12] Arleo F. and Peigne' S., *JHEP* **1303** (2013) 122
[13] Arleo F., Gossiaux P.-B., Gousset T. and Aichelin J., *Phys. Rev.* C **61** (2000) 054906
[14] McGlinchey D., Frawley A.D and Vogt R., *Phys. Rev.* C **87** (2013) 054910
[15] V. Khachatryan et al., (CMS Collaboration) *Phys. Rev. Lett.* **113** (2014) 262301

Journal of Physics: Conference Series **612** (2015) 012034 doi:10.1088/1742-6596/612/1/012034

Measurements of non-photonic electrons with the STAR experiment

Olga Rusňáková for the STAR Collaboration
Czech Technical University in Prague
Faculty of Nuclear Sciences and Physical Engineering
Břehová 7, 11519, Prague 1, Czech Republic

E-mail: olga.hajkova@fjfi.cvut.cz

Abstract. Non-photonic electrons (NPE), produced by semileptonic decays of D and B mesons, are good probes to study the properties of hot and dense medium created in relativistic heavy ion collisions at RHIC. Studies of heavy quark production in p+p collisions can test the validity of perturbative QCD. They also provide a baseline to study the effects of nuclear matter on the production of heavy quarks in heavy ion collisions. In this paper, we present recent results of NPE spectra measured in p+p collisions at $\sqrt{s} = 200$ GeV in mid-rapidity. We also report NPE nuclear modification factor R_{AA} and elliptic flow v_2 in Au+Au collisions at $\sqrt{s_{NN}} = 200$ GeV.

1. Introduction

Heavy quarks are good probes to study the properties of strongly interacting matter, the Quark-Gluon Plasma (QGP). Due to their large masses, they are produced mainly during the initial hard parton scatterings at RHIC, before the QGP is formed. They are expected to interact with the medium differently from the light quarks. Hot nuclear matter effects, such as heavy quark inteactions with the QGP, change the heavy quark kinematic distribution but do not change the total heavy quark yield. Due to energy loss of heavy quarks in the QGP, their transverse momentum p_T distributions in heavy ion collisions may fall steeper than those in p+p collisions. At small p_T, heavy quarks may thermalize with the medium and exhibit collective flow effects [1].

At RHIC, heavy quarks could be studied by measuring NPE which are produced from semileptonic decays of D and B mesons. Measurements of NPE provide information on heavy quarks energy loss and elliptic flow in the hot and dense nuclear matter created in relativistic heavy ion collisions.

Hot nuclear matter effects can be quantified by the nuclear modification factor (R_{AA}). R_{AA} is defined as the ratio of the number of particles produced in nucleus-nucleus collisions to that in proton-proton collisions, scaled by the average number of binary collisions for a given centrality. $R_{AA} < 1$ may indicate heavy quark energy loss. In the case there are no medium effects the nuclear modification factor will be equal to unity.

2. Analysis

Data reported in this proceedings were collected in p+p collisions at $\sqrt{s} = 200$ GeV in 2009, and in Au+Au collisions at $\sqrt{s_{NN}} = 200$ GeV in 2010 with Minimum Bias (MB) and High Tower (HT) triggers, where MB triggered data are used for $p_T < 2$ GeV/c results and High Tower triggered data for $p_T > 2$ GeV/c results.

The Time Projection Chamber (TPC) is the primary tracking device for charged particle momentum determination at STAR. Information from the TPC was used for particle identification together with information from the Barrel Electromagnetic Calorimeter (BEMC), and the Time Of Flight (TOF) detector. The BEMC was used for high p_T electron energy measurement and online trigger. The TOF detector was used to reduce hadron contamination at low p_T.

Electron candidates are identified via ionization energy loss measured by the TPC combined with TOF velocity information, which together provide good PID capability at low-p_T region. Electrons at high-p_T are selected using the ratio of the track momentum and the energy deposited in the BEMC, the BSMD shower profile, and the distance between TPC track projected position at BEMC and reconstructed BEMC cluster position. The raw NPE yield is calculated as: $N_{NPE} = N_{Inc} * \epsilon_{pur} - N_{PHE}/\epsilon_{PHE}$, where N_{NPE} is NPE yield, N_{Inc} represents inclusive electron candidate yield, ϵ_{pur} is the purity of inclusive electron sample, N_{PHE} is the yield of reconstructed photonic electron background, which mainly originates from photon conversion in the detector material and from Dalitz decay of π^0 or η mesons, and ϵ_{PHE} represents the photonic electron reconstruction efficiency. ϵ_{PHE} is estimated by full GEANT simulation of the detector response. Finally, NPE yield is corrected by detector acceptance and efficiencies.

3. NPE results in p+p collisions at $\sqrt{s} = 200$ GeV

The recent results of STAR NPE measurement in p+p colisions at $\sqrt{s} = 200$ GeV are shown in Fig. 1. Left panel presents the measured NPE yield as a function of p_T, where black marks show results obtained in the 2009 analysis and blue points show former analysis results [3]. The new data extends to lower p_T region compared to the previous STAR measurement. Green points represent data from PHENIX (pseudorapidity range $|\eta| < 0.35$) [4]. STAR and PHENIX results are consistent in the overlappping p_T range. STAR results are also compared with pQCD FONLL calculation (Fixed Order plus Next-to-Leading Logarithms [2]), where the FONLL central result is presented by the blue solid line and its upper and lower uncertainties by black lines. Experimental results are in good agreement with the FONLL calculation and lie between its central value and the upper limit.

Data and FONLL calculation are also compared in Fig. 1 (right panel) where a ratio to FONLL central values is plotted.

4. NPE results in Au+Au collisions at $\sqrt{s_{NN}} = 200$ GeV

In this section the recent results of NPE measurements in Au+Au collisions at $\sqrt{s_{NN}}$=200 GeV will be discussed. The NPE invariant yields were compared with FONLL calculations scaled by the average number of binary collisions in five centrality bins. In central and semicentral Au+Au collisions we observed a suppresion of NPE production compared to the FONLL calculation [2]. Figure 2 (left) shows the NPE R_{AA} for the 0-10% most central collisions compared to several theoretical models of heavy quark energy loss [5-9]. Gluon radiation scenario alone [5] (dashed green line) fails to explain the large NPE suppression at high p_T. When the collisional energy loss is added model calculations describe the data better. The collisional dissociation model [8] (red line) and the AdS/CFT calculation [9] (blue line) also describe the data well. Note that the baseline for nuclear modification factor calculation is produced from a combination of NPE spectra measured in the years 2005 and 2008 [3]. By including the aforementioned new p+p results from 2009 data, the R_{AA} uncertainties will be reduced.

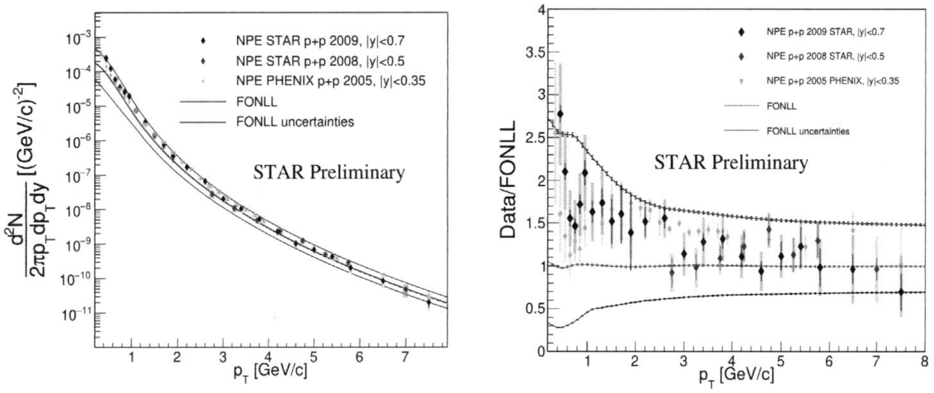

Figure 1. Non-photonic electron yield (left) and ratio of data over FONLL calculation (right) [2], STAR results from p+p collisions at $\sqrt{s}$ = 200 GeV from the year 2009 (black) compared with STAR data from the year 2008 (blue) [3] and PHENIX p+p results [4].

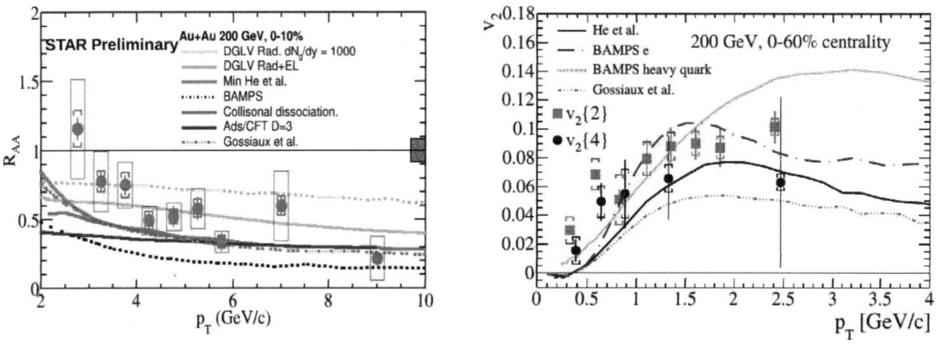

Figure 2. Non-photonic electron results from Au+Au collisions at $\sqrt{s}$ = 200 GeV from 2010. On the left plot is the NPE nuclear modification factor and on the right plot is the NPE elliptic flow. Both results are compared with theoretical models [5-14].

The right panel of Fig. 2 shows a measurements of NPE v_2 in Au+Au collisions at $\sqrt{s_{NN}}$=200 GeV. These results are obtained using 2-particle ($v_2\{2\}$) and 4-particle ($v_2\{4\}$) correlations [10]. These results are compared with theoretical models [11-14]. We observe finite v_2 at low p_T and at high p_T we observe increasing of v_2 which can arise from non-flow effects such as jet-like correlations.

5. Summary

The NPE invariant yield in p+p collisions at $\sqrt{s}$ = 200 GeV has been measured by the STAR experiment using high statistics data. The result can be described by FONLL calculations. Compared to the previously published STAR results, the new measurement using 2009 data extends the NPE spectrum to lower p_T region. Results of NPE measurements in Au+Au collisions at $\sqrt{s_{NN}}$=200 GeV from STAR show large suppression of NPE production in central

Au+Au collisions which cannot be explained by gluon radiation energy loss alone. Finite NPE v_2 is observed at low p_T which together with strong NPE suppresion in central collisions indicates strong charm-medium interaction.

6. Outlook

The Heavy Flavor Tracker (HFT), STAR's new sub-detector, which started its operation in 2014, extends the STAR particle identification capability to heavy flavor particles (i.e. particles containing heavy quarks). The HFT is able to topologically reconstruct charm mesons and baryons. This is made possible by reconstruction of secondary vertices with high precision of the order of 50 μm. For NPE measurements it will be possible to distinguish between electrons from D and B mesons. Such studies will improve our understanding of heavy quarks interaction with QGP. Projected R_{CP} results for electrons from these two sources are shown in Fig. 3, where R_{CP} is a central to peripheral nuclear modification factor, define as a ratio of yield in central collisions to yield in peripheral collisions scaled by number of binary collisions.

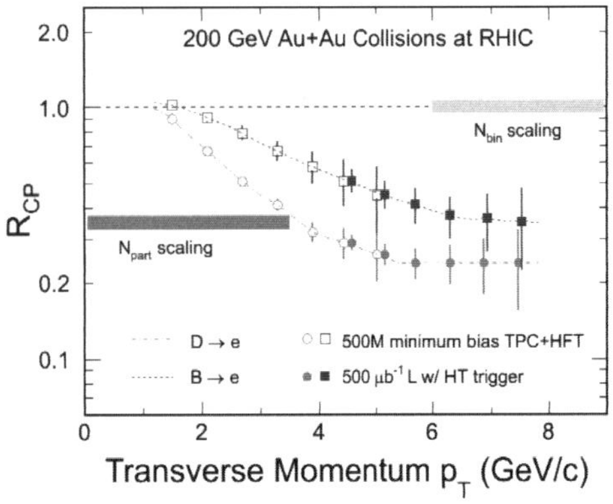

Figure 3. Projection of R_{CP} of non-photonic electrons from D (red) and B (blue) mesons separately [15]. Open symbols are for 500M Au+Au simulated minimum-bias events, and filled symbols are for HT data with 500 μb^{-1} sampled luminosity.

Acknowledgments

This work was supported by Grant Agency of the Czech Technical University in Prague, grant No. SGS13/215/OHK4/3T/14 and by Grant Agency of the Czech Republic, grant No.13-20841S.

References

[1] A. D. Frawley, T. Ullrich, and R. Vogt, Phys. Rept. 462, 125 (2008).
[2] M. Cacciari, P. Nason and R. Vogt, Phys. Rev. Lett. 95, 122001 (2005).
[3] H. Agakishiev et al. [STAR Collaboration], Phys. Rev. D83, 052006,(2011).
[4] A.Adare et al. [PHENIX Collaboration, Phys. Rev. C. 84, 044905 (2011).
[5] M. Djordjevic, M. Gyulassy, R. Vogt, and S. Wicks, Phys. Lett. B 632, 81 (2006).
[6] A. Buzzatti, M. Gyulassy, Nucl. Phys. A 910-911, 490 (2013).
[7] H. Van Hees, M. Mannarelli, V. Greco, and R. Rapp, Phys. Rev. Lett. 100, 192301 (2008).
[8] R. Sharma, I. Vitev, B. Zhang, Phys. Rev. C 80, 054902 (2009).
[9] W. A. Horowitz, J.Phys. G35, 044025 (2008).
[10] L.Adamczyk et al. [STAR Collaboration], arXiv:1405.6348 (2014).
[11] J. Uphoff, O. Fochler, Z. Xu, C. Greiner, Phys.Rev. C84, 024908 (2011).
[12] J. Uphoff, O. Fochler, Z. Xu, C. Greiner, Phys.Lett. B717, 430435 (2012).
[13] M. He, R. J. Fries, and R. Rapp, Phys.Rev. C86, 014903 (2012).
[14] Gossiaux, J. Aichelin, T. Gousset, and V. Guiho, J.Phys. G37, 094019 (2010).
[15] The STAR Heavy Flavor Tracker, Technical Design Report, (2011).

Journal of Physics: Conference Series **612** (2015) 012035 doi:10.1088/1742-6596/612/1/012035

Heavy flavours in high-energy nuclear collisions: quenching, flow and correlations

A. Beraudo[†], A. De Pace, M. Monteno, M. Nardi, F. Prino
INFN, sezione di Torino, Via Pietro Giuria 1, I-10125 Torino
E-mail: [†]`beraudo@to.infn.it`

Abstract. We present results for the quenching, elliptic flow and azimuthal correlations of heavy flavour particles in high-energy nucleus-nucleus collisions obtained through the POWLANG transport setup, developed in the past to study the propagation of heavy quarks in the Quark-Gluon Plasma and here extended to include a modeling of their hadronization in the presence of a medium. Hadronization is described as occurring via the fragmentation of strings with endpoints given by the heavy (anti-)quark $Q(\overline{Q})$ and a thermal parton $\overline{q}(q)$ from the medium. The flow of the light quarks is shown to affect significantly the $R_{\rm AA}$ and v_2 of the final D mesons, leading to a better agreement with the experimental data.

Heavy quarks – indirectly accessible through D-mesons, heavy-flavour decay electrons and muons and J/ψ's from B decays – have been used for a long time as probes of the medium formed in heavy-ion collisions. In a series of papers [1, 2, 3] over the last few years we developed a complete setup (referred to as POWLANG) for the study of heavy flavour observables in high-energy nucleus-nucleus collisions, describing the initial hard production of the $Q\overline{Q}$ pairs and the corresponding parton-shower stage through the POWHEG-BOX package [4, 5] and addressing the successive evolution in the plasma through the relativistic Langevin equation. Here we present a brief summary of our recent efforts aimed at supplementing the above numerical tool by modeling the hadronization of the heavy quarks accounting for the presence of a surrounding medium made of light thermal partons feeling the collective flow of the fluid: a comprehensive exposition can be found in [6].

In order to simulate the hadronization of heavy quarks in the medium at the end of their propagation in the QGP we proceed as follows. Once a heavy quark Q, during its stochastic propagation in the fireball, has reached a fluid cell below the decoupling temperature $T_{\rm dec}$, it is forced to hadronize. One extracts then a light antiquark $\overline{q}_{\rm light}$ (up, down or strange, with relative thermal abundances dictated by the ratio $m/T_{\rm dec}$) from a thermal momentum distribution corresponding to the temperature $T_{\rm dec}$ in the Local Rest Frame (LRF) of the fluid; information on the local fluid four-velocity $u^{\mu}_{\rm fluid}$ provided by hydrodynamics allows one to boost the momentum of $\overline{q}_{\rm light}$ from the LRF to the laboratory frame. A string is then constructed joining the endpoints given by Q and $\overline{q}_{\rm light}$ and is then passed to PYTHIA 6.4 [7] to simulate its fragmentation into hadrons (and their final decays). In agreement with PYTHIA, in evaluating their momentum distribution, light quarks are taken as "dressed" particles with the effective masses $m_{u/d} = 0.33$ GeV and $m_s = 0.5$ GeV. Concerning $T_{\rm dec}$ the values 0.155 and 0.17 GeV are explored. In the following some representative results obtained with the new hadronization procedure are displayed and compared to experimental data (when available).

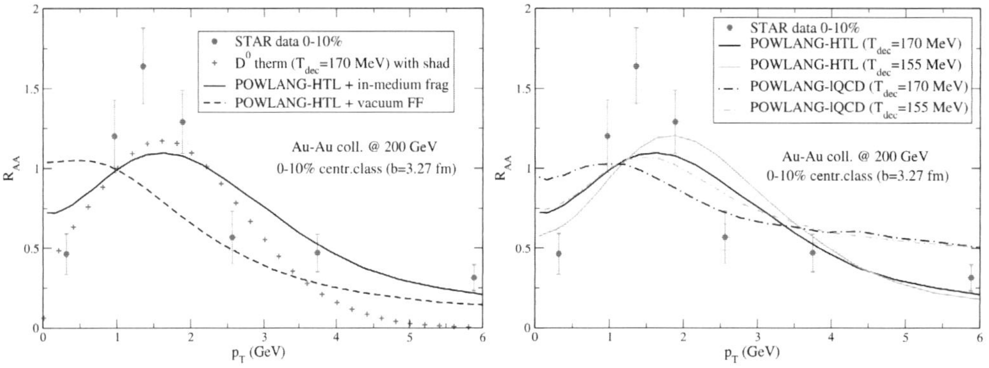

Figure 1. R_{AA} of D^0 mesons in central Au-Au collisions at $\sqrt{s_{NN}} = 200$ GeV. Theory curves are compared to STAR data [8]. Left panel: POWLANG results obtained with HTL transport coefficients and $T_{dec} = 170$ MeV are displayed, the new in-medium hadronization mechanism leading to a characteristic bump in the R_{AA} due to the radial flow of light quarks; also shown for comparison are the limiting case of full kinetic thermalization of D mesons and the transport result with in-vacuum fragmentation. Right panel: results for two different choices of transport coefficients (weak-coupling HTL, continuous curves, and lattice-QCD, dot-dashed curves) and decoupling temperatures ($T_{dec} = 170$ MeV, in black, and 155 MeV, in orange) are compared.

We start addressing the D meson R_{AA} at RHIC. In Fig. 1 we show the results of the POWLANG + in-medium fragmentation setup compared to STAR data [8] in central Au-Au collisions at $\sqrt{s_{NN}} = 200$ GeV. One can appreciate how, at variance with the case of vacuum Fragmentation Functions (black dashed curve), due to the additional radial flow inherited from the light thermal parton the theory curves (various choices of transport coefficients and decoupling temperatures are explored) display a bump around $p_T \approx 1.5$ GeV in qualitative agreement with the data.

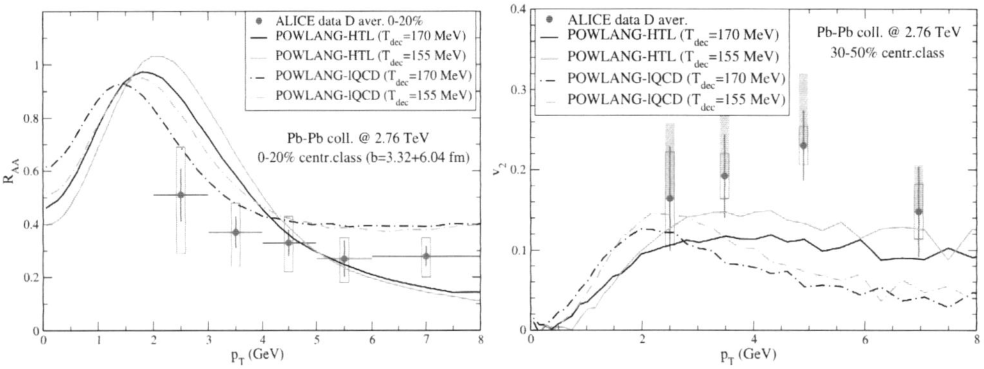

Figure 2. The R_{AA} (left panel) and v_2 (right panel) of D mesons in central (left) and semi-central (right) Pb-Pb collisions at $\sqrt{s_{NN}} = 2.76$ TeV. POWLANG results with in-medium hadronization and different choices of transport coefficients and decoupling temperatures are compared to ALICE data [9, 10].

In Fig. 2 we consider the nuclear modification factor and elliptic flow of D mesons in central (for the $R_{\rm AA}$, left panel) and semi-central (for the v_2, right panel) Pb-Pb collisions at $\sqrt{s_{\rm NN}}$ = 2.76 at the LHC; theory results are compared to experimental data from the ALICE Collaboration [9, 10]. Notice that, with the present broader binning and accessible p_T domain, ALICE data for the D meson $R_{\rm AA}$ don't show evidence of the bump from radial flow found by STAR at lower center-of-mass energy. Concerning the D meson v_2, the additional elliptic flow acquired from the light thermal parton picked-up at hadronization moves the theory curves significantly closer to the data with respect to past results with vacuum Fragmentation Functions, which underestimated the experimental findings.

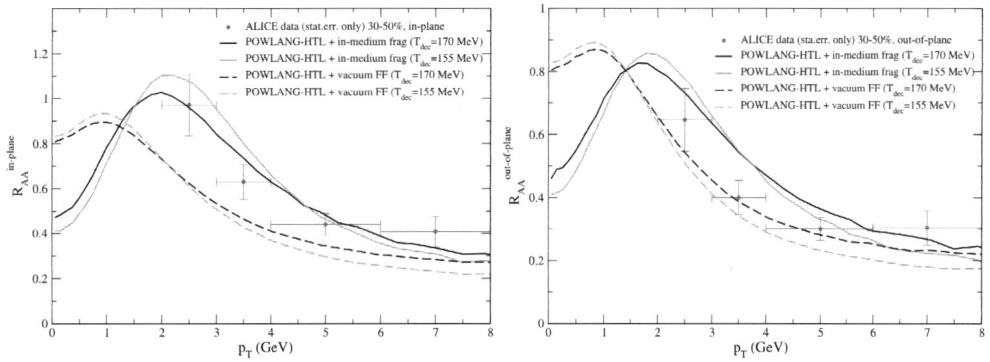

Figure 3. $R_{\rm AA}$ in-plane (left panel) and out-of-plane (right panel) of D mesons. ALICE data in the 30−50% centrality class [11] are compared to POWLANG results obtained with different hadronization mechanisms (in-medium, solid curves, and vacuum fragmentation, dashed curves) and decoupling temperatures ($T_{\rm dec}$ = 170 MeV, black curves, and 155 MeV, orange curves).

A possible way of combining the information contained in the $R_{\rm AA}$ and v_2 data (the latter reflecting the collective flow of the medium at low p_T and the path-length dependence of the energy-loss at high p_T) consists in studying the azimuthal dependence (with respect to the reaction plane) of the nuclear modification factor of hard particles. The ALICE Collaboration has recently released the data for the in-plane/out-of-plane $R_{\rm AA}$ of D mesons in semi-central (30-50%) Pb-Pb collisions [11]. Experimental findings are displayed in Fig. 3 and compared to the transport predictions of the POWLANG setup: also in this case model results with in-medium fragmentation look in slightly better agreement with the experimental data.

Finally, we decide to address, although at the limit of the present experimental capabilities, more differential observables like azimuthal angular correlations between heavy flavour particles (or their decay products) and the charged hadrons produced in the same collision. In Figs. 4 and 5 we display then our results for $D-h$ and $e-h$ correlations in the rapidity window $|\eta| > 1$. All figures are obtained with weak-coupling HTL transport coefficients in the QGP phase. In general, one observes a strong suppression of the away-side peak around $\Delta\phi = \pi$. Depending on the cuts imposed on the trigger particles (either D-mesons or heavy-flavour decay electrons) and on the associated hadrons this can be mostly due either to the energy loss (moving particles below the p_T-cut) or to the angular decorrelation (moving particles away from $\Delta\phi = \pi$) of the parent heavy quark. For $e-h$ correlations we have plotted separately the charm and beauty contributions. While in the case of electrons from charm decays we have always found an almost complete suppression of the away-side peak independently on the centrality of the collision and on the kinematic cuts, beauty decay electrons turn out to be less decorrelated, allowing one in principle to extract more information on the heavy-quark interaction with the medium.

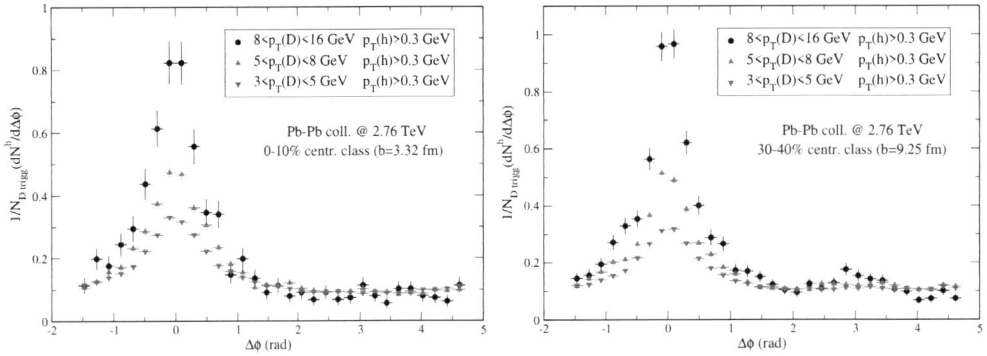

Figure 4. $D-h$ correlations in Pb-Pb collisions at $\sqrt{s_{\rm NN}}=2.76$ TeV for various p_T cuts on the trigger particle and different centrality classes: $0-10\%$ (left panel) and $30-40\%$ (right panel).

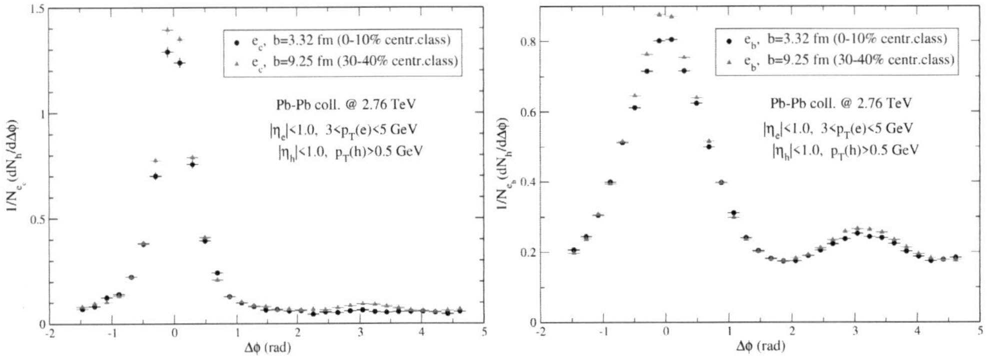

Figure 5. e–h correlations in Pb-Pb collisions at $\sqrt{s_{\rm NN}}=2.76$ TeV for various centrality classes: $0-10\%$ (black circles) and $30-40\%$ (red triangles). The separate charm (left panel) and beauty (right panel) results are shown.

In summary, the new in-medium fragmentation model for heavy quark hadronization we developed, once interfaced to our POWLANG transport setup, allows us to get results for the D meson $R_{\rm AA}$ and v_2 in better agreement with the experimental data. We have also tried to extend our predictions to more differential observables, like heavy flavour azimuthal angular correlations: this has to be considered just an exploratory study, leaving the issue of the information one can extract from such a kind of measurements for future work.

[1] A. Beraudo, A. De Pace, W.M. Alberico and A. Molinari 2009 *Nucl. Phys.* A **831** 59
[2] W.M. Alberico *et al.* 2011 *Eur. Phys. J.* C **71** 1666
[3] W.M. Alberico *et al.* 2013 *Eur. Phys. J.* C **73** 2481
[4] S. Frixione, P. Nason, G. Ridolfi 2007 *JHEP* **0709** 126
[5] S. Alioli, P. Nason, C. Oleari and E. Re 2010 *JHEP* **1006** 043
[6] A Beraudo *et al.* 2014 arXiv:1410.6082 [hep-ph]
[7] T. Sjostrand, S. Mrenna and P.Z. Skands 2006 *JHEP* **0605** 026
[8] STAR Collaboration (L. Adamczyk *et al.*) 2014 *Phys. Rev. Lett.* **113** 142301
[9] ALICE Collaboration (B. Abelev *et al.*) 2012 *JHEP* **1209** 112
[10] ALICE Collaboration (B. Abelev *et al.*) 2013 *Phys. Rev. Lett.* **111** 102301
[11] ALICE Collaboration (B. Abelev *et al.*) 2014 *Phys. Rev.* C **90** 034904

Υ production in hadron collisions at forward rapidity with ALICE at the LHC

Massimiliano Marchisone, for the ALICE Collaboration
Dipartimento di Fisica Sperimentale dell'Università di Torino and Sezione INFN di Torino
E-mail: marchiso@to.infn.it

Abstract. Ultrarelativistic heavy-ion collisions produce strongly interacting matter at high temperature and energy density. Under these extreme conditions, a deconfined partonic state, called Quark-Gluon Plasma, is formed. The measurement of charmonia and bottomonia in AA collisions is expected to provide essential information about the QGP properties. In pp collisions high precision data serve as crucial test for several competing models of quarkonium hadroproduction and provide the reference for the measurements in heavy-ion collisions, while pA collisions are useful to disentangle hot and cold nuclear matter effects. In the ALICE Muon Spectrometer, bottomonium production can be measured at forward rapidity ($2.5 < y < 4$) and down to $p_{\mathrm{T}} = 0$ via the dimuon decay channel. The latest results on Υ production in pp, Pb–Pb and p–Pb collisions are discussed and compared to various theoretical predictions.

1. Introduction

ALICE [1] is the LHC experiment dedicated to the study of heavy-ion collisions, but it has also a strong physics program in pp and p–Pb collisions. At forward rapidity ($2.5 < y < 4$) the quarkonia ($c\bar{c}$ and $b\bar{b}$ mesons) are reconstructed with the Muon Spectrometer down to a transverse momentum (p_{T}) equal to zero, exploiting their decay into two muons.

The study of bottomonium production in hadron collisions is important for several reasons. New measurements in pp collisions help to constrain the various models describing the quarkonium production mechanisms and the theoretical calculations for bottomonium are more robust than those of charmonium due to the higher mass. Furthermore, according to the colour-screening model [2], the dissociation probability due to the Quark-Gluon Plasma (QGP) of the different quarkonium states is expected to provide essential information about the properties of the system produced in AA collisions. A competing effect such as the quarkonium regeneration is much smaller for the bottomonium than for the charmonium [3]. Other important mechanisms are the cold nuclear matter (CNM) effects, which can modify the quarkonium production even in absence of the QGP [4]. Data from pA collisions are necessary to disentangle these effects from the hot ones.

2. Υ production in pp collisions

The $\Upsilon(1S)$ and $\Upsilon(2S)$ inclusive cross sections have been measured at forward rapidity in pp collisions at an energy of $\sqrt{s} = 7$ TeV [5].

The $\Upsilon(1S)$ and $\Upsilon(2S)$ rapidity differential cross sections presented in Fig. 1 (left) show a good agreement with the values reported by LHCb [6] in the same rapidity range and complement

Journal of Physics: Conference Series **612** (2015) 012036 doi:10.1088/1742-6596/612/1/012036

the results obtained by CMS at midrapidity [7, 8].

In Fig. 1 (right) the inclusive $\Upsilon(1S)$ differential production cross section as a function of p_{T} is compared to three Color Singlet Model (CSM) calculations which account for the feed down from higher mass states [9]. The LO calculation underestimates the data for $p_{\mathrm{T}} > 4$ GeV/c and falls too rapidly with increasing p_{T}. The p_{T} dependence of the NLO calculation is closer to the measurements, but the prediction still underestimates the cross section over the full p_{T} range. A good agreement is achieved at NNLO* (it includes the leading-p_{T} contributions appearing at NNLO [10]), but over a limited p_{T} range and with large theoretical uncertainties.

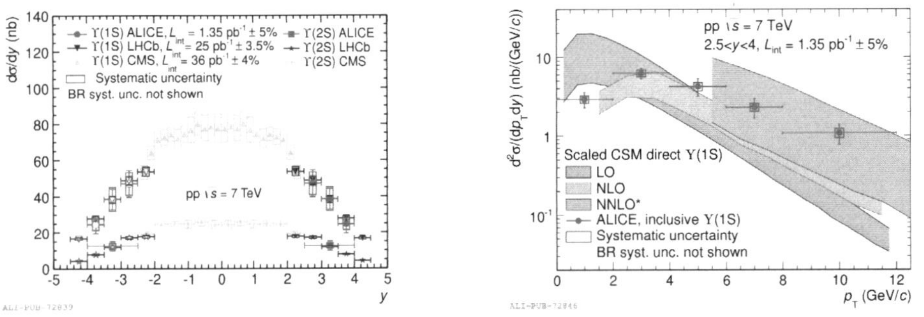

Figure 1. On the left, rapidity differential cross sections of $\Upsilon(1S)$ and $\Upsilon(2S)$ measured by ALICE [5], LHCb [6] and CMS [7, 8]. On the right p_{T} differential cross section of $\Upsilon(1S)$ compared to three theoretical calculations [9].

3. Υ production in Pb–Pb collisions

The effects of the hot and dense medium on the $\Upsilon(1S)$ production in Pb–Pb collisions at $\sqrt{s_{\mathrm{NN}}} = 2.76$ TeV are quantified by means of the nuclear modification factor $R_{\mathrm{AA}} = \frac{Y^{\Upsilon}_{\mathrm{Pb-Pb}}}{T_{\mathrm{AA}} \cdot \sigma^{\Upsilon}_{\mathrm{pp}}}$, defined as the meson yield in Pb–Pb collisions, divided by the nuclear overlap function and the $\Upsilon(1S)$ production cross section in pp collisions at the same energy [11].

The R_{AA} shown in Fig. 2 (left) shows a more pronounced suppression in central collisions. Moreover, the rapidity dependence in the right panel suggests a stronger suppression at forward than at midrapidity as it appears from the comparison with the CMS measurement in $|y| < 2.4$ [12].

In the same figures, the measurements are also compared to the predictions of a dynamical model which does not include CNM nor regeneration effects [13]. The predictions are made assuming two possible initial temperature profiles in rapidity: a boost-invariant plateau and a Gaussian profile. For each of them, three values of QGP shear viscosity over entropy density (η/s) are used in the calculations. The predictions underestimate the measured suppression, for every temperature profiles and η/s values considered. All sets used in the model predicts a rapidity dependence opposite to the measured one.

Predictions from a transport model are shown in Fig. 3. These calculations are based on a kinetic rate-equation approach in an evolving QGP and include both suppression and regeneration effects [14]. CNM effects are calculated by varying an effective absorption cross section between 0 and 2 mb, resulting in an uncertainty band. This model underestimates the observed suppression, even if the centrality dependence is fairly reproduced. The model predicts also an almost constant R_{AA} as a function of the rapidity which is in disagreement with the trend observed by ALICE and CMS.

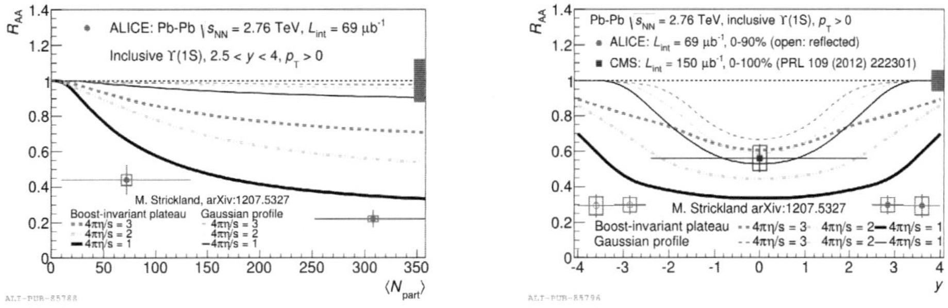

Figure 2. Inclusive $\Upsilon(1S)$ R_{AA} as a function of the average number of participants (left) and rapidity (right) compared to predictions from a dynamical model [13]. Horizontal bars represent the uncertainty associated to $\langle N_{\mathrm{part}} \rangle$ (from a Glauber model) in the former plot and the rapidity bin width in the latter.

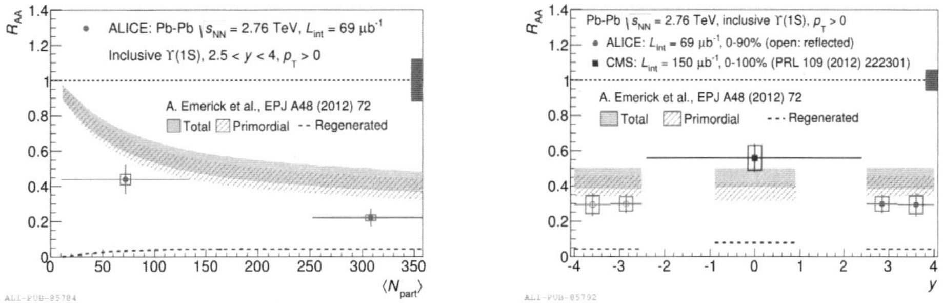

Figure 3. Inclusive $\Upsilon(1S)$ R_{AA} as a function of the average number of participants (left) and rapidity (right) compared to predictions from a transport model [14].

4. Υ production in p–Pb collisions

The nuclear modification factor $R_{\mathrm{pPb}} = \frac{\sigma_{\mathrm{pPb}}}{A_{\mathrm{Pb}} \cdot \sigma_{\mathrm{pp}}}$ (i.e. the ratio between the Υ production cross sections in p–Pb and pp collisions, normalized to the Pb mass number) is used to determine the CNM effects in p–Pb collisions at $\sqrt{s_{\mathrm{NN}}} = 5.02$ TeV [15]. It is equivalent to the R_{AA} used to investigate the QGP effects in Pb–Pb collisions.

As shown in Fig. 4, the inclusive $\Upsilon(1S)$ production is suppressed at forward rapidity (p-going direction), while at backward rapidity (Pb-going direction) the measurement is compatible with unity within uncertainties, disfavoring a strong gluon anti-shadowing.

In the left plot, the ALICE results are compared to a NLO Color Evaporation Model (CEM) calculation with shadowing parametrized by EPS09 at NLO [16] which tends to overestimate the observed $\Upsilon(1S)$ R_{pPb}. Coherent parton energy loss calculations [17] with or without EPS09 are also shown: the former reproduces the data at forward rapidity, while the latter is in better agreement with our measurements at backward rapidity.

In the right panel, the results are compared to a LO calculation of a $gg \to \Upsilon g$ production with shadowing parametrization [18]. The two bands show the uncertainties related to EPS09 LO in the shadowing region and in the EMC region. A calculation based on the CGC framework coupled with a CEM production is also shown for positive values of y_{cms} [19]. Although this prediction only slightly underestimates the $\Upsilon(1S)$ R_{pPb}, it is not able to reproduce the analogous J/ψ measurement in the same rapidity range [20].

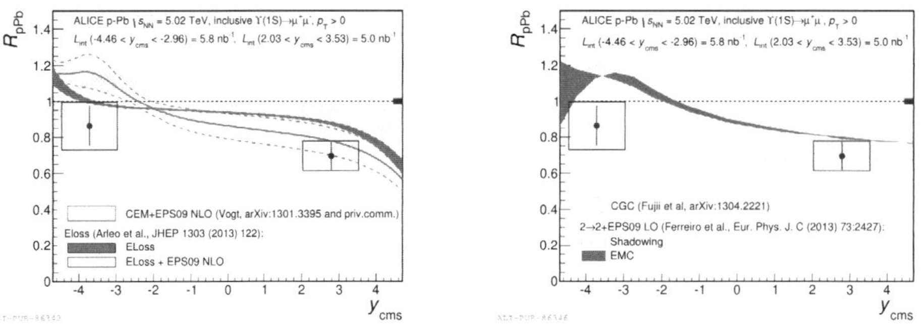

Figure 4. Nuclear modification factor of inclusive $\Upsilon(1S)$ in p–Pb collisions as a function of rapidity compared to several model calculations: CEM with EPS09 shadowing at NLO [16] and parton energy loss with and without EPS09 shadowing at NLO [17] on the left; CSM with EPS09 shadowing at LO [18] and CGC [19] on the right.

5. Conclusions

The measured inclusive production cross sections of $\Upsilon(1S)$ and $\Upsilon(2S)$ at forward rapidity in pp collisions are in good agreement with measurements by LHCb and complement those by CMS at midrapidity. Both CSM LO and NLO calculations underestimate the data at large transverse momentum, while the addition of the leading-p_T NNLO contributions helps to reduce this disagreement, but with larger theoretical uncertainties.

The observed suppression of inclusive $\Upsilon(1S)$ in Pb–Pb collisions is stronger in central than in semiperipheral collisions and shows a pronounced rapidity dependence over the large domain covered by ALICE and CMS. The suppression, larger than that predicted by the models considered, might point to a significant dissociation of direct $\Upsilon(1S)$.

Finally, the $\Upsilon(1S)$ production in p–Pb collisions is suppressed at forward rapidity, while it is consistent with unity at backward rapidity. Models including the nuclear modification of the gluon PDF or a contribution from coherent parton energy loss tend to overestimate the measured $R_{\rm pPb}$ and cannot simultaneously describe the forward and backward rapidity measurements.

References

[1] K. Aamodt et al. (ALICE Collaboration), JINST 3 (2008) S08002
[2] T. Matsui and H. Satz, Phys. Lett. B 178 (1986) 416
[3] L. Grandchamp et al., Phys. Rev. C 73 (2006) 064906
[4] R. Vogt, Phys. Rev. C 81 (2010) 044903
[5] B. Abelev et al. (ALICE Collaboration), Eur. Phys. J. C 74 (2014) 2974
[6] R. Aaij et al. (LHCb Collaboration), Eur. Phys. J. C 72 (2012) 2025
[7] V. Khachatryan et al. (CMS Collaboration), Phys. Rev. D 83 (2011) 112004
[8] S. Chatrchyan et al. (CMS Collaboration), Phys. Lett. B 727 (2013) 101
[9] J.-P. Lansberg, Nucl. Phys. A 470 (2013) 910
[10] J.-P. Lansberg, J. Phys. G 38 (2011) 124110
[11] B. Abelev et al. (ALICE Collaboration), Phys. Lett. B 738 (2014) 361–372
[12] S. Chatrchyan et al. (CMS Collaboration), Phys. Rev. Lett. 109 (2012) 222301
[13] M. Strickland, Phys. Rev. Lett. 107 (2011) 132301
[14] A. Emerick et al., Eur. Phys. J. A 48 (2012) 72
[15] B. Abelev et al. (ALICE Collaboration), Phys. Lett. B 740 (2015) 105–117
[16] J. L. Albacete et al., Int. J. Mod. Phys. E 22 (2013) 1330007
[17] F. Arleo et al., JHEP 1303 (2013) 122
[18] E. G. Ferreiro, Eur. Phys. J. C 73 (2011) 2427
[19] H. Fuji et al., Nucl. Phys. A 915 (2013) 1–23
[20] B. Abelev et al. (ALICE Collaboration), JHEP 1402 (2014) 073

Measurement of beauty-hadron decay electrons in Pb–Pb collisions at $\sqrt{s_{NN}} = 2.76$ TeV with ALICE

Martin Völkl for the ALICE Collaboration
Physikalisches Institut der Universität Heidelberg, Heidelberg, Germany
E-mail: `voelkl@physi.uni-heidelberg.de`

Abstract. The ALICE Collaboration at the LHC studies heavy-ion collisions to investigate the properties of the Quark-Gluon Plasma (QGP). Heavy quarks (charm and beauty) are effective probes for this purpose. Both their energy loss in the medium as well as their possible thermalization yield information about the medium properties. In ALICE, the reconstruction of exclusive decays is currently accessible for charm, but not for beauty hadrons. For hadrons with beauty valence quarks a promising strategy is the measurement of their decay electrons. To separate these from the background electrons (mainly from charm hadron decays, photon conversions or light-meson decays) the large decay length of beauty hadrons can be utilized. It leads to a relatively large typical impact parameter of the decay electrons. By comparing the impact parameter distribution of the signal electrons with those from the background sources, the signal can be statistically separated from the background. For this purpose a maximum likelihood fit is employed using impact parameter distribution templates from simulations. The resulting nuclear modification factor for electrons from beauty-hadron decays shows a sizeable suppression for $p_{\mathrm{T}} > 3$ GeV/c, albeit still with large uncertainties.

1. Introduction

The measurement of hadrons with heavy valence quarks (charm or beauty) via their decay products in heavy-ion collisions can give insight into the properties of the Quark-Gluon Plasma. Of particular interest are the energy loss the quarks experience in the medium and their possible thermalization. Theoretical calculations predict a lower energy loss for quarks than for gluons and a lower energy loss for heavy quarks compared with lighter ones making a flavor-separated measurement interesting [1]. Such a measurement is possible through the observation of hadrons containing heavy valence quarks.

For charm hadrons a full reconstruction from hadronic decay products was carried out [2]. Due to small branching ratios this approach is not used for the decays of beauty hadrons. To access beauty quarks, use can be made of the large branching ratios ($\approx 10\%$) of decays with an electron (or positron) in the final state. A comparison of the p_{T} distributions of electrons from beauty-hadron decays in Pb–Pb and pp collisions based on the nuclear modification factor (R_{AA}) is sensitive to the medium effects on b-quarks. R_{AA} is the ratio of the p_{T}-differential yield in Pb–Pb collisions to the corresponding cross section in pp collisions scaled with the nuclear overlap function. Figure 1 shows a measurement of the nuclear modification factor of electrons from heavy-flavor hadron decays for the 10% most central Pb–Pb collisions at $\sqrt{s_{NN}} = 2.76$ TeV [3]. The value is significantly below unity, pointing to an energy loss of heavy quarks in the hot and dense medium. As shown by the model expectations [4, 5], it is interesting to separate the contributions from beauty and charm.

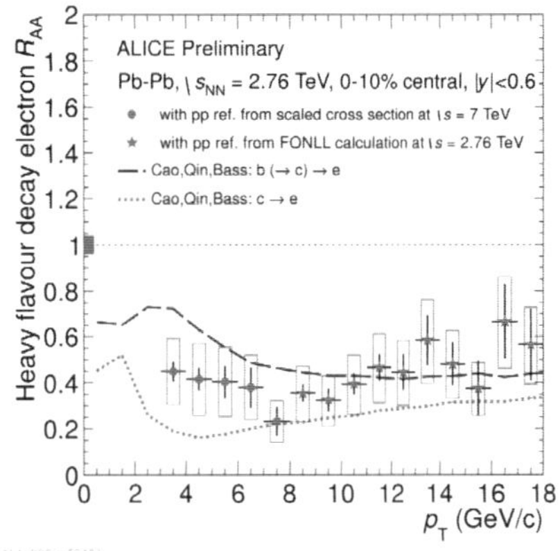

Figure 1. R_{AA} of electrons from heavy-flavor decays in central Pb–Pb collisions [3].

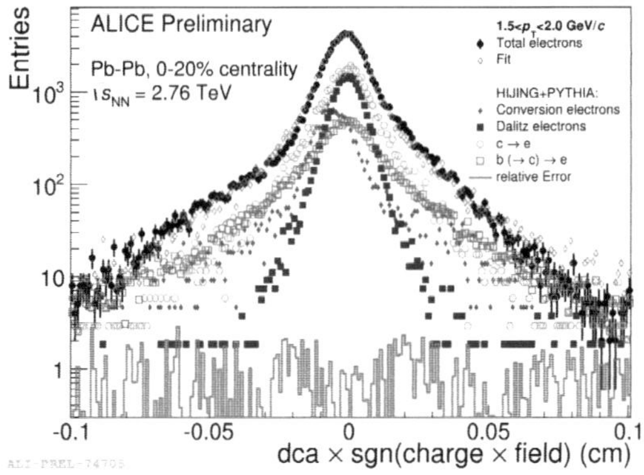

Figure 2. Impact parameter distribution of electron candidates in data and electrons from different sources in simulations.

2. Electron Identification and Background Sources

Electron identification on the data set of Pb–Pb collisions recorded in 2010 is done using the ALICE Time Projection Chamber (TPC) and the Time of Flight detector (TOF) [6]. Particle identification with the TPC is based on the specific energy loss of charged particles in its large gas volume. Electrons are identified by requiring the measured time-of-flight to be within 3σ of the expected value for electrons and the $\mathrm{d}E/\mathrm{d}x$ in the TPC to be within 0 and $+3\sigma$ with respect to the expected $\mathrm{d}E/\mathrm{d}x$ of electrons. This asymmetric cut keeps the contamination due to pions low. The remaining hadron contamination was studied and was found not to contribute strongly to the systematic uncertainty of the measured spectrum of beauty-hadron decay electrons.

The number of electrons from sources other than heavy-flavor decays can be estimated using other measurements performed by the ALICE collaboration. The background is dominated by Dalitz decays of light mesons and photon conversions in the detector material.

Knowledge of this background allows for a measurement of the number of electrons from heavy-flavor decays based on the subtraction of background electrons from the total number of electrons [7]. The measured D-meson p_{T}-differential cross sections allow for the subtraction of the charm-decay contribution by simulating the decay of the charm hadrons to electrons. However, a measurement of the beauty contribution with this approach leads to large uncertainties because a large background is subtracted from a large electron sample. Thus, additional information is needed to improve the precision of the measurement.

3. The Impact Parameter

The typically large decay length of beauty hadrons [8] can be used for the separation of the different electron sources. Due to the fact that only one decay particle is measured (the electron), the decay vertex cannot be measured directly. The proxy quantity is the track impact parameter. A track propagated backwards from a secondary vertex will typically not go through the primary vertex. The distance of closest approach to the primary vertex of this backwards propagated track is called impact parameter. A positive or negative sign is assigned depending on whether the reconstructed track passes to the left or right of the primary vertex. An accurate measurement of the impact parameter is possible due to the excellent resolution of the tracks and the primary vertex provided by the Inner Tracking System (ITS) [6]. Four types of electron sources can be distinguished via their impact parameter distributions as shown in Figure 2:

Journal of Physics: Conference Series **612** (2015) 012037 doi:10.1088/1742-6596/612/1/012037

Electrons from beauty-hadron decays. Due to the large decay length ($c\tau \approx 500$ μm) of beauty hadrons this distribution is the widest among the different electron sources.

Electrons from charm-hadron decays. The smaller decay length of the charm hadrons ($c\tau \approx 100 - 300$ μm) leads to a narrower distribution as compared to beauty-hadron decay electrons.

Electrons from the primary vertex. All electrons produced at or very close to the primary vertex have the same impact parameter distribution which is determined by the resolution of the track parameters. These electrons mostly come from Dalitz decays of π^0 and will be referred to as Dalitz electrons.

Electrons from photon conversions in the detector material. Due to the small opening angle of photon conversions a non-zero impact parameter is due to the influence of the magnetic field. It has proven useful to multiply the impact parameter by the signs of the particle charge and field polarity. This results in an asymmetric distribution.

There are two approaches of using the impact parameter information to measure electrons from beauty-hadron decays. The first approach is based on a cut on the minimum absolute value of the impact parameter in order to preferentially select beauty-hadron decay electrons. Afterwards the remaining background electrons are subtracted. The impact parameter cut allows for a substantial improvement of the signal-to-background ratio leading to a smaller uncertainty on the beauty-hadron decay electron yield with respect to the case in which the impact parameter selection is not applied. The pp reference at $\sqrt{s} = 7$ TeV for the R_{AA} [9] was measured with this approach. The second approach is based on the statistical separation of the electron sources. It was used for the Pb–Pb analysis as described in the next section.

4. The Impact Parameter Fit Method

The basic idea is to compare the impact parameter distributions of the various electron sources from simulation (templates) to the impact parameter distribution of all measured electrons. The simulations use the particles from event generators (HIJING [10], PYTHIA [11]) which are transported through the ALICE apparatus with GEANT3 [12]. The accuracy of the simulated impact parameter distributions was shown to not be a dominant source of systematic uncertainty. The fit of the impact parameter distributions is performed separately for each of the six p_{T}-intervals considered in this analysis. With the template distributions $a_{source,bin}$, the expectation values of the counts in each impact-parameter bin are introduced as the free fit parameters $A_{source,bin}$. The expectation value of the data counts is the fit function f_{bin}. It can be written as [13]:

$$f_{bin} = \sum_{source} p_{source} \cdot A_{source,bin} \; .$$

The p_{source} are the strength factors of the fit. Assuming Poissonian fluctuations in the data bins d_{bin} and template bins $a_{source,bin}$, this leads to the likelihood:

$$\log L = \sum_{bin} d_{bin} \log f_{bin} - f_{bin} + \sum_{bin} \sum_{source} a_{source,bin} \log A_{source,bin} - A_{source,bin} \; ,$$

neglecting constant terms. The first term incorporates the fluctuations of the data, as in a fit with a smooth function. The second term includes the fluctuations of the template bin contents. The minimization of the $A_{source,bin}$ can be done fairly efficiently [13], leaving only the amplitudes p_{source} as free parameters for numerical minimization. An example of such a fit can be found in Figure 2.

To estimate the uncertainties from the maximum likelihood method, toy simulation and data distributions were created from the weighted sums of the templates. Taking the measured results

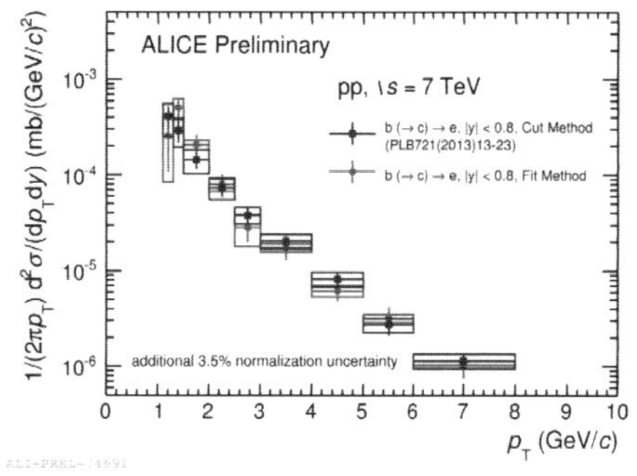

Figure 3. Comparison of beauty-decay electron cross sections obtained from fit and cut methods in pp collisions at $\sqrt{s} = 7$ TeV.

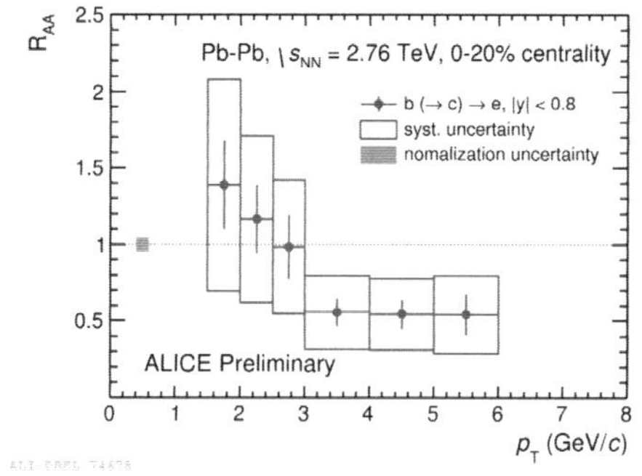

Figure 4. R_{AA} of electrons from beauty-hadron decays measured in central Pb–Pb collisions as a function of p_{T}.

as a starting point the fit was repeatedly simulated. The spread of the resulting amplitudes was identified as the statistical uncertainty while the difference of the mean of the results to the true value of the toy model was interpreted as a systematic bias of the maximum likelihood method.

5. Results

The method was tested on the data sample of pp collisions at $\sqrt{s} = 7$ TeV and compared with the results from the impact parameter cut method. The results from the two methods are in good agreement within their systematic uncertainties as shown in Figure 3. The systematic uncertainties of the two approaches are of a similar size. The advantage of the fit method is its independence from other measurements. The fit method was applied to the 20% most central Pb–Pb collisions at $\sqrt{s_{NN}} = 2.76$ TeV recorded in 2010. The uncertainties are currently dominated by the strong selection criteria of the particle identification in Pb–Pb collisions with smaller contributions from the accuracy of the simulated impact parameter distributions, the pp reference and its energy scaling and the fit method. The resulting R_{AA} is shown in Figure 4. It shows a suppression ($R_{\mathrm{AA}} < 1$) for $p_{\mathrm{T}} > 3$ GeV/c - albeit with sizeable uncertainties - with the indication of a rise towards lower p_{T}. This is consistent with in-medium energy loss of beauty quarks causing a redistribution of the beauty-hadron decay electrons towards lower p_{T} in Pb–Pb collisions compared to pp collisions. Due to the large uncertainties no clear indication of a quark-mass dependent in-medium energy loss is observed at the current stage of the analysis.

References

[1] Dokshitzer Y and Kharzeev D 2001 *Phys. Lett. B* **519** 199 – 206 ISSN 0370-2693
[2] Abelev B *et al.* (ALICE Collaboration) 2012 *J. High Energy Phys.* **2012** 128
[3] Sakai S (ALICE Collaboration) 2013 *Nuc. Phys. A* **904–905** 661c – 664c ISSN 0375-9474
[4] Cao S, Qin G Y and Bass S 2013 *Phys. Rev. C* **88**(4) 044907
[5] Cao S, Qin G Y and Bass S Private communication
[6] Abelev B *et al.* (ALICE Collaboration) 2014 *Int. J. Mod. Phys. A* **29** 1430044
[7] Abelev B *et al.* (ALICE Collaboration) 2012 *Phys. Rev. D* **86**(11) 112007
[8] Particle Data Group 2010 *J. Phys. G Nuc. Partic.* **37** 075021
[9] Abelev B *et al.* (ALICE Collaboration) 2013 *Phys. Lett. B* **721** 13 – 23 ISSN 0370-2693
[10] Wang X N and Gyulassy M 1991 *Phys. Rev. D* **44**(11) 3501–3516
[11] Sjöstrand T, Mrenna S and Skands P 2006 *Journal of High Energy Physics* **2006** 026
[12] Brun R *et al.* 1994 *CERN program library long write-up* **W5013**
[13] Barlow R and Beeston C 1993 *Comput. Phys. Commun.* **77** 219 – 228 ISSN 0010-4655

Journal of Physics: Conference Series **612** (2015) 012038 doi:10.1088/1742-6596/612/1/012038

J/ψ and $\psi(2S)$ measurement in p+p collisions at $\sqrt{s}$ = 200 and 500 GeV in the STAR experiment

Barbara Trzeciak[1] **for the STAR Collaboration**

[1]Faculty of Nuclear Sciences and Physical Engineering, Czech Technical University in Prague, Brehova 7, 115 19 Praha 1, Czech Republic

E-mail: trzecbar@fjfi.cvut.cz

Abstract. In this paper, results on the J/ψ cross section and polarization measured via the dielectron decay channel at mid-rapidity in $p + p$ collisions at $\sqrt{s}$ = 200 and 500 GeV in the STAR experiment are discussed. The first measurement of $\psi(2S)$ to J/ψ ratio at $\sqrt{s}$ = 500 GeV is also reported.

1. Introduction

J/ψ and $\psi(2S)$ are bound states of charm (c) and anti-charm ($\bar{c}$) quarks. Charmonium physical states have to be colorless, however they can be formed via a color-singlet (CS) or color-octet (CO) intermediate $c\bar{c}$ state. One of the first models of the charmonium production, the Color Singlet Model (CSM) [1], assumed that J/ψ is created through the color-singlet state only. This early prediction failed to describe the measured charmonium cross section which has led to the development of new models. For example, Non-Relativistic QCD (NRQCD) [1] calculations were proposed in which a $c\bar{c}$ color-octet intermediate states, in addition to a color-singlet states, can bind to form charmonia.

However, the charmonium production mechanism in elementary particle collisions is not yet exactly known. For many years measurements of the J/ψ cross section have been used to test different J/ψ production models. While many models can describe relatively well the experimental data on the J/ψ cross section in $p + p$ collisions [2–9], they have different predictions for the J/ψ polarization. Therefore, measurements of the J/ψ polarization may allow to discriminate among different models and provide new insight into the J/ψ production mechanism.

2. Charmonium measurements in STAR

In STAR, charmonia have been measured so far via the dielectron decay channel. The STAR detector [10] is a multi-purpose detector that has large acceptance at mid-rapidity, $|\eta| < 1$ with a full azimuthal coverage. Electrons can be identified using the Time Projection Chamber (TPC) [11] through ionization energy loss (dE/dx) measurement. The Time Of Flight (TOF) detector [12] greatly enhances the electron identification capability at low momenta where the dE/dx bands for electrons and hadrons cross each other. At high p_T, electron identification can be improved by the Barrel Electromagnetic Calorimeter (BEMC) [13] which measures electron energy and shower shape. The BEMC is also used to trigger on high-p_T electrons (HT trigger). Minimum Bias (MB) events are triggered by the Vertex Position Detectors (VPD) [14].

3. J/ψ measurements in p+p at $\sqrt{s}$ = 200 GeV

STAR has measured inclusive J/ψ p_T spectra and polarization in $p + p$ collisions at $\sqrt{s} = 200$ GeV via the dielectron decay channel ($B_{ee} = 5.9\%$) at mid-rapidity ($|y| < 1$). These results are compared to different model predictions to understand J/ψ production mechanism in elementary collisions.

Left panel of Fig. 1 shows STAR low and high-p_T measurements of J/ψ p_T spectra [3, 15] compared to model predictions. The Color Evaporation Model (CEM) [16] for prompt J/ψ can describe the p_T spectrum reasonably well, except the region around $p_T \approx 3$ GeV/c where it over-predicts the data. NLO NRQCD calculations with color-singlet and color-octet transitions [17] for prompt J/ψ match the data for $p_T > 4$ GeV/c. NNLO* CS model [18] for direct J/ψ production under-predicts the STAR data, but the prediction does not include contributions from $\psi(2S)$, χ_C and B-meson decays to J/ψ.

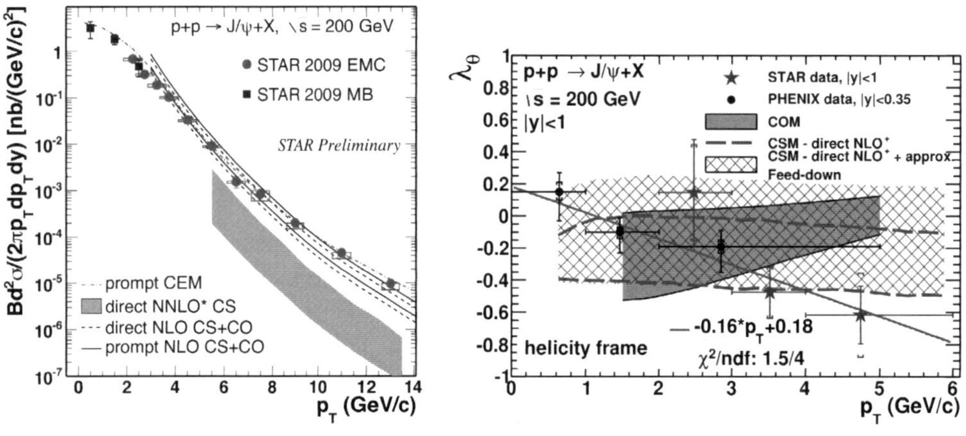

Figure 1. Left: J/ψ invariant cross section vs p_T in p+p collisions at $\sqrt{s} = 200$ GeV at mid-rapidity at low [15] and high p_T [3] shown as blue squares and red circles, respectively, compared to different model predictions [16–18]. Right: Polarization parameter λ_θ vs J/ψ p_T for $|y| < 1$ [19] compared to the PHENIX measurement [20] and two model predictions [21, 22].

In p+p collisions at $\sqrt{s} = 200$ GeV STAR has also measured J/ψ polarization parameter λ_θ in the helicity frame at mid-rapidity and $2 < p_T < 6$ GeV/c [19]. J/ψ polarization is analyzed via the angular distribution of the decay electrons that is described by: $\frac{d^2N}{d(\cos\theta)d\phi} \propto 1 + \lambda_\theta \cos^2\theta + \lambda_\phi \sin^2\theta \cos 2\phi + \lambda_{\theta\phi} \sin 2\theta \cos\phi$, where θ and ϕ are polar and azimuthal angles, respectively; λ_θ, λ_ϕ and $\lambda_{\theta\phi}$ are the angular decay coefficients. The p_T dependence of λ_θ is shown on the right panel of Fig. 1 with low-p_T PHENIX results [20] and compared to NRQCD calculations [21] and the NLO$^+$ CSM prediction [22]. A trend observed in the RHIC data is towards longitudinal polarization as p_T increases and, within experimental and theoretical uncertainties, the result is consistent with the NLO$^+$ CSM model.

The inclusive J/ψ production is a combination of prompt and non-prompt J/ψ. The prompt J/ψ production consists of the direct one (~60%) and feed-down from excited states $\psi(2S)$(~10%) and χ_C(~30%), while non-prompt J/ψ originate from B-hadron decays. STAR has estimated the contribution from B-meson decays using a measurement of azimuthal angular correlation between high-p_T J/ψ and charged hadrons [2, 3]. The relative contribution of B-hadron decays to inclusive J/ψ yield is strongly p_T dependent and it is 10-25% for $4 < p_T < 12$ GeV/c, as it is shown on the left panel of Fig. 2. The measurement is consistent with the FONLL+CEM prediction [23, 24].

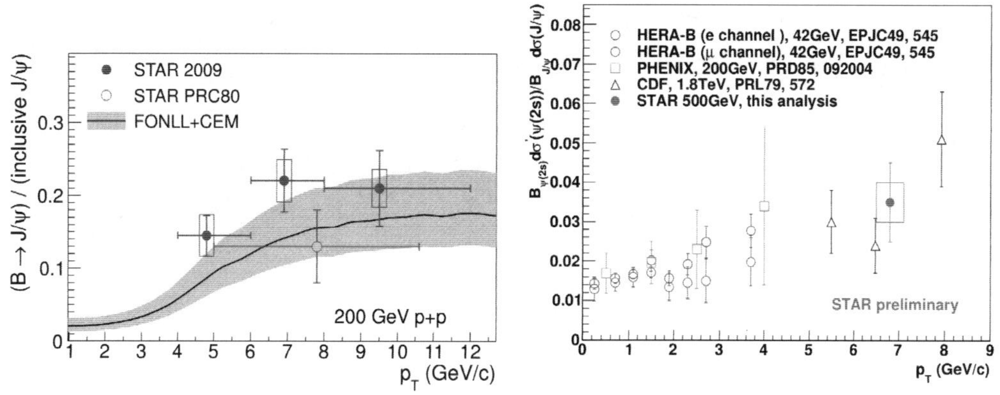

Figure 2. Left: relative contribution from B-meson decays to inclusive J/ψ production in p+p at $\sqrt{s} = 200$ GeV [3] compared to FONLL+CEM calculations [23, 24]. Right: ratio of $\psi(2S)$ to J/ψ in $p+p$ collisions at $\sqrt{s} = 500$ GeV from STAR (red circle) compared to results from other experiments at different energies.

4. J/ψ and $\psi(2S)$ measurements in p+p at $\sqrt{s} = 500$ GeV

In order to further test the charmonium production mechanism and constrain the feed-down contribution from the excited states to the inclusive J/ψ production, the J/ψ and $\psi(2S)$ signals were extracted in $p + p$ collisions at $\sqrt{s} = 500$ GeV at mid-rapidity. The J/ψ p_T spectrum is shown on the left panel of Fig. 3. The STAR results at $\sqrt{s} = 500$ GeV (full circles) are compared to those at $\sqrt{s} = 200$ GeV (open circles) and with measurements of other experiments in p+$\bar{p}$ collisions at different energies. The STAR measurements cover p_T range of 4 - 20 GeV/c with a good precision. It was also observed that J/ψ cross section follows the x_T scaling: $\frac{d^2\sigma}{2\pi p_T dp_T dy} = g(x_T)/(\sqrt{s})^n$, where $x_T = 2p_T/\sqrt{s}$, with $n = 5.6 \pm 0.2$ at mid-rapidity and $p_T >$ 5 GeV/c for a wide range of colliding energies [2]. At $\sqrt{s} = 500$ GeV the same x_T scaling of high-p_T J/ψ production is seen, as shown on the right panel of Fig. 3.

Right panel of Fig. 2 shows $\psi(2S)/J/\psi$ ratio from STAR (red full circle) compared to measurements of other experiments at different colliding energies, in $p + p$ and p+A collisions. The STAR data point is consistent with the observed trend, and no collision energy dependence of the $\psi(2S)$ to J/ψ ratio is seen with the current precision.

The statistics available at $\sqrt{s} = 500$ GeV will allow us to extract the frame invariant polarization parameter, also in different reference frames, providing model independent information about the J/ψ polarization [25]. It will be possible to measure the azimuthal polarization parameter, λ_ϕ, and improve precision of the λ_θ measurement. Analysis of J/ψ polarization at $\sqrt{s} = 500$ GeV is ongoing.

5. Summary

In summary, STAR has measured the inclusive J/ψ cross section and polarization in p+p collisions at $\sqrt{s} = 200$ GeV as a function of p_T. The measurements are compared to different model predictions of the J/ψ production. The p_T spectrum is described well by the NRQCD calculations while the measured polarization parameter λ_θ is consistent with the NLO$^+$ CSM prediction. STAR new result for J/ψ at $\sqrt{s} = 500$ GeV extends p_T reach up to 20 GeV/c. The first measurement of $\psi(2S)/J/\psi$ ratio in p+p collisions at $\sqrt{s} = 500$ GeV is reported and compared with results from other experiments. No collision energy dependence is observed.

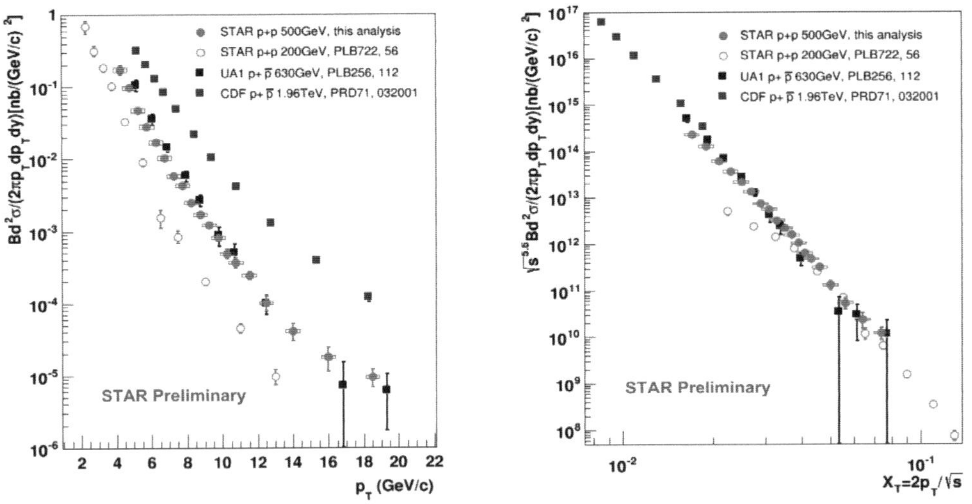

Figure 3. J/ψ invariant cross section vs p_T, left panel, and invariant cross section multiplied by $\sqrt{s}^{5.6}$ vs x_T, right panel, in p+p collisions at $\sqrt{s}$ = 500 GeV at mid-rapidity shown as full circles compared to measurements at different energies.

Acknowledgements
This publication was supported by the European social fund within the framework of realizing the project „Support of inter-sectoral mobility and quality enhancement of research teams at Czech Technical University in Prague", CZ.1.07/2.3.00/30.0034.

References

[1] Braaten E, Fleming S and Yuan T C 1996 *Ann. Rev. Nucl. Part. Sci.* **46** 197–235 (*Preprint* hep-ph/9602374)
[2] Abelev B *et al.* (STAR Collaboration) 2009 *Phys. Rev.* **C 80** 041902 (*Preprint* 0904.0439)
[3] Adamczyk L *et al.* (STAR Collaboration) 2013 *Phys. Lett.* **B722** 55–62 (*Preprint* 1208.2736)
[4] Adare A *et al.* (PHENIX Collaboration) 2012 *Phys.Rev.* **D85** 092004 (*Preprint* 1105.1966)
[5] Abe F *et al.* (CDF Collaboration) 1997 *Phys. Rev. Lett.* **79**(4) 572–577
[6] Acosta D *et al.* (CDF Collaboration) 2005 *Phys.Rev.* **D71** 032001 (*Preprint* hep-ex/0412071)
[7] Aad G *et al.* (ATLAS Collaboration) 2011 *Nucl.Phys.* **B850** 387–444 (*Preprint* 1104.3038)
[8] Khachatryan V *et al.* (CMS Collaboration) 2011 *Eur.Phys.J.* **C71** 1575 (*Preprint* 1011.4193)
[9] Aaij R *et al.* (LHCb Collaboration) 2011 *Eur.Phys.J.* **C71** 1645 (*Preprint* 1103.0423)
[10] Ackermann K *et al.* (STAR Collaboration) 2003 *Nucl. Instrum. Meth.* **A 499** 624–632
[11] Anderson M *et al.* 2003 *Nucl. Instrum. Meth.* **A 499** 659–678 (*Preprint* nucl-ex/0301015)
[12] Llope W J *et al.* 2012 *Nucl. Instrum. Meth.* **A 661** 110–113
[13] Beddo M *et al.* (STAR Collaboration) 2003 *Nucl. Instrum. Meth.* **A 499** 725–739
[14] Llope W J *et al.* 2004 *Nucl. Instrum. Meth.* **A 522** 252–273 (*Preprint* nucl-ex/0308022)
[15] Kosarzewski L (STAR Collaboration) 2012 *Acta Phys.Polon.Supp.* **5** 543–548
[16] Frawley A D, Ullrich T and Vogt R 2008 *Phys.Rept.* **462** 125–175 (*Preprint* 0806.1013)
[17] Ma Y Q, Wang K and Chao K T 2011 *Phys.Rev.* **D84** 114001 (*Preprint* 1012.1030)
[18] Artoisenet P *et al.* 2008 *Phys.Rev.Lett.* **101** 152001 (*Preprint* 0806.3282)
[19] Adamczyk L *et al.* (STAR Collaboration) 2013 *Phys.Lett.* **B739** 180 (*Preprint* 1311.1621)
[20] Adare A *et al.* (PHENIX Collaboration) 2010 *Phys. Rev.* **D 82**(1) 012001
[21] Chung H S, Yu C, Kim S and Lee J 2010 *Phys. Rev.* **D 81**(1) 014020
[22] Lansberg J 2011 *Phys. Lett.* **B 695** 149–156 (*Preprint* 1003.4319)
[23] Bedjidian M, Blaschke D, Bodwin G T, Carrer N, Cole B *et al.* 2004 (*Preprint* hep-ph/0311048)
[24] Cacciari M, Nason P and Vogt R 2005 *Phys.Rev.Lett.* **95** 122001 (*Preprint* hep-ph/0502203)
[25] Faccioli P, Lourenco C, Seixas J and Wohri H K 2010 *Eur. Phys. J.* **C 69** 657–673 (*Preprint* 1006.2738)

Dilepton Production in Transport-based Approaches

Janus Weil[1], Stephan Endres[1], Hendrik van Hees[1], Marcus Bleicher[1], Ulrich Mosel[2]

[1]Frankfurt Institute for Advanced Studies , Ruth-Moufang-Str. 1, 60438 Frankfurt, Germany
[2]Institut für Theoretische Physik, JLU Giessen, Heinrich-Buff-Ring 16, 35392 Giessen, Germany

E-mail: weil@fias.uni-frankfurt.de

Abstract. We investigate dilepton production in transport-based approaches and show that the baryon couplings of the ρ meson represent the most important ingredient for understanding the measured dilepton spectra. At SIS energies, the baryon resonances naturally play a major role and affect already the vacuum spectra via Dalitz-like contributions, which can be captured well in transport simulations. Recent pion-beam measurements at GSI will help to constrain the properties of the involved resonances further.

1. Introduction

Lepton pairs are an ideal probe for studying phenomena at high densities and temperatures. They are created at all stages of a heavy-ion collision, but unlike hadrons they can escape the hot and dense zone almost undisturbed (since they only interact electromagnetically) and thus can carry genuine in-medium information out to the detector. Dileptons are particularly well-suited to study the in-medium properties of vector mesons, since the latter can directly convert into a virtual photon, and thus a lepton pair [1, 2]. One of the groundbreaking experiments in this field was NA60 at the CERN SPS, which revealed that the ρ spectral function is strongly broadened in the medium. Calculations by Rapp et al. have shown that this collisional broadening is mostly driven by baryonic effects, i.e., the coupling of the ρ meson to baryon resonances (N^*, Δ^*) [3]. In the low-energy regime, the data taken by the DLS detector have puzzled theorists for years and have recently been confirmed and extended by new measurements by the HADES collaboration [4, 5]. At such low energies, it is expected that not only the in-medium properties are determined by baryonic effects, but that already the production mechanism of vector mesons is dominated by the coupling to baryons (even in the vacuum).

2. The model: hadronic transport + VMD

Already our previous investigations [6] based on the GiBUU transport model [7] have shown that the baryonic N^* and Δ^* resonances give important Dalitz-like contributions to dilepton spectra at SIS energies for pp as well as AA collisions. This finding is based on the assumption that these resonances decay into a lepton pair exclusively via an intermediate ρ meson (i.e., strict vector-meson dominance). In the transport simulation, the Dalitz decays $R \to e^+e^-N$ are treated as a two-step process, where the first part is an $R \to \rho N$ decay, followed by a subsequent conversion of the ρ into a lepton pair ($\rho \to e^+e^-$). The branching ratios for the $R \to \rho N$ decays are taken from the partial-wave analysis by Manley et al. [8], while the decay width for the second part is calculated under the strict-VMD assumption. Recently we have extended the VMD assumption also to the $\Delta(1232)$ state [9].

In principle it is clear that all the dilepton Dalitz decays of the baryonic resonances ($R \to Ne^+e^-$) must involve an electromagnetic transition form factor. From an experimental point of view this form factor is essentially unknown in the time-like region. In consequence it is frequently neglected in transport simulations. Our approach is to use strict VMD as a sensible ansatz for this form factor, which is easy to implement in a transport scheme.

3. Dilepton spectra from p+p and A+A collisions

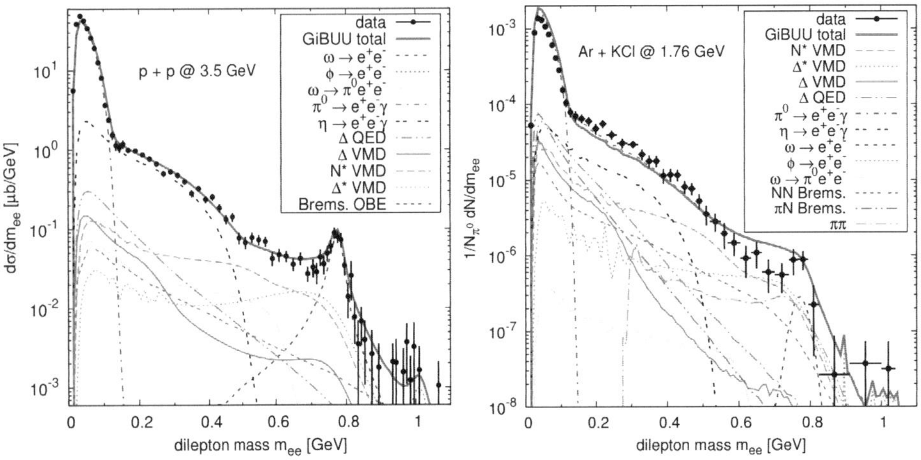

Figure 1. Dilepton mass spectra for p+p and Ar+KCl collisions, in comparison to the data from [4, 5].

In Fig. 1 we present two examples of results obtained with our model [9]. The left-hand side shows the simulated dilepton mass spectrum from elementary p+p collisions at $E_{\rm kin} = 3.5\,\rm GeV$ (inside the detector acceptance) compared to the data measured by the HADES collaboration.

While the low-mass regime is dominated by the Dalitz decays of the π^0 and η mesons, the region of higher masses shows a clear ω peak (whose width is dictated by detector resolution) and a hint of a ϕ peak at the very end of the spectrum. The most interesting region, however, is the intermediate-mass regime below the ω peak, which according to our model is dominated by baryonic contributions from N^* (and Δ^*) resonances. There is a whole cocktail of such resonances which are known to couple to the ρ meson, but in particular the light ones, such as the $D_{13}(1520)$, give important contributions here, without which the data in the intermediate-mass regime could not be described. The $\Delta(1232)$ only gives subdominant contributions in our model (both in the QED and VMD approaches), which is in conflict with other results [10], which claim a much more dominant role of the Δ.

The right panel of Fig. 1 shows the dilepton spectrum from Ar+KCl collisions at an energy of 1.76 AGeV compared to the HADES data. There are additional channels in this medium-sized nucleus-nucleus system, which do not occur in pure pp collisions, such as πN bremsstrahlung or the contribution from $\pi\pi \to \rho \to e^+e^-$, but they are not dominant. As in the elementary case, the contributions from the baryonic resonances are significant. Overall we observe an underestimation at intermediate masses and a slight excess in the vector-meson pole region. Reasons for these deviations could be uncertainties in the used resonance parameters or a missing dynamical treatment of density-dependent spectral functions. Such a treatment may be provided by the so-called "coarse-graining" approach, which is subject of ongoing investigations [11]. It could help to improve the description of heavy systems and to make the connection to dilepton measurements at higher energies.

4. Dilepton spectra from pion-induced collisions

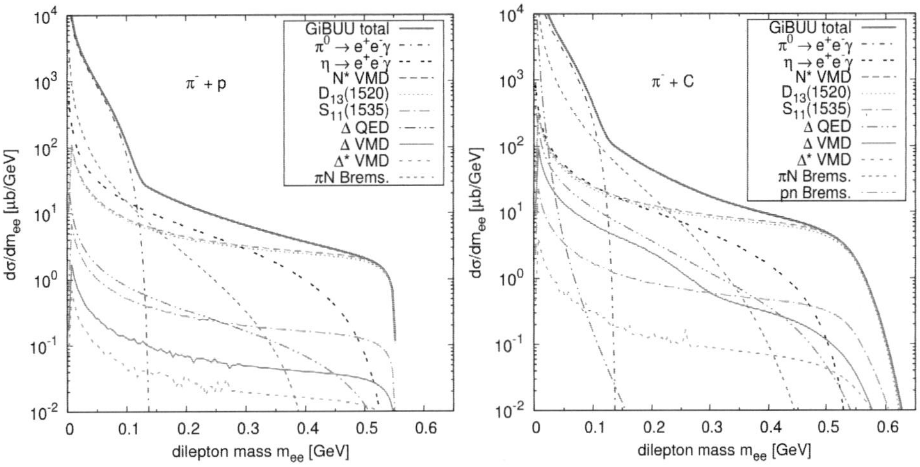

Figure 2. Dilepton mass spectra for $\pi^- p$ and $\pi^- C$ collisions at $\sqrt{s_{\pi N}} = 1.49\,\text{GeV}$.

Pion-induced reactions offer a much cleaner testing ground for dilepton production than AA or pp reactions, since the latter always involve excitation of baryonic resonances at various energies, while πN collisions operate at a fixed $\sqrt{s_{\pi N}}$, corresponding to a fixed resonance mass. This makes it much easier to disentangle contributions from different resonances, as pointed out e.g. in [12].

Fig. 2 shows the prediction of our model for dilepton mass spectra from pion-induced reactions on proton and carbon targets at a beam kinetic energy of $E_{\text{kin}} = 0.566\,\text{GeV}$ (corresponding to $p_{\text{lab}} = 0.690\,\text{GeV}$ or $\sqrt{s} = 1.49\,\text{GeV}$) which have recently been measured at GSI. In contrast to Fig. 1, no detector acceptance and resolution effects are included here.

The considered energy is just above the η threshold and close to the pole masses of resonance states like $D_{13}(1520)$ and $S_{11}(1535)$. In particular the former is believed to have a strong coupling to the ρ meson and thus should have a large influence on dilepton spectra. In our model, the $D_{13}(1520)$ plays a very dominant role at this energy. When disregarding the η Dalitz contribution, which can be subtracted in the experimental spectra, essentially all of the high-mass yield is due to the $D_{13}(1520)$, while contributions from other resonances like the $S_{11}(1535)$ or the $\Delta(1232)$ are suppressed by roughly an order of magnitude in the elementary $\pi^- p$ reaction. The Δ contribution is again shown in two different approaches, and there also is a πN bremsstrahlung contribution. However, the latter should be taken with a grain of salt, since it is only calculated in soft-photon approximation, which may not be very reliable in the πN case, in particular for large dilepton masses.

When moving to the carbon target, as shown on the right-hand side of Fig. 2, one can see that the Δ and bremsstrahlung contributions gain additional strength from secondary collisions. Yet, the relative strength of the η contribution drops a bit, since it is produced only in the primary collisions, but can be absorbed afterwards. The dominance of the $D_{13}(1520)$ in the high-mass regions persists.

We stress that our model currently relies on the assumption of a strict-VMD form factor for the $D_{13}(1520)$ (as for all other N^* resonances). However, the strong dominance of this resonance in the reaction considered here might make it possible to obtain a rather direct measurement of its electromagnetic transition form factor. It will be very interesting to see whether the experimental data will reveal any deviation from the VMD assumption.

In any case, it should also be mentioned that some models claim rather strong destructive interference effects at this energy [13] (which are neglected in our transport simulation, since they are very hard to handle). However, those effects are subject to controversy and do not show up so strongly at this energy in other models [14].

5. Conclusions

We have shown that the HADES dilepton data from pp and AA collisions can be described rather well with a combination of a resonance-model-based transport approach with a strict-VMD coupling of the baryons to the electromagnetic sector, where a mix of different baryonic resonances contributes to the total dilepton yield. Unfortunately not all parameters of these resonances are known sufficiently well. However, recent pion-beam measurements at GSI could help to shed new light on the baryonic resonances and possibly even fix the transition form factor of their electromagnetic Dalitz decays.

Acknowledgments

This work was supported by the Hessian Initiative for Excellence (LOEWE) through the Helmholtz International Center for FAIR and by the Federal Ministry of Education and Research (BMBF). J.W. acknowledges funding of a Helmholtz Young Investigator Group VH-NG-822 from the Helmholtz Association and GSI.

References

[1] S. Leupold, V. Metag, U. Mosel, Int. J. Mod. Phys. E 19, 147 (2010)
[2] R. Rapp, J. Wambach, H. van Hees, Landolt-Börnstein I/23, 4 (2010)
[3] H. van Hees, R. Rapp, Phys. Rev. Lett. 97, 102301 (2006)
[4] G. Agakishiev et al. (HADES Collaboration), Eur. Phys. J. A 48, 64 (2012)
[5] G. Agakishiev et al. (HADES Collaboration), Phys. Rev. C 84, 014902 (2011)
[6] J. Weil, H. van Hees, U. Mosel, Eur. Phys. J. A 48, 111 (2012)
[7] O. Buss, T. Gaitanos, K. Gallmeister, H. van Hees, M. Kaskulov et al., Phys. Rep. 512, 1 (2012)
[8] D. Manley, E. Saleski, Phys. Rev. D 45, 4002 (1992)
[9] J. Weil, S. Endres, H. van Hees, M. Bleicher, U. Mosel, arXiv:1410.4206
[10] E. Bratkovskaya, J. Aichelin, M. Thomere, S. Vogel, M. Bleicher, Phys. Rev. C 87(6), 064907 (2013)
[11] S. Endres, H. van Hees, J. Weil, M. Bleicher, J. Phys. Conf. Ser. 503, 012039 (2014)
[12] J. Weil, PhD thesis, 2013
[13] M. Lutz, B. Friman, M. Soyeur, Nucl. Phys. A 713, 97 (2003)
[14] M. Zetenyi, G. Wolf, Phys. Rev. C 86, 065209 (2012)

Hot Quarks 2014 IOP Publishing
Journal of Physics: Conference Series **612** (2015) 012040 doi:10.1088/1742-6596/612/1/012040

Multiplicity dependence of transverse momentum correlations in p-Pb and Pb-Pb collisions in ALICE

Prabhat Pujahari (for the ALICE collaboration)
Department of Physics and Astronomy, Wayne State University, Detroit, USA
E-mail: p.pujahari@cern.ch

Abstract. Two-particle correlations provide information about particle production mechanisms in heavy-ion collisions. We report on the study of two particle number correlations (R_2) and transverse momentum correlations ($\Delta p_T \Delta p_T$) in Pb-Pb collisions at $\sqrt{s_{NN}} = 2.76$ TeV and in p-Pb collisions at $\sqrt{s_{NN}} = 5.02$ TeV measured with ALICE detector at the LHC. We study the evolution of both correlation functions with collision centrality and particle multiplicity and find they exhibit a strong dependence on either the centrality in Pb-Pb collisions or the multiplicity in p-Pb collisions. We present studies of pseudorapidity dependent Fourier decompositions of these correlation functions. Given that the details of the collision dynamics may depend on the charge of the particles, we analyze like-sign and unlike-sign particle pairs separately. We combine the correlation functions of like-sign and unlike-sign to obtain a charge independent function. The positive correlation of $\Delta p_T \Delta p_T$ indicates that both particles from a pair are more likely to have momentum above or below the average transverse momentum in an event ensemble.

1. Introduction

Two-particle correlation functions measured at RHIC and the LHC energies have shown unanticipated structures both on the near- and away-side [1, 2, 3, 4]. There remains ambiguity, however. In an effort to lift some of these ambiguities, it is interesting to consider additional observables and types of correlation functions. In this work, we present measurements of R_2, a number correlation function, which we complement with measurements of a transverse momentum correlation function, $\Delta p_T \Delta p_T$ [5]. These two correlation functions are studied in p-Pb collisions at $\sqrt{s_{NN}} = 5.02$ TeV and Pb-Pb collisions at $\sqrt{s_{NN}} = 2.76$ TeV as a function of charged particle pair relative pseudorapidity, $\Delta\eta$, and azimuthal angle, $\Delta\varphi$, as well as collision centrality. The observable $\Delta p_T \Delta p_T$ features an explicit dependence on the transverse momentum of the produced particles. As such, it provides sensitivity to the correlation "hardness," i.e. how much high momentum transfer collisions contribute to the correlation dynamics. The measurements presented in this work will thus provide additional quantitative constraints on existing models.

2. Observable definition

Let $\rho_1(\eta_1, \varphi_1)$ and $\rho_2(\eta_1, \varphi_1, \eta_2, \varphi_2)$ be one- and two-particle densities, respectively. The number correlation function is written in terms of the cumulant, i.e. $C_2 = \rho_2 - \rho_1\rho_1$, as: $R_2 = \frac{C_2}{\rho_1\rho_1}$. R_2 is by construction independent to first approximation of the detection efficiency, i.e. the detector efficiency cancels out in the ratio. The transverse momentum correlation is defined as:

$$\langle \Delta p_t \Delta p_t \rangle(\eta_1, \varphi_1, \eta_2, \varphi_2) = \frac{\int \rho_2(\eta_1, \varphi_1, p_{t,1}, \eta_2, \varphi_2, p_{t,2}) \Delta p_{t,1} \Delta p_{t,2} dp_{t,1} dp_{t,2}}{\rho_2(\eta_1, \varphi_1, \eta_2, \varphi_2)} \tag{1}$$

The difference, $\Delta p_{t,i} = p_{t,i} - \langle p_{t,i} \rangle$, taken for each particle, measures deviations from the inclusive average momentum defined as

$$\langle p_{t,i} \rangle = \frac{\int \rho_1(\mathbf{p}_i) p_{t,i} dp_{t,i}}{\int \rho_1(\mathbf{p}_i) dp_{t,i}} \tag{2}$$

The observables in this analysis are discussed in detail in [5].

3. Data analysis

The data presented were acquired by the ALICE detector [6] in Pb-Pb collisions at $\sqrt{s_{\rm NN}} =$ 2.76 TeV and p-Pb collisions at $\sqrt{s_{\rm NN}} = 5.02$ TeV in the years 2010 and 2013, respectively. For this analysis, the ALICE Inner Tracking System (ITS) and the Time Projection Chamber (TPC) are used to reconstruct charged particle tracks. The VZERO scintillator hodoscopes and the Silicon Pixel Detector (SPD) are used for the trigger definition. Track candidates in the TPC are selected in the pseudorapidity range $|\eta| < 1.0$. Track quality cuts in the TPC are based on the number of reconstructed space points (at least 70 out of a maximum of 159). Tracks are rejected from the final sample if their distance of closest approach to the reconstructed vertex in the longitudinal (d_Z) and radial (d_{xy}) directions are greater than 3.2 cm and 2.4 cm, respectively. Contamination from HBT effects in like-sign pairs is not removed in this measurement. An electron rejection cut is applied to suppress contribution from photon conversions. Four different charge pairs $(+-, -+, ++, --)$ are measured and the information is combined into a charge independent (CI) correlation function.

$$CI = \frac{1}{4}\{(+-) + (-+) + (++) + (--)\} \tag{3}$$

4. Results

Figures 1 and 2 display the correlation function R_2 and $\Delta p_{\rm T} \Delta p_{\rm T}$, respectively, in three different centrality classes in Pb-Pb collisions. Both correlation functions exhibit decreasing amplitude with increasing number of particles, which reflects correlation dilution with large multiplicity. While an away-side doubly peaked structure is observed in $\Delta p_{\rm T} \Delta p_{\rm T}$ correlation in the 0-5% centrality bin, there is no such feature for the R_2 correlation in the same centrality bin and it emerges only in more central (0-2%) collisions [7]. For the 30-40% centrality bin, the distributions are dominated by a large $\cos(2\Delta\varphi)$ modulation.

To characterize the various features observed in the correlation functions and to disantangle them, we proceed to fit the data with a Fourier decomposition in relative azimuth, $\Delta\varphi$, with coefficients that depend on $\Delta\eta$:

$$F(\Delta\eta, \Delta\varphi) = c_0(\Delta\eta) + 2 \times \sum_{n=1}^{6} c_n \cos(n\Delta\varphi). \tag{4}$$

The left and right panel of Figure 3 show the c_1 and c_2 coefficients, respectively, in selected centralities in Pb-Pb collisions at $\sqrt{s_{\rm NN}} = 2.76$ TeV. However, given the correlation functions have been symmetrized in $\Delta\eta$, we show the results only for $\Delta\eta > 0$. The first order coefficient, $c_1(\Delta\eta)$, is influenced by the presence of the near-side peak in the correlation functions. One consequently observes positive values at all collision centralities for $\Delta\eta < 0.7$ and negative values

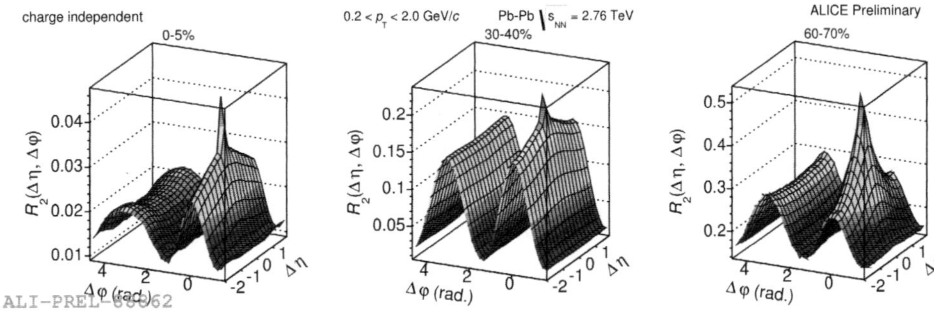

Figure 1. Correlation function R_2 for charge independent (CI) pairs ($0.2 < p_{\rm T} < 2.0$ GeV/c) in Pb-Pb collisions at $\sqrt{s_{\rm NN}} = 2.76$ TeV for selected collision centrality ranges.

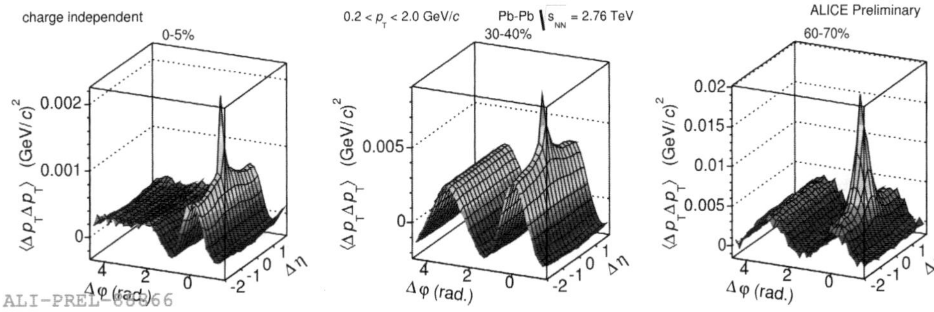

Figure 2. Correlation function $\Delta p_{\rm T} \Delta p_{\rm T}$ for charge independent (CI) pairs ($0.2 < p_{\rm T} < 2.0$ GeV/c) in Pb-Pb collisions at $\sqrt{s_{\rm NN}} = 2.76$ TeV for selected collision centrality ranges.

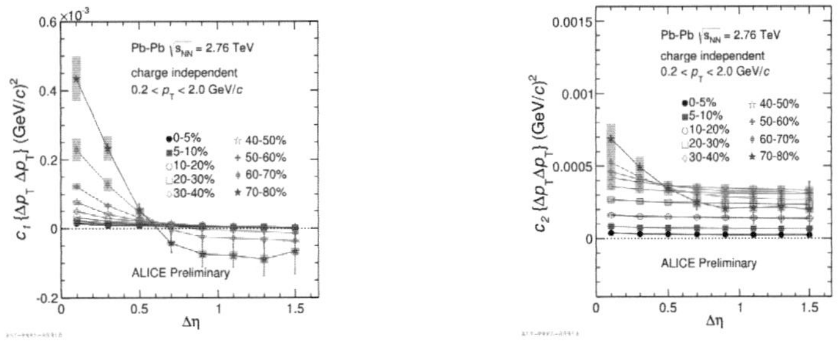

Figure 3. Harmonic coefficient $c_1\{\Delta p_{\rm T} \Delta p_{\rm T}\}$ (left) and $c_2\{\Delta p_{\rm T} \Delta p_{\rm T}\}$ (right) in Pb-Pb collisions at $\sqrt{s_{\rm NN}} = 2.76$ TeV.

otherwise. Negative values reflect, in part, the fact that momentum conservation requires that particles should be emitted back-to-back in azimuth. The second order coefficient, c_2, depends on the collision centrality. We note that the values of c_2 exhibit a smaller dependence on $\Delta\eta$ for all results from centrality bins between 0 and 50% in Pb-Pb collisions.

While comparing the coefficients, $c_n\{R_2\}$ vs. $c_n\{\Delta p_{\rm T} \Delta p_{\rm T}\}$ in p-Pb collisions, we observe

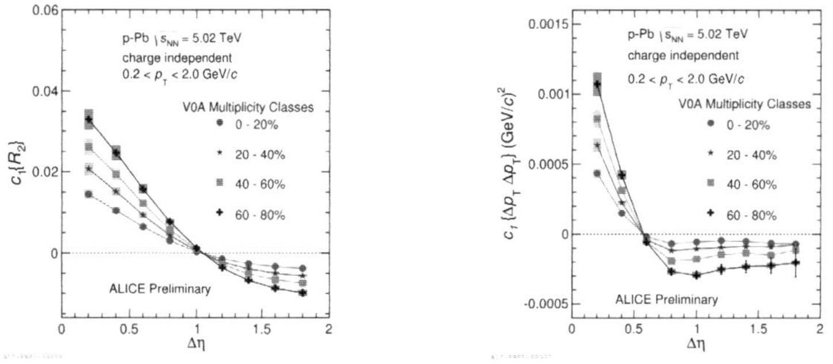

Figure 4. Harmonic coefficient $c_1\{R_2\}$ (left) and $c_1\{\Delta p_{\rm T}\Delta p_{\rm T}\}$ (right) in p-Pb collisions at $\sqrt{s_{\rm NN}} = 5.02$ TeV.

different $\Delta\eta$ dependence for the correlation observables. As shown in Figure 4, $c_1\{\Delta p_{\rm T}\Delta p_{\rm T}\}$ changes sign earlier ($\Delta\eta \approx 0.6$) than $c_1\{R_2\}$ ($\Delta\eta \approx 1.2$). Figure 5 displays $c_2\{R_2\}$ (left panel) and $c_2\{\Delta p_{\rm T}\Delta p_{\rm T}\}$ (right panel). It is observed that $c_2\{\Delta p_{\rm T}\Delta p_{\rm T}\}$ saturates earlier in $\Delta\eta$ than $c_2\{R_2\}$.

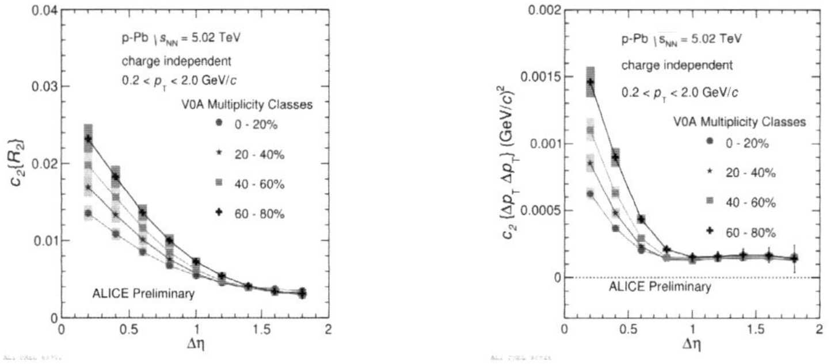

Figure 5. Harmonic coefficient $c_2\{R_2\}$ (left) and $c_2\{\Delta p_{\rm T}\Delta p_{\rm T}\}$ (right) in p-Pb collisions.

5. Summary

The first measurement of differential $\Delta p_{\rm T}\Delta p_{\rm T}$ correlations at the LHC energies are discussed in p-Pb and Pb-Pb collisions. $\Delta p_{\rm T}\Delta p_{\rm T}$ correlation shows an away-side doubly peaked structure in 0-5% centrality which in contrast to what is observed in R_2 correlation function in the same centrality in Pb-Pb collisions at $\sqrt{s_{\rm NN}} = 2.76$ TeV. The harmonic coefficients obtained from $\Delta p_{\rm T}\Delta p_{\rm T}$ show a different evolution vs. $\Delta\eta$ as compared to those from R_2.

6. References

[1] J. Adams *et al.*, [STAR Collaboration], Phys. Rev. C **73** (2006) 064907.
[2] M. P. McCumber *et al.*, [PHENIX Collaboration], J. Phys. G **35** (2008) 104081.
[3] S. Chatrchyan *et al.*, [CMS Collaboration], Phys. Lett. B**724** (2013) 213.
[4] B. Abelev *et al.*, [ALICE Collaboration], Phys. Lett. B**719** (2013) 29.
[5] M. Sharma, C. A. Pruneau, Phys. Rev. **C79**, 024905 (2009)
[6] K. Aamodt *et al.*, [ALICE Collaboration], JINST, 3, (2008) S08002.
[7] B. Abelev *et al.*, [ALICE Collaboration], Phys. Lett. B**708** (2012) 249-264.

Parametrization for chemical freeze-out conditions from net-charge fluctuations measured at RHIC

M Bluhm[1], P Alba[2], W Alberico[2], R Bellwied[3], V Mantovani Sarti[2], M Nahrgang[4], C Ratti[2,3]

[1] Department of Physics, North Carolina State University, Raleigh, NC 27695, USA
[2] Department of Physics, Torino University and INFN, Sezione di Torino, via P. Giuria 1, 10125 Torino, Italy
[3] Department of Physics, University of Houston, Houston, TX 77204, USA
[4] Department of Physics, Duke University, Durham, NC 27708, USA

E-mail: mbluhm@ncsu.edu

Abstract. We discuss details of our thermal model applied to extract chemical freeze-out conditions from fluctuations in the net-electric charge and net-proton number measured at RHIC. A parametrization for these conditions as a function of the beam energy is given.

1. Introduction

Over the past decade, statistical hadronization model (SHM) approaches, cf. e.g. [1], were quite successful in describing the particle multiplicities measured in heavy-ion collisions at various beam energies $\sqrt{s}$, allowing one to deduce the chemical freeze-out parameters as unique functions of the beam energy. As a complementary method for determining the conditions at freeze-out, fluctuations in the conserved charges of QCD, i.e. baryon number B, electric charge Q and strangeness S, were proposed [2]. The latter can be calculated from first-principles in lattice QCD [3, 4] and compared with corresponding measurements of the moments of multiplicity distributions. Experimental data on fluctuations in the net-electric charge and the net-proton number (not a conserved charge but often assumed to be a good proxy for net-baryon number) became recently available from the beam energy scan program at RHIC [5, 6]. Thermal models provide another possibility to extract the freeze-out parameters from fluctuation observables. In the approach discussed here, we consider a grandcanonical ensemble of hadrons and resonances as listed in [7]. This implies that charge conservation is respected in the means but not in the higher-order moments. Given that this as well as other non-included fluctuation sources, e.g. critical fluctuations, have a negligible influence on the observables we study, we show that an extraction of the chemical freeze-out parameters is feasible within our approach.

2. Freeze-out conditions from moments of net-charge distributions

In this work, we concentrate on the lowest moments of the measured multiplicity distributions and determine the conditions at chemical freeze-out such that the experimental data [5, 6] for mean M and variance σ^2 are reproduced within error bars. For the net-electric charge one has $M_{\text{net}-Q} = \sum_i Q_i \langle N_i \rangle$, where $\langle N_i \rangle$ is the average of the number N_i of particles of type i with electric charge Q_i, and $\sigma^2_{\text{net}-Q} = \sum_i \sum_j Q_i Q_j \langle \Delta N_i \Delta N_j \rangle$ with $\Delta N = N - \langle N \rangle$. The sums run over all charged hadrons contained in the measurement. As electric charge is dominated by π^+, π^-, K^+, K^-, p and $\bar{p}$, we restrict our model to this ensemble. For the net-proton number one has $M_{p-\bar{p}} = \langle N_p \rangle - \langle N_{\bar{p}} \rangle$ and $\sigma^2_{p-\bar{p}} = \langle (\Delta N_p)^2 \rangle + \langle (\Delta N_{\bar{p}})^2 \rangle - 2\langle \Delta N_p \Delta N_{\bar{p}} \rangle$, where the covariance between protons and anti-protons $\langle \Delta N_p \Delta N_{\bar{p}} \rangle$ vanishes in our approach. To take the actual experimental situation more accurately into account, several physically relevant aspects are included in our study as explained in the following.

2.1. *Influence of kinematic cuts and resonance decays*

The experimental coverage in rapidity (or pseudo-rapidity) and transverse momentum is limited by the detector design and the demands from particle reconstruction and identification. To account for this acceptance limitation, it was proposed in [8] to restrict the phase-space integrals in a thermal model correspondingly. In [8], the sensitivity of moment ratios to kinematic cuts was found to be small for net-protons and more pronounced for the net-electric charge. We follow this proposal and apply kinematic cuts in line with the experimental analysis [5, 6].

Resonance decays can, event by event, significantly influence the final multiplicities of stable (with respect to strong and electromagnetic decays) particles. A contamination from secondaries, e.g. weak decay contributions from Λ to p or spallation protons from the beam pipe, was reported to be suppressed through the experimental analysis [5]. In our model, contributions from resonance decays to the averages $\langle N_i \rangle$ and $\langle \Delta N_i \Delta N_j \rangle$ in M and σ^2 are included via

$$\langle N_i \rangle = \langle N_i^* \rangle_T + \sum_R \langle N_R^* \rangle_T \langle n_i \rangle_R \,, \tag{1}$$

$$\langle \Delta N_i \Delta N_j \rangle = \langle \Delta N_i^* \Delta N_j^* \rangle_T + \sum_R \langle (\Delta N_R^*)^2 \rangle_T \langle n_i \rangle_R \langle n_j \rangle_R + \sum_R \langle N_R^* \rangle_T \langle \Delta n_i \Delta n_j \rangle_R \,. \tag{2}$$

In Eqs. (1) and (2), N_i^* and N_R^* denote the primordial (before resonance decay) numbers of stable particles i and resonances R, while $\langle n_i \rangle_R = \sum_r b_r^R n_{i,r}^R$ and $\langle \Delta n_i \Delta n_j \rangle_R = \sum_r b_r^R n_{i,r}^R n_{j,r}^R - \sum_r b_r^R n_{i,r}^R \cdot \sum_{r'} b_{r'}^R n_{j,r'}^R$ are decay-channel averages, with r-th channel branching ratio b_r^R and $n_{i,r}^R$ the number of i produced in that decay-channel of R. As the decay of a resonance is a probabilistic process, which itself causes event-by-event fluctuations in the final hadron multiplicities, the averages in M and σ^2 involve an averaging over the thermal ensemble and over resonance decays. The thermal, primordial means and (co)variances are obtained in our model by derivatives of the pressure P with respect to the particle chemical potentials μ_l as (indices i and j stand for hadrons or resonances here)

$$\langle N_i^* \rangle_T = VT^3 \frac{\partial (P(T, \{\mu_l\})/T^4)}{\partial (\mu_i/T)} \,, \quad \langle \Delta N_i^* \Delta N_j^* \rangle_T = VT^3 \frac{\partial^2 (P(T, \{\mu_l\})/T^4)}{\partial (\mu_i/T) \partial (\mu_j/T)} \,. \tag{3}$$

Although in the grandcanonical ensemble the primordial covariances $\langle \Delta N_i^* \Delta N_j^* \rangle_T$ between different particle species vanish, resonance decays can induce correlations such that the covariances between different hadrons affect the moments of the net-electric charge distribution. Due to baryon-number conservation this is, however, not the case for net protons. Interestingly, the form of the factors $\langle \Delta n_i \Delta n_j \rangle_R$ implies that the fluctuation contributions associated with the probabilistic nature of resonance decays can be small or zero depending on the charge under consideration: if the number of produced charges is the same in each decay channel of R, they will exactly vanish. In the case of the net-electric charge this means that although individual factors such as $\langle \Delta n_{\pi^+} \Delta n_{K^-} \rangle_R$ can be non-zero, the probabilistic contributions cancel each other exactly in the double-sum over all charged hadrons in $\sigma^2_{\text{net}-Q}$ because electric charge is conserved in each decay. Since we restrict the ensemble of considered hadrons, however, we expect some but sub-dominant contributions from resonances decaying, in addition, into charged hyperons. In the case of the net-proton number, probabilistic fluctuation contributions are significant and vanish only for the Δ^{++}-resonances. A systematic study for net protons was presented in [9].

2.2. Influence of the isospin randomization of nucleons

Final state effects may significantly influence fluctuations, in particular, in the net-proton number. The regeneration and subsequent decay of Δ-resonances via $p(n) + \pi \to \Delta \to n(p) + \pi$ can change the isospin identity of the nucleons p and n (similar for anti-nucleons) and, thus, influence their multiplicity distributions. As the electric charge is conserved in these reactions, they do not affect the fluctuations in the net-electric charge. Also, the mean $M_{p-\bar{p}}$ is not altered, but the higher-order moments of the net-proton distribution are influenced. For high enough pion densities and a long enough hadronic phase (before free streaming), such that a full isospin randomization can be assumed, correction expressions for the moments of the net-proton distribution based on the moments of baryon and anti-baryon distributions can be derived, cf. [10, 11]. It was argued in [10, 11] that the necessary conditions are satisfied for $\sqrt{s} \gtrsim 10$ GeV.

When we exclude weak decay contributions from hyperons explicitly, these correction expressions may be formulated in terms of the moments of the nucleon and anti-nucleon distributions. Accordingly, the effect of isospin randomization is included in our model as

$$\sigma^2_{p-\bar{p}} = \langle (r \cdot \Delta N_N - \bar{r} \cdot \Delta N_{\bar{N}})^2 \rangle + \langle r(1-r)N_N + \bar{r}(1-\bar{r})N_{\bar{N}} \rangle \tag{4}$$

with $r = \langle N_p \rangle / \langle N_N \rangle$, $\bar{r} = \langle N_{\bar{p}} \rangle / \langle N_{\bar{N}} \rangle$ and $N_N = N_p + N_n$, $N_{\bar{N}} = N_{\bar{p}} + N_{\bar{n}}$. The entering quantities are calculated via Eqs. (1) - (3). As shown in [9], the main effect of these corrections is to make the net-proton distribution look Poissonian independent of whether the probabilistic fluctuation contributions from resonance decays are taken into account or not.

3. Parametrization for the chemical freeze-out conditions

With the abovementioned refinements in our thermal model, one can determine the values for temperature T and baryo-chemical potential μ_B at chemical freeze-out by comparing the measured moment ratios $\sigma^2_{\text{net}-Q}/M_{\text{net}-Q}$ and $\sigma^2_{p-\bar{p}}/M_{p-\bar{p}}$ with the model results. In addition, physical constraints on $M_{\text{net}-Q}/M_{\text{net}-B}$ and $M_{\text{net}-S}/M_{\text{net}-B}$ have to be fulfilled. In the moment ratios, the volume V which appears in Eq. (3) cancels when fluctuations in V are ignored. The corresponding parameter values for T and μ_B were reported in [12] and contrasted with different results from lattice QCD and with SHM results in [13].

In the left panel of Fig. 1 we show the quality of the description of the double-ratio $R^{\text{net}-Q}_{12}/R^{p-\bar{p}}_{12} = (\sigma^2_{p-\bar{p}}/M_{p-\bar{p}})/(\sigma^2_{\text{net}-Q}/M_{\text{net}-Q})$ of the measured moment ratios with our model obtained for the freeze-out parameters [12]. Since we assume full isospin randomization independent of the beam energy, we restrict our study to $\sqrt{s} \geq 11.5$ GeV. The ratio μ_B/T of our freeze-out parameters (with error bars based on the reported errors in the experimental data [5, 6]) is shown in the right panel of Fig. 1 (symbols). A suitable parametrization of these results as a function of $\sqrt{s}$ within $11.5 \leq \sqrt{s}/\text{GeV} \leq 200$ can be given in the form

$$\frac{\mu_B}{T} = \frac{a_0}{(\sqrt{s}/\text{GeV})^{a_1}} + a_2 + a_3(\sqrt{s}/\text{GeV}) \tag{5}$$

with $a_0 = 57.24$, $a_1 = 1.345$, $a_2 = 0.276$ and $a_3 = -0.00080$, which describes accurately our central values for μ_B/T (cf. solid curve in the right panel of Fig. 1).

4. Conclusion

We described in detail the physically relevant aspects taken into account in our thermal model in order to successfully deduce chemical freeze-out parameters from a comparison of model results with the moment ratios σ^2/M of the measured net-electric charge and net-proton distributions. We provided a suitable parametrization for the ratio μ_B/T of extracted freeze-out conditions as a function of $\sqrt{s}$. As discussed in [13], other sources of fluctuations that are not included in

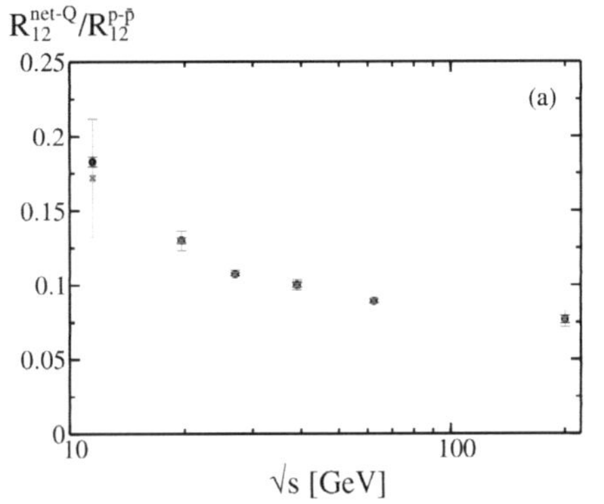

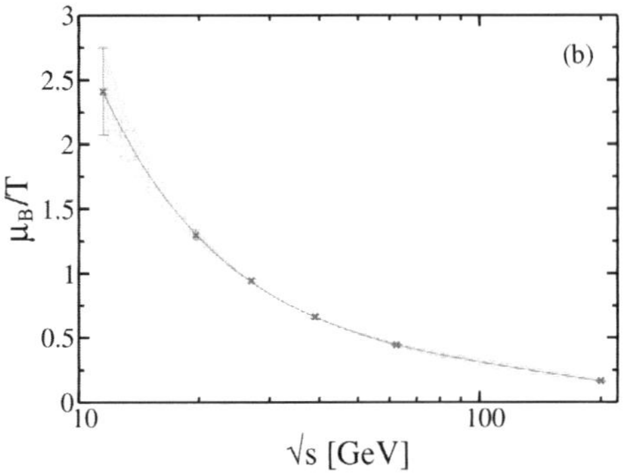

Figure 1. (Color online) (a): double-ratio of measured σ^2/M-ratios for net protons and net-electric charge [5, 6] (circles) as a function of $\sqrt{s}$, compared with our thermal model results (crosses) using the chemical freeze-out parameters from [12]. (b): ratio μ_B/T of these freeze-out parameters (crosses) as a function of $\sqrt{s}$ together with a suitable parametrization thereof (solid curve) of the form Eq. (5) with parameter values $a_0 = 57.24\pm^{73.78}_{35.91}$, $a_1 = 1.345\pm^{0.298}_{0.386}$, $a_2 = 0.276\pm^{0.116}_{0.251}$ and $a_3 = -0.00080\mp^{0.00040}_{0.00082}$ (errors define the shaded band).

our approach, such as critical fluctuations, volume fluctuations or the influence of exact charge conservation, have a negligible impact on these results for the highest beam energies considered in our analysis.

Acknowledgments
This work is supported by the Italian Ministry of Education, Universities and Research under the FIRB Research Grant RBFR0814TT, by the DAAD and by the U.S. Department of Energy grants DE-FG02-03ER41260, DE-FG02-05ER41367 and DE-FG02-07ER41521.

References

[1] Cleymans J, Oeschler H, Redlich K and Wheaton S 2006 *Phys. Rev.* C **73** 034905
[2] Karsch F 2012 *Central Eur. J. Phys.* **10** 1234
[3] Bazavov A *et al* 2012 *Phys. Rev. Lett.* **109** 192302
[4] Borsanyi S, Fodor Z, Katz S D, Krieg S, Ratti C and Szabo K K 2013 *Phys. Rev. Lett.* **111** 062005; 2014 *Phys. Rev. Lett.* **113** 052301
[5] Adamczyk L *et al* [STAR Collaboration] 2014 *Phys. Rev. Lett.* **112** 032302
[6] Adamczyk L *et al* [STAR Collaboration] 2014 *Phys. Rev. Lett.* **113** 092301
[7] Eidelman S *et al* [Particle Data Group] 2004 *Phys. Lett.* B **592** 1
[8] Garg P, Mishra D K, Netrakanti P K, Mohanty B, Mohanty A K, Singh B K and Xu N 2013 *Phys. Lett.* B **726** 691-696
[9] Nahrgang M, Bluhm M, Alba P, Bellwied R and Ratti C 2014 Impact of resonance regeneration and decay on the net-proton fluctuations in a hadron resonance gas *Preprint* arXiv:1402.1238 [hep-ph]
[10] Kitazawa M and Asakawa M 2012 *Phys. Rev.* C **85** 021901
[11] Kitazawa M and Asakawa M 2012 *Phys. Rev.* C **86** 024904; 2012 *Phys. Rev.* C **86** 069902
[12] Alba P, Alberico W, Bellwied R, Bluhm M, Mantovani Sarti V, Nahrgang M and Ratti C 2014 *Phys. Lett.* B **738** 305-310
[13] Bluhm M, Alba P, Alberico W, Bellwied R, Mantovani Sarti V, Nahrgang M and Ratti C 2014 Determination of freeze-out conditions from fluctuation observables measured at RHIC *Preprint* arXiv:1408.4734 [hep-ph]

Journal of Physics: Conference Series **612** (2015) 012042 doi:10.1088/1742-6596/612/1/012042

Overview of fluctuations at the QCD phase transition in heavy-ion collisions

Marlene Nahrgang

Department of Physics, Duke University, Durham, North Carolina 27708-0305, USA

E-mail: `marlene.nahrgang@phy.duke.edu`

Abstract. I will summarize our current understanding of the QCD phase diagram with special emphasis on what we can expect to learn from heavy-ion collision experiments. With the upcoming Beam Energy Scan Phase II at RHIC and the experimental facilities FAIR, GSI, and NICA, Dubna, several challenges need to be met both experimentally and theoretically. I will discuss in which aspects theoretical advances are necessary in order to be able to interpret the experimental results.

1. Introduction

Heavy-ion collisions have proved to be an excellent experimental tool to study strongly interacting matter under extreme conditions such as high temperatures and high densities for a couple of decades. While the available beam energies increased steadily from one generation of accelators to the next with the LHC soon to operate at $\sqrt{s_{\rm NN}} = 5$ TeV, a revived interest in lower beam energies is seen. This is predominently due to the possibility of discovering a critical point in the phase diagram of QCD.

In the chiral limit, where the up and down quarks are massless, the phase transition between a quark-gluon plasma (QGP) and the hadronic phase at vanishing baryochemical potential $\mu_B = 0$ is of second order if the strange quark has mass $m_s = \infty$ and turns into first-order at some finite value of m_s [1, 2]. It has nowadays been established by lattice QCD calculations that the phase transition at $\mu_B = 0$ and at the physical quark masses is an analytic crossover [3, 4]. Being the only available exact method to solving QCD in the nonperturbative regime, current lattice QCD methods are only feasible at $\mu_B = 0$. Experimentally this corresponds to LHC and top-RHIC energies. Due to the stopping of the incoming baryon currents of the two colliding nuclei, one can study strongly-interacting matter at finite baryonic densities in the laboratory by decreasing the beam energy. This fascinating idea is the main motivation behind the beam energy scan (BES) at RHIC. Phase I has been completed by runs in 2010 and 2011 with data taken at at $\sqrt{s} = 7.7$, 11.5, 19.6, 27, 39, 62.4 and 200 GeV and in 2014 with an additional run at $\sqrt{s} = 14.5$ GeV. Phase II with larger statistics and additional energies is planned for 2018/19.

This brief overview is organised as follows: in the next section I will turn to theoretical studies that conjecture the existence of a critical point and an adjacent line of first-order phase transition at finite μ_B. Sections 3 and 4 discuss the potentials to discover the critical point and the first-order phase transition in heavy-ion collisions via possible signatures and point out the necessity of dynamical modeling of the phase transition. In section 5 some challenges for the upcoming BES phase II will be highlighted.

2. Indications for the existence of a critical point

As there is not yet any exact method to solve QCD thermodynamics at finite baryochemical potential, our current knowledge is limited to model studies or asymptotic behavior. While lattice QCD gives very reliable results at $\mu_B = 0$, a finite baryochemical potential makes it impossible to perform Monte-Carlo importance sampling due to the complex fermionic determinant, often called the "sign problem".

Functional methods, including Dyson-Schwinger equations (DSE) [5, 6] and functional renormalization group (FRG) [7, 8], do not suffer from the sign problem and can be applied in the entire phase diagram and in the nonperturbative regime. In the DSE approach the temperature-dependent quenched gluon propagator is usually taken as input from lattice QCD calculations and the back reaction of the quark loops is systematically included. For the quark-gluon vertex a general ansatz has to be made with parameters fixed to match the gluon input, the quark condensate and to obtain the correct asymptotic behaviors in the infrared and ultraviolet. Solving the gluon and quark DSE excellent agreement with lattice calculations for the order parameters (chiral condensate and Polyakov-loop) in the crossover transition at $\mu_B = 0$ is found. At finite baryochemical potential a critical point and a first-order phase transition are found [5, 6]. In the FRG the thermodynamic potential is solved momentum shell by momentum shell, thus obtaining a solution for the QCD flow equation. This method achieves excellent agreement with lattice calculations for vacuum QCD [9] and for pure glue at finite temperature [10] and has widely been applied to low-energy effective models, see below. Full QCD calculations in the (T, μ_B) plane are work in progress.

From both functional methods combined one can so far conclude that there is no critical point for $T/\mu_B \lesssim 2$. For larger values a more thorough treatment of baryonic degrees of freedom needs to be included.

The phase diagram of low-energy effective models has extensively been studied in chiral mesonic models with constituent quarks like the Quark-Meson (QM) or the Nambu-Jona-Lasino (NJL) model, and show the expected structures of crossover, critical point and first-order phase transition with increasing baryonic density. These models can further be extended to include the temporal gauge field A_0 in terms of the Polyakov-loop

$$\ell = \frac{1}{N_c}\langle tr_c \mathcal{P}\rangle_\beta \, , \quad \bar{\ell} = \frac{1}{N_c}\langle tr_c \mathcal{P}^\dagger\rangle_\beta \, , \tag{1}$$

for QCD with N_c number of colors and with the operator $\mathcal{P}$

$$\mathcal{P} = P\exp\left(ig_s \int_0^\beta \mathrm{d}\tau A_0 \right) \, , \tag{2}$$

where g_s is the strong coupling and $\beta = 1/T$ the inverse temperature. The Lagrangian of the Polyakov-Quark-Meson (PQM) model [11] reads

$$\mathcal{L} = \bar{q}\left[i\left(\gamma^\mu \partial_\mu - ig_s\gamma^0 A_0\right) - g\left(\sigma + i\gamma_5\vec{\tau}\cdot\vec{\pi}\right)\right] q + \frac{1}{2}\left(\partial_\mu \sigma\right)^2 + \frac{1}{2}\left(\partial_\mu \vec{\pi}\right)^2 - U\left(\sigma, \vec{\pi}\right) - \mathcal{U}_T(\ell, \bar{\ell}) \, . \tag{3}$$

Here, the potential of the mesonic fields $U(\sigma, \vec{\pi})$ is the classical Mexican hat potential and the temperature-dependent effective potential for the Polyakov loop is a parametrization of quenched QCD calculations from the lattice

$$\frac{\mathcal{U}_T}{T^4}\left(\ell, \bar{\ell}\right) = -\frac{b_2(T)}{4}\left(|\ell|^2 + |\bar{\ell}|^2\right) - \frac{b_3}{6}\left(\ell^3 + \bar{\ell}^3\right) + \frac{b_4}{16}\left(|\ell|^2 + |\bar{\ell}|^2\right)^2 \, , \tag{4}$$

for which the parameters can for example be found in [11].

Journal of Physics: Conference Series **612** (2015) 012042 doi:10.1088/1742-6596/612/1/012042

The inclusion of the Polyakov-loop leads to the suppression of one- and two-quark states in the thermodynamic potential of the PQM model, the so-called statistical confinement. This is due to the change of the numerical value of the Polyakov-loop. It vanishes below the transition temperature and has a finite value at larger temperatures. It is thus an order parameter of the confinement/deconfinement transition. The exact position of the critical point in these models depends among other aspects on the way fluctuations are treated, on the mean-field level or beyond [11, 12].

The existence of a first-order phase transition is further backed by work on color superconductivity at low T and very large μ_B, where a rich phase structure of color-flavor locking and 2SC phases is expected [13, 14, 15], and on neutron stars [16, 17].

3. Potential to discover the critical point in heavy-ion collisions

A critical point is a very special point in the phase diagram of any system. Here, fluctuations grow very large and the thermodynamic characteristics of the systems are governed by only some macroscopic parameters, like e.g. the dimension. Microscopically, very different systems show a common, universal, behavior at their respective critical point. One of the particular features is the divergence of the correlation length of the order parameter, $\xi \to \infty$, which gives rise to the divergence of the ensemble fluctuations themselves.

By coupling the order parameter, e.g. the sigma field σ for chiral symmetry, to measurable particles like pions, ie. $g_\pi \sigma\pi\pi$, or protons, ie. $g_p \bar{p}\sigma p$, the diverging fluctuations of the sigma field should translate into event-by-event fluctuations of pion or proton multiplicities in heavy-ion collisions [18, 19]. This was realized not without noting that in small systems the correlation length can obviously not become infinitely large beyond the size of the system. In a heavy-ion collision one can assume that due to spatial limits the correlation length can maximally grow up to $\xi = 6 - 10$ fm. First fluctuations were measured at the NA49 experiment at the SPS. It was specifically looked for nonmonotonic behavior of the fluctuation measure as a function of the beam energy. Results based on second-order moments did, however, not show the expected nonmonotonic behavior [20].

More important than finiteness in space is actually the finiteness in time. It is not only the correlation length, which diverges at a critical point, but also the relaxation time. Even if a system is in equilibrium above the critical point it is necessarily driven out of equilibrium as it evolves through the critical point at any finite time. It was shown in [21] by using a phenomenological evolution equation for the mass of the sigma field $m_\sigma = 1/\xi$ that the correlation length only grows up to $\xi = 1.5-2.5$ fm. Experimentally, it thus turns out to be more favorable to look at higher-order moments of the fluctuations since they diverge proportionally to larger exponents of the correlation length [22].

At finite baryochemical potential the net-baryon density mixes with the sigma field. Due to its diffusive dynamics it becomes the true critical mode in terms of long-time dynamics [23, 24]. Therefore one expects more generally that at the critical point a non-monotonic behavior in the fluctuations of conserved charges such as net-baryon number and net-electric charge can be seen [13, 2, 18, 19, 25]. The fluctuations can be quantified by the susceptibilities, which are given by

$$\chi_l = \left.\frac{\partial^l (P/T^4)}{\partial(\mu/T)^l}\right|_T \tag{5}$$

where the derivative of the pressure P is taken with respect to the chemical potential corresponding to the conserved charge of interest. These susceptibilities can naturally be related to the cumulants of measured multiplicity distributions of the net quantities via

$$\chi_1 = \frac{1}{VT^3}\langle N\rangle_c = \frac{1}{VT^3}\langle N\rangle\,, \tag{6}$$

$$\chi_2 = \frac{1}{VT^3}\langle(\Delta N)^2\rangle_c = \frac{1}{VT^3}\langle(\Delta N)^2\rangle\,, \tag{7}$$

$$\chi_3 = \frac{1}{VT^3}\langle(\Delta N)^3\rangle_c = \frac{1}{VT^3}\langle(\Delta N)^3\rangle\,, \tag{8}$$

$$\chi_4 = \frac{1}{VT^3}\langle(\Delta N)^4\rangle_c \equiv \frac{1}{VT^3}\left(\langle(\Delta N)^4\rangle - 3\langle(\Delta N)^2\rangle^2\right)\,, \tag{9}$$

where the first three cumulants are simply given by the corresponding central moments. The fourth cumulant is a combination of fourth and second central moments. The fluctuations are the event-wise deviation from the event-by-event mean: $\Delta N = N - \langle N\rangle$.

In the above equations for χ_l experimentally unknown quantities like the temperature and the volume of the system appear. It turns out that suitable ratios of the susceptibilities can be expressed in terms of the mean $M = \langle N\rangle$, the variance $\sigma^2 = \langle(\Delta N)^2\rangle$, the skewness $S = \langle(\Delta N)^3\rangle/\langle(\Delta N)^2\rangle^{3/2}$ and the kurtosis $\kappa = \langle(\Delta N)^4\rangle/\langle(\Delta N)^2\rangle^2 - 3$, by

$$\frac{\chi_2}{\chi_1} = \frac{\sigma^2}{M}, \quad \frac{\chi_3}{\chi_2} = S\sigma, \quad \frac{\chi_4}{\chi_2} = \kappa\sigma^2\,. \tag{10}$$

The unknown volume and temperature and the simple dependence on the particle multiplicity according to the central limit theorm of statistics cancel in these ratios.

In the years 2010 and 2011, the beam energy scan phase I was run successfully at the RHIC facility. Data taken at $\sqrt{s}$ = 7.7, 11.5, 19.6, 27, 39, 62.4 and 200 GeV has been analysed and the first results of higher-order cumulants of the net-proton distribution [26] and net-charge [27] were reported.

In order to make a better connection between experimental data and the theoretical expectation for critical phenomena it will become increasingly important to develop dynamical models of the phase transition. One approach that can effectly treat fluctuations at the critical point dynamically is nonequilibrium chiral fluid dynamics (NχFD) [28, 29, 30, 31, 32, 33] where fluctuations of the sigma field and the Polyakov-loop are propagated explicitly via a coupling to a fluid dynamical expansion of heavy-ion collisions. After the discussion of some aspects of the first order phase transition in the next section I will highlight some challenges for dynamical models in section 5.

4. ... and the first order phase transition

An indirect discovery of the critical point could go via a disappearance of signals of a first-order phase transition toward higher $\sqrt{s}$. The first-order phase transition is characterized by a coexistence region where the two phases have the same pressure. Above and below the transition temperature, meta-stable states exist, in which the system can be trapped when nonequilibrium effects become important. For example the fast collective expansion of matter produced in a heavy-ion collision can lead to supercooling [34, 35, 36, 30]. If the nucleation rates are low enough, this phase will decay via spinodal instabilities [37, 38, 39]. It was shown in recent works that this effect can lead to domain formation in the net-baryon density [40, 41, 32]. In [40, 41] a nonequilibium equation of state constructed by joining a QGP equation of state via a spline with a hadron gas equation of state was applied in a fluid dynamical simulation of heavy-ion collisions. This is different from the usual equations of state with a first-order phase transition, as they are taken in the equilibrium mode, which corresponds to a Maxwell construction over the coexistence and spinodal region. It was seen that initial inhomogeneities, present in the fluctuating initial conditions from UrQMD, get amplified in the spinodal region. In NχFD fluctuations are generated dynamically which means that even for smooth initial conditions spinodal instabilities may arise during the evolution through the phase transition, as was shown in [32].

If fluctuations are to be seen from a critical point scenario or a first-order phase transition is a question if the system is sufficiently in or out of equilibrium. One can perform a similar analysis of susceptibilities in Eq. (5) in the spinodal region instead of restricting oneselves to the thermodynamically stable states. In [42] it was found that the quark number susceptibility diverges along the isothermal spinodal lines due to the convex structure of the pressure in the presence of mechanically instable regions.

5. Challenges for the Beam Energy Scan Phase II

The following points are some important challenges for experiment and theory for current and future investigation of the QCD phase diagram.

- *Net-baryon versus net-proton fluctuations*: If net-baryon number is the driving force behind the critical fluctuations then the critical signals in the experimentally measured fluctuations in net-proton number are significantly obscured due to isospin randomization. Assuming that the number of (anti-)protons and neutrons among a certain number of (anti)baryons are distributed according to a binomial distribution, the net-baryon number fluctuations can be reconstructed from the measured (anti-)proton distributions via formulas in [43, 44]. The isospin randomization can also be included in benchmark calculations starting from a hadron resonance gas model [45].
- *Efficiency corrections*: Additional fluctuations can come from the experimental setup itself. Limited reconstruction efficiency will affect all higher-order moments of measured distributions [46]. It turns out that if the momentum coverage is extended the efficiency corrections need to be treated momentum dependent [47] and they will also affect the experimental errors [48].
- *Global charge conservation*: In heavy-ion collisions the net-baryon number, net-electric charge and net-strangeness are conserved not on average but exactly in each collision. Due to the limited experimental acceptance fluctuations are observed. It remains a challenging question if the system can at all be described in a grandcanonical ensemble or if a canoncial ensemble is more suitable. Experimental cuts can strongly affect the fluctuations [49, 50]. In transport model studies, that account for the microcanonical character of the individual scatterings, it was shown that net-proton fluctuations (in contrast to net-baryon number) are affected only at lower $\sqrt{s}$ [49]. Observing only a small fraction of all net-baryon in each event leads also to a strong bias toward observing a Poisson distribution although the underlying distribution might be very different from Poisson [50].
- *Initial state and final state*: In theoretical modeling of the fluctuations at the phase transition it cannot be forgotten that there are other sources of fluctuations coming from the initial and the final state. Especially the initial state at high baryonic densities is only very poorly understood so far. Since we are interested in fluctuations in net-baryon number, the initial fluctuations due to baryon stopping at lower beam energies need to be treated more carefully.
- *Equation of state and transport coefficients*: It is an open question if the QCD phase transition at high baryonic densities is predominantly a liquid-gas or a hadron-quark phase transition, for which the (pseudo-)critical pressure either increases or decreases as a function of the temperature [51, 52]. This will have significant phenomenological consequences. In addition to the equation of state, any realistic simulation at finite baryochemical potential requires transport coefficients as an input.

- *Phase transition and freeze-out conditions*: It needs to be emphasized that an interplay of the location of freeze-out conditions [53] and variations in $\sqrt{s}$ can generate various nonmonotonic fluctuation patterns, which need to be investigated thoroughly. In order to distinguish between fluctuations at the crossover [54], critical point and first-order phase transition a combined analysis of all higher-order cumulants needs to be performed.

Acknowledgments
The author acknowledges support from the U.S. Department of Energy under grant DE-FG02-05ER41367 and a fellowship within the Postdoc-Program of the German Academic Exchange Service (DAAD). I thank the organisers of Hot Quarks 2014 for this inspiring conference.

References
[1] Brown F R, Butler F P, Chen H, Christ N H, Dong Z h *et al.* 1990 *Phys.Rev.Lett.* **65** 2491–2494
[2] Halasz A M, Jackson A, Shrock R, Stephanov M A and Verbaarschot J 1998 *Phys.Rev.* **D58** 096007 (*Preprint* hep-ph/9804290)
[3] Aoki Y, Endrodi G, Fodor Z, Katz S and Szabo K 2006 *Nature* **443** 675–678 (*Preprint* hep-lat/0611014)
[4] Borsanyi S *et al.* (Wuppertal-Budapest Collaboration) 2010 *JHEP* **1009** 073 (*Preprint* 1005.3508)
[5] Fischer C S and Luecker J 2013 *Phys.Lett.* **B718** 1036–1043 (*Preprint* 1206.5191)
[6] Fischer C S, Luecker J and Welzbacher C A 2014 *Phys.Rev.* **D90** 034022 (*Preprint* 1405.4762)
[7] Pawlowski J M 2007 *Annals Phys.* **322** 2831–2915 (*Preprint* hep-th/0512261)
[8] Pawlowski J M 2011 *AIP Conf.Proc.* **1343** 75–80 (*Preprint* 1012.5075)
[9] Braun J, Fister L, Pawlowski J M and Rennecke F 2014 (*Preprint* 1412.1045)
[10] Fister L and Pawlowski J M 2011 (*Preprint* 1112.5440)
[11] Schaefer B J, Pawlowski J M and Wambach J 2007 *Phys.Rev.* **D76** 074023 (*Preprint* 0704.3234)
[12] Herbst T K, Pawlowski J M and Schaefer B J 2011 *Phys.Lett.* **B696** 58–67 (*Preprint* 1008.0081)
[13] Berges J and Rajagopal K 1999 *Nucl.Phys.* **B538** 215–232 (*Preprint* hep-ph/9804233)
[14] Rischke D, Son D and Stephanov M A 2001 *Phys.Rev.Lett.* **87** 062001 (*Preprint* hep-ph/0011379)
[15] Muller D, Buballa M and Wambach J 2013 *Eur.Phys.J.* **A49** 96 (*Preprint* 1303.2693)
[16] Lattimer J and Prakash M 2004 *Science* **304** 536–542 (*Preprint* astro-ph/0405262)
[17] Buballa M, Dexheimer V, Drago A, Fraga E, Haensel P *et al.* 2014 *J.Phys.* **G41** 123001 (*Preprint* 1402.6911)
[18] Stephanov M A, Rajagopal K and Shuryak E V 1998 *Phys.Rev.Lett.* **81** 4816–4819 (*Preprint* hep-ph/9806219)
[19] Stephanov M A, Rajagopal K and Shuryak E V 1999 *Phys.Rev.* **D60** 114028 (*Preprint* hep-ph/9903292)
[20] Alt C *et al.* (NA49 Collaboration) 2008 *Phys.Rev.* **C78** 034914 (*Preprint* 0712.3216)
[21] Berdnikov B and Rajagopal K 2000 *Phys.Rev.* **D61** 105017 (*Preprint* hep-ph/9912274)
[22] Stephanov M 2009 *Phys.Rev.Lett.* **102** 032301 (*Preprint* 0809.3450)
[23] Fujii H and Ohtani M 2004 *Phys.Rev.* **D70** 014016 (*Preprint* hep-ph/0402263)
[24] Son D and Stephanov M 2004 *Phys.Rev.* **D70** 056001 (*Preprint* hep-ph/0401052)
[25] Hatta Y and Ikeda T 2003 *Phys.Rev.* **D67** 014028 (*Preprint* hep-ph/0210284)
[26] Adamczyk L *et al.* (STAR Collaboration) 2014 *Phys.Rev.Lett.* **112** 032302 (*Preprint* 1309.5681)
[27] Adamczyk L *et al.* (STAR Collaboration) 2014 *Phys.Rev.Lett.* **113** 092301 (*Preprint* 1402.1558)
[28] Nahrgang M, Leupold S, Herold C and Bleicher M 2011 *Phys.Rev.* **C84** 024912 (*Preprint* 1105.0622)
[29] Nahrgang M, Leupold S and Bleicher M 2012 *Phys.Lett.* **B711** 109–116 (*Preprint* 1105.1396)
[30] Nahrgang M, Herold C, Leupold S, Mishustin I and Bleicher M 2013 *J.Phys.* **G40** 055108 (*Preprint* 1105.1962)
[31] Herold C, Nahrgang M, Mishustin I and Bleicher M 2013 *Phys.Rev.* **C87** 014907 (*Preprint* 1301.1214)
[32] Herold C, Nahrgang M, Mishustin I and Bleicher M 2014 *Nucl.Phys.* **A925** 14–24 (*Preprint* 1304.5372)
[33] Herold C, Nahrgang M, Yan Y and Kobdaj C 2014 *J.Phys.* **G41** 115106 (*Preprint* 1407.8277)
[34] Csernai L and Mishustin I 1995 *Phys.Rev.Lett.* **74** 5005–5008
[35] Zabrodin E, Bravina L, Csernai L, Stoecker H and Greiner W 1998 *Phys.Lett.* **B423** 373–378 (*Preprint* hep-ph/9806207)
[36] Keranen A, Manninen J, Csernai L and Magas V 2003 *Phys.Rev.* **C67** 034905 (*Preprint* nucl-th/0205019)
[37] Mishustin I 1999 *Phys.Rev.Lett.* **82** 4779–4782 (*Preprint* hep-ph/9811307)
[38] Randrup J 2009 *Phys.Rev.* **C79** 054911 (*Preprint* 0903.4736)
[39] Randrup J 2010 *Phys.Rev.* **C82** 034902 (*Preprint* 1007.1448)
[40] Steinheimer J and Randrup J 2012 *Phys.Rev.Lett.* **109** 212301 (*Preprint* 1209.2462)
[41] Steinheimer J and Randrup J 2013 *Phys.Rev.* **C87** 054903 (*Preprint* 1302.2956)
[42] Sasaki C, Friman B and Redlich K 2007 *Phys.Rev.Lett.* **99** 232301 (*Preprint* hep-ph/0702254)

[43] Kitazawa M and Asakawa M 2012 *Phys.Rev.* **C85** 021901 (*Preprint* 1107.2755)
[44] Kitazawa M and Asakawa M 2012 *Phys.Rev.* **C86** 024904 (*Preprint* 1205.3292)
[45] Nahrgang M, Bluhm M, Alba P, Bellwied R and Ratti C 2014 (*Preprint* 1402.1238)
[46] Bzdak A and Koch V 2012 *Phys.Rev.* **C86** 044904 (*Preprint* 1206.4286)
[47] Bzdak A and Koch V 2013 (*Preprint* 1312.4574)
[48] Luo X 2014 (*Preprint* 1410.3914)
[49] Nahrgang M, Schuster T, Mitrovski M, Stock R and Bleicher M 2012 *Eur.Phys.J.* **C72** 2143 (*Preprint* 0903.2911)
[50] Bzdak A, Koch V and Skokov V 2013 *Phys.Rev.* **C87** 014901 (*Preprint* 1203.4529)
[51] Steinheimer J, Randrup J and Koch V 2014 *Phys.Rev.* **C89** 034901 (*Preprint* 1311.0999)
[52] Hempel M, Dexheimer V, Schramm S and Iosilevskiy I 2013 *Phys.Rev.* **C88** 014906 (*Preprint* 1302.2835)
[53] Alba P, Alberico W, Bellwied R, Bluhm M, Mantovani Sarti V *et al.* 2014 *Phys.Lett.* **B738** 305–310 (*Preprint* 1403.4903)
[54] Redlich K 2012 *Central Eur.J.Phys.* **10** 1254–1257 (*Preprint* 1207.2610)

Journal of Physics: Conference Series **612** (2015) 012043 doi:10.1088/1742-6596/612/1/012043

Final source eccentricity measured by HBT interferometry with the event shape selection

Takafumi Niida for the PHENIX Collaboration
University of Tsukuba, 1-1-1 Tennoudai, Tsukuba, Ibaraki 305-8571, Japan
E-mail: niida@bnl.gov

Abstract. Azimuthal angle dependence of the pion source radii has been measured applying the event shape engineering technique at the PHENIX experiment. When events with higher magnitude of second-order flow vector are selected, the oscillation of the source radii is enhanced as well as v_2 which leads to the enhancement of the measured final source eccentricity. The event twist effect in the spatial source distribution in the final state has been also explored with AMPT model. Results indicate a possible twisted source due to the initial longitudinal fluctuations.

1. Introduction

Higher-order flow coefficients v_n are useful observables to constrain the properties of the quark-gluon plasma, such as a shear viscosity over entropy density ratio, in a heavy ion collision [1, 2]. The v_2 is mainly caused by the almond shape of the nuclear overlap region, but higher-order flow, v_3, v_4, $\cdots$, especially their odd components are originating from the initial spatial fluctuations of participant nucleons. Recently ATLAS experiment has presented results of event-by-event v_n [3], where v_n values show large variations even in the fixed centrality bin due to initial fluctuations. To control such an initial fluctuation, the event shape engineering was suggested [4]. This could be a useful tool to study the response of initial state to the system evolution and connect the initial and final states. In these proceedings, we present results on HBT measurements using charged pions and applying the event shape engineering technique for Au+Au collisions at $\sqrt{s_{NN}}$=200 GeV recorded with the PHENIX experiment.

The event shape engineering focuses on the fluctuations in the transverse plane, but the presence of fluctuations in longitudinal direction could cause a twisted source along that direction [5]. The number of participants going to the forward and backward directions is not necessarily the same, and also participant eccentricities and participant planes might be different at both angles, which leads to different event plane angles between forward and backward rapidities. Thus the initial twist may survive as a twisted flow in the final state [8, 9]. In these proceedings, we examine the possibility of a spatially twisted source in the final state using HBT interferometry in AMPT model.

2. HBT measurements with event shape selection at PHENIX

The event shape selection was performed by selecting the magnitude of 2$^{\rm nd}$-order flow vectors, Q_2, which were measured by the Reaction Plane Detector (RXN, $1 < |\eta| < 2.8$). The Q_2 is

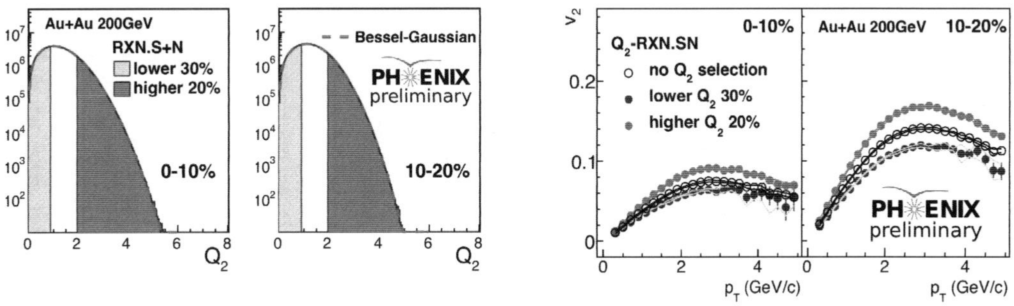

Figure 1. (Left) Q_2 distributions measured with the RXN in Au+Au 200 GeV collisions, indicating 20%(30%) higher(lower) Q_2 events. (Right) Charged hadron v_2 as a function of p_T with and without event shape selection.

defined as $Q_2 = \sqrt{Q_{2,x}^2 + Q_{2,y}^2}/\sqrt{\Sigma w_i}$, where $Q_{2,x} = \Sigma w_i \cos(2\phi)$, $Q_{2,y} = \Sigma w_i \sin(2\phi)$, and w_i reflects the multiplicity within the element i of the RXN. Figure 1(left) shows Q_2 distributions for two centrality bins, where the higher 20% and the lower 30% Q_2 events are shown filled. We have tested the effect of the Q_2 selection on charged hadron v_2 as shown in Fig. 1(right), where the v_2 is measured at mid-rapidity ($|\eta| < 0.35$) and the systematic uncertainty from different event planes of south, north, and both combined RXN is included. It is confirmed that the Q_2 selection enhances or decreases the strength of v_2. Then we have applied this technique to the HBT measurement using charged pion pairs. Charged pions were identified by the electromagnetic calorimeter (EMCal, $|\eta| < 0.35$) and the pair selection cuts at the drift chamber and EMCal were applied to remove the effects of mis-reconstructed tracks and detector inefficiency. For the HBT analysis, pion pairs were analyzed with the out-side-long parameterization [10, 11] in the longitudinally co-moving system. The effect of the event plane resolution was also corrected for both cases with and without Q_2 selection. Figure 2(left) shows the extracted pion HBT radii, R_s and R_o, as a function of azimuthal pair angle ϕ relative to the second-order event plane Ψ_2. Results show that the higher Q_2 selection increases the oscillation strength compared to the case without Q_2 selection. These oscillations of HBT radii are supposed to be sensitive to the final source eccentricity at freeze-out, $\varepsilon_{\text{final}}$. Blast-wave studies suggest that the quantity of $2R_{s,2}^2/R_{s,0}$ would be a good probe to $\varepsilon_{\text{final}}$ in the limit of $k_T = 0$, where k_T denotes a mean pair transverse momentum. The oscillation amplitudes of R_s^2 and R_o^2 in a form of the final eccentricity are plotted as a function of the number of participants calculated by Glauber model in Fig. 2(right). The higher Q_2 selection enhances the measured $\varepsilon_{\text{final}}$ as well as v_2. It could be originating from a larger initial eccentricity, although there should be a contribution from the selected flow itself because the radii modulations also depend on the anisotropy in the momentum space [6, 7].

3. HBT measurements with event twist selection in AMPT model

To study the twisted source in the final state, we used the data of Pb+Pb 2.76 TeV collisions simulated using AMPT model (v2.25 with string melting), where the impact parameter was fixed to 8 fm. For the HBT study, the interference effect between two identical particles, $1 + \cos(\mathbf{\Delta r} \cdot \mathbf{\Delta p})$, was calculated and weighted to the relative pair momentum distributions. Then the correlation functions were reconstructed by taking a ratio of the distributions with and without the weight. Also, all charged pions were allowed to make a pair with each other including $\pi^+\pi^-$ to increase the statistics (the consistency between results for positive and negative pairs was checked). The event plane was determined using particles in $4 < |\eta| < 6$, where particles

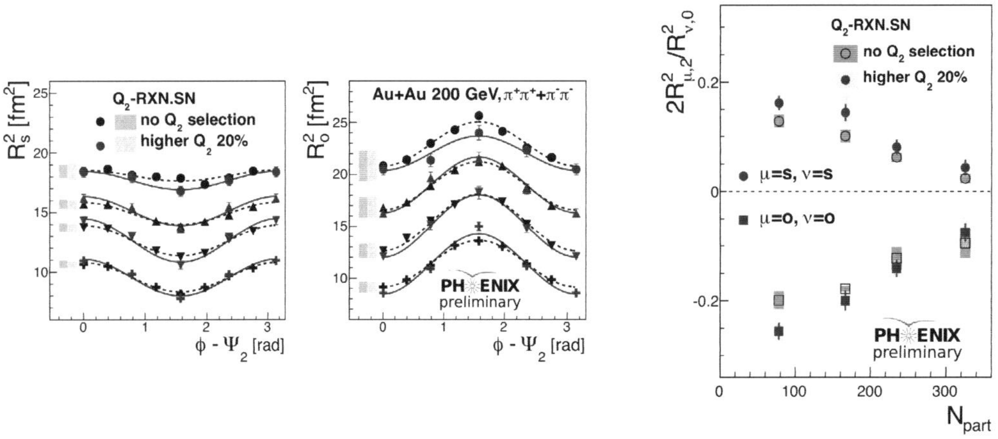

Figure 2. (Left and center) Azimuthal angle dependence of R_s^2 and R_o^2 relative to Ψ_2. (Right) 2nd-order azimuthal oscillation on R_s^2 and R_o^2. In both panels, results with and without higher Q_2 selection are shown.

were divided into two sub-groups. A set of forward and backward event planes (Ψ_2^F, Ψ_2^B) were used for the event cut which requires finite difference between Ψ_2^F and Ψ_2^B as a event twist selection, whereas the other set of events was used for a reference angle of azimuthal HBT measurement.

Figure 3(left) shows R_s^2, R_o^2, and R_{os}^2 as a function of azimuthal pair angle relative to the Ψ_2^B, $\Delta\phi$, for four η regions, where $(\Psi_2^B - \Psi_2^F) > 0.6$ was required. The oscillations of three HBT radii measured in positive η regions are shifted to negative direction, which is the direction of Ψ_2^F in the current event cut. This phase shift can be understood to be a possible twist effect in the final source distribution. These oscillations were fitted with the following functions:

$$R_\mu^2(\Delta\phi) = R_{\mu,0}^2 + 2R_{\mu,2}^2\cos(2\Delta\phi + \alpha) \ \ (\text{for } \mu = o, s), \quad (1)$$
$$R_\mu^2(\Delta\phi) = R_{\mu,0}^2 + 2R_{\mu,2}^2\sin(2\Delta\phi + \alpha) \ \ (\text{for } \mu = os), \quad (2)$$

to extract the magnitude of the phase shift, which is taken into account with α. The phase

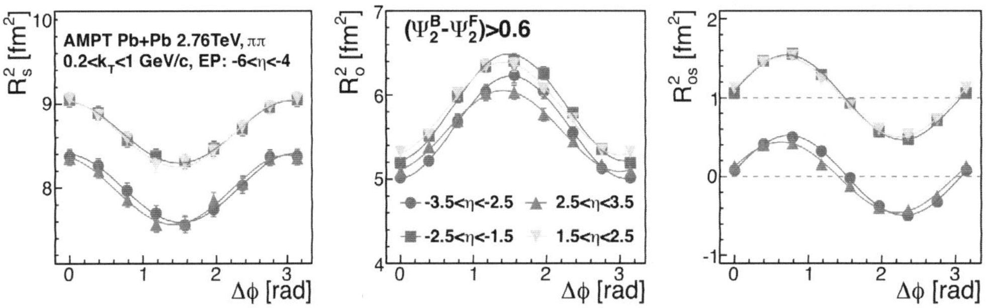

Figure 3. Azimuthal angle dependence of R_s^2, R_o^2, and R_{os}^2 relative to the backward Ψ_2 with the event cut of $(\Psi_2^B - \Psi_2^F) > 0.6$, where the dashed lines show $R_{os}^2 = 0$.

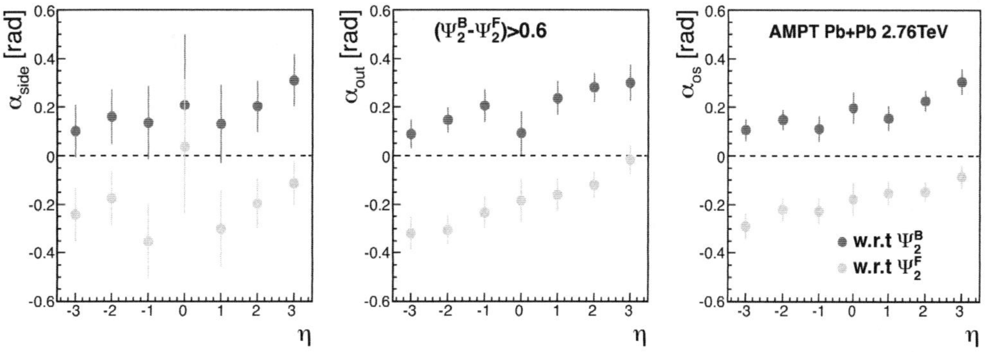

Figure 4. Phase shift parameters α obtained from the R_s^2, R_o^2, and R_{os} as a function of η. Results measured with respect to forward and backward event planes (Ψ_2^F, Ψ_2^B) are shown, with event cut of $(\Psi_2^B - \Psi_2^F) > 0.6$.

shift parameter α obtained from the results with respect to Ψ_2^F and Ψ_2^B is plotted as a function of η in Fig. 4. The α increases with going from backward to forward angle in all cases. The variation of α in the η dependence is comparable to the difference between results relative to Ψ_2^F and Ψ_2^B at the same η. These results indicate that the source in the final state is also twisted due to longitudinal fluctuations in the initial state, as well as the twisted event plane and flow as discussed in Ref. [8, 9].

4. Summary

We presented the results of HBT measurements using event shape engineering for collisions recorded with the PHENIX experiment. We found that the higher Q_2 selection enhances the measured final source eccentricity as well as v_2. Although the model comparison is needed to disentangle both spatial and dynamical effects on the HBT radii, this study clarifies the relation between initial and final eccentricity and constrains better the system dynamics.

We have also studied the event twist effect using the AMPT model. When selecting events with finite difference between forward and backward event plane angles, the oscillations of HBT radii are shifted in the phase and the phase shift increases with η. The results indicate a possible twisted source in the final state preserving the initial twist due to the longitudinal fluctuations. This effect could be measured in experiments at RHIC and the LHC.

Both techniques could be useful to probe and control initial fluctuations in transverse plane and longitudinal directions, as well as to study the response of the system to the space-time evolution.

References

[1] A. Adare *et al.* (PHENIX Collaboration), Phys. Rev. Lett. **107**, 252301 (2011).
[2] C. Gale, S. Jeon, B. Schenke, P. Tribedy, and R. Venugopalan, Phys. Rev. Lett. **100**, 012302 (2013).
[3] G. Aad *et al.* (ATLAS Collaboration), JHEP **11** (2013) 183.
[4] J. Schukraft, A. Timmins, and S. A. Voloshin, Phys. Lett. B **719**, 394 (2013).
[5] P. Bożek, W. Broniowski, and J. Moreira, Phys. Rev. C **83**, 034911 (2011).
[6] F. Retière and M. A. Lisa, Phys. Rev. C **70**, 044907 (2004).
[7] C. J. Plumberg, C. Shen, and U. Heinz, Phys. Rev. C **88**, 044914 (2013).
[8] J. Jia and P. Huo, Phys. Rev. C **90**, 034915 (2014).
[9] J. Jia and P. Huo, Phys. Rev. C **90**, 034905 (2014).
[10] S. Pratt, Phys. Rev. D **33**, 72 (1986).
[11] G. Bertsch, M. Gong, and M. Tohyama, Phys. Rev. C **37**, 1896 (1988).

Journal of Physics: Conference Series **612** (2015) 012044 doi:10.1088/1742-6596/612/1/012044

Charge dependence of identified two-hadron correlation relative to the reaction plane in Pb-Pb collisions measured with ALICE

Jacobus Onderwaater (for the ALICE Collaboration)

Institut für Kernphysik, Technische Universität Darmstadt, Schlossgrabenstr. 9, 64289 Darmstadt

Research Division and ExtreMe Matter Institute, GSI Helmholtzzentrum für Schwerionenforschung, Planckstr. 1, 64291 Darmstadt

E-mail: jacobus.onderwaater@cern.ch

Abstract. Recently a non-zero charge dependence of two-particle correlation relative to the reaction plane in relativistic heavy-ion collisions was observed by RHIC and LHC experiments. The interpretation of these results is a hot topic of debate in the heavy-ion community because of its possible implication for our understanding of parity violation in strong interactions. We extend the ALICE measurement of the charge dependent two-particle correlation in Pb-Pb collisions at $\sqrt{s_{\rm NN}} = 2.76$ TeV with one identified hadron (pion, kaon, or proton). Correlations are reported as a function of the identified hadron transverse momentum. These new results are important for disentangling contributions from a number of competing physics effects, such as local charge conservation coupled with strong anisotropic flow, flow fluctuations, and possible contribution from parity violation coupled with strong magnetic fields, the so-called chiral magnetic effect.

1. Introduction

Heavy ion collisions offer an opportunity to study possible parity violation in interactions between quarks. The magnitude of global parity violation in strong interactions is constrained to a very small value through measurements of the neutron electric dipole moment [1,2]. On a local scale, parity symmetry can be spontaneously broken via QCD transitions known as instantons and sphalerons [3]. In small regions with non-zero topological charge, these interactions can couple to quarks, leading to a change of chirality and a violation of P- and CP-symmetry. It has been suggested that in the presence of a strong magnetic field, such as generated in a heavy-ion collision, the spin and momentum of produced quarks align with the field lines. In combination with the parity violating effects described above this results in a charge separation relative to the reaction plane called the Chiral Magnetic Effect (CME). This can, in principle, be measured in existing relativistic heavy-ion experiments [4].

One of the main observables to probe the charge separation effect, as proposed by Voloshin [5], is a two particle correlation relative to the reaction plane,

$$c_{\alpha\beta} = \langle \cos(\varphi_\alpha + \varphi_\beta - 2\Psi_{\rm RP}) \rangle, \tag{1}$$

where φ denotes the azimuthal angle, α, β the electric charges of two correlated particles and $\Psi_{\rm RP}$ the reaction plane angle. The reaction plane, which is defined by the impact-parameter vector and the beam direction, is oriented on average perpendicular to the magnetic field. The angular brackets denote the average over all pairs in all collisions. The parity-even correlation in Eq. 1 is sensitive to parity-odd symmetry breaking while cancelling out background sources independent to the reaction plane. This correlation has been extensively studied at RHIC and LHC [6–9], where a significant charge separation was observed. The interpretation of these results is subject to discussion, because various physical reaction plane dependent sources other than CME are likely to contribute to the observed correlation, like local charge conservation and flow fluctuations [10–12]. It is therefore important to develop not only a qualitative but a quantitative understanding of the contributing sources through additional measurements and detailed model calculations. Correlations with identified particles help to elucidate the effect of different particle properties on the charge dependent correlation. Local charge conservation may contribute to the observed charge seperation through modulation by elliptic flow, making it reaction plane dependent. Blast-wave fits have shown that this can explain a significant fraction of the measured correlations [13]. In this case, correlations with identified particles that exhibit different elliptic flow [14] are expected to have a different charge correlation. The focus of this study is the correlation

$$c^{ij}_{\alpha\beta}(p_{\rm T}) = \langle \cos(\varphi^i_\alpha(p_{\rm T}) + \varphi^j_\beta - 2\Psi_{\rm RP})\rangle, \tag{2}$$

where i is a charged pion, kaon or proton and j is an unidentified charged hadron. The observable is studied as a function of $p_{\rm T}$, the transverse momentum of the identified hadron.

2. Analysis details

Around 12 million minimum bias Pb-Pb events at $\sqrt{s_{\rm NN}} = 2.76$ TeV from the 2010 LHC run, collected by the ALICE detector, were analyzed. Details on the ALICE detector and collision trigger can be found in [15]. Tracks reconstructed with the Time Projection Chamber (TPC) in the kinematic range of $0.2 < p_{\rm T} < 5.0$ GeV/c and $|\eta| < 0.8$ are used for the analysis. Particles are identified with the method described in [14] using energy loss in the TPC ($\mathrm{d}E/\mathrm{d}x$) and the speed measurement (β) in the Time of Flight detector. The orientation of the reaction plane is estimated with the second order event plane angle $\Psi_{\rm EP,2}$. The measured correlation relative to the event plane is corrected for the event plane resolution [16]:

$$c^{ij}_{\alpha\beta}(p_{\rm T})\{\rm EP\} = \langle \cos(\varphi^i_\alpha(p_{\rm T}) + \varphi^j_\beta - 2\Psi_{\rm EP,2})\rangle / R_2. \tag{3}$$

The event plane is estimated with the V0-A and V0-C forward multiplicity detectors which respectively cover the pseudo-rapidity region $-3.7 < \eta < -1.7$ and $2.8 < \eta < 5.1$. After applying recentering corrections to the event plane measurement, the 3-subevent method is employed to extract the resolution correction factor R_2 [16]:

$$R^A_2 = \sqrt{\frac{\langle \cos 2(\Psi^A_{\rm EP,2} - \Psi^B_{\rm EP,2})\rangle \langle \cos 2(\Psi^A_{\rm EP,2} - \Psi^C_{\rm EP,2})\rangle}{\langle \cos 2(\Psi^B_{\rm EP,2} - \Psi^C_{\rm EP,2})\rangle}}. \tag{4}$$

In Fig. 1 the resolution correction factor is shown versus collision centrality for a number of ALICE subdetectors. For different limits of the resolution correction, the estimated elliptic flow is [17, 18]

$$v_2\{\rm EP\} = \begin{cases} \langle v_2 \rangle, & \text{for } R_2 \to 1 \\ \sqrt{\langle v_2^2 \rangle}, & \text{for } R_2 \to 0 \end{cases}. \tag{5}$$

Because of flow fluctuations, the estimate of v_2 with the event plane method can significantly differ depending on the event plane resolution correction. The right side of Fig. 1 shows that

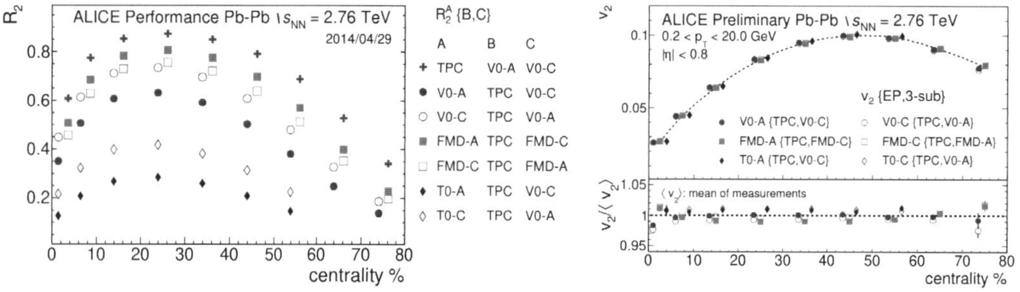

Figure 1. (left) Second order event plane resolution correction factor R_2 for the TPC, V0-A/C, FMD-A/C and T0-A/C subdetectors estimated with the 3-subevent method. (right) Estimation of $v_2\{\mathrm{EP}\}$ with six different event plane estimators.

measurements of v_2 with different estimations from ALICE subdetectors are consistent within $\sim 4\%$ [19], and the ordering of the magnitude of the v_2 measurements is consistent with Eq. 5. The two particle correlation in Eq. 2 is measured with respect to V0-A and V0-C event plane estimates in 10% centrality bins. For the final result the two event plane measurements were combined, as well as the 10% centrality ranges to a 10-30% and 30-50% centrality range.

3. Results

In Fig. 2 the correlation Eq. 2 is shown for the 10-30% and 30-50% centrality range. The different charge combinations of correlated particle pairs are combined in $\alpha = \beta$ (same) and $\alpha = -\beta$ (opp.) charge pairs. For the proton correlation at $p_{\mathrm{T}} < 1$ GeV/c only anti-protons are used in order to avoid contamination from knock-out protons from detector material. The trends for hadron, pion, kaon and proton correlations for like-sign and unlike-sign are similar in both centrality ranges, while the magnitude of the observed correlations are approximately three times stronger in more peripheral collisions. A clear particle species dependence is observed in the unlike-sign correlations, which is close to zero for pions, but increases towards negative

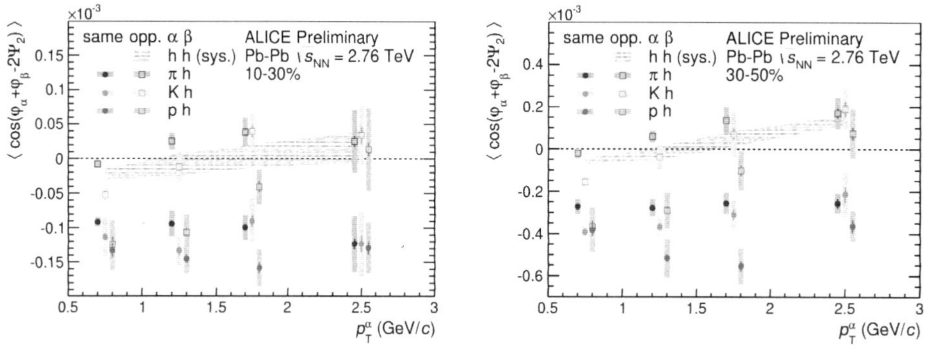

Figure 2. Results for $c^{ij}_{\alpha\beta}$ defined by Eq. 2 in the 10-30% (left) and the 30-50% (right) centrality range. The fine hatched areas show the unlike-sign hadron correlation and the coarse hatched areas the like-sign hadron correlation. The unlike-sign correlations for identified hadrons are shown in open squares, and the like-sign correlations in full squares. Pions are shown in blue, kaons in green, and protons in red. The systemetic error is indicated with the colored bands.

values at low transverse momentum for kaons and protons. The like-sign correlations do not have a significant slope as a function of p_T. Although an ordering appears from pions to protons in the magnitude of the correlations, the differences are not significant as for the unlike-sign correlations.

Several checks are made to evaluate the systematic uncertainty of the measurement. Two track reconstruction approaches, using TPC standalone and a combined TPC and Inner Tracking System information, are considered. Requirements on the track reconstruction quality and particle identification efficiency and purity were studied similar to [14]. The effect of magnet polarity, which determines the direction of curvature for charged particles, was checked for both configurations, as well as individual charge combinations (++,--,+-,-+). Residual effects from detector non-uniformity were checked by replacing the cosine in Eq. 2 by a sine, and different event plane estimators (V0-A/C, FMD-A/C). All contribute significantly to the final systematic uncertainty, which has been calculated for each p_T bin with deviations of the mentioned systematic checks added in quadrature. In the intermediate p_T region the total relative systemetic uncertainty is around 5% for unindentified hadron correlations and 20% for the correlations with an identified hadron. Finally a weigthed average was made of the identified particle correlations using identified particle spectra and demonstrated consistency with the unidentified hadron correlation.

4. Summary

In this study a new charge-dependent measurement with identified particles is presented. The charge-dependent correlation $\langle \cos(\varphi^i_\alpha(p_T) + \varphi^j_\beta - 2\Psi_{RP}) \rangle$ (Eq. 2) is measured with a charged pion, kaon or proton (i) and an unidentified charged hadron (j). In the opposite sign correlations a significant dependence on particle species is observed at low p_T. This new result signals that effects contributing to the previously observed charge separation with unidentified hadrons have a different contribution depending on particle species. Model calculations, in particular of the local charge conservation with identified particles, and further measurements are required to quantify and describe the observed charge separation effects, and achieve a detailed understanding of the charge-dependent correlations and possible contributions from CME.

References

[1] Pospelov M and Ritz A 1999 *Phys.Rev.Lett.* **83** 2526–2529
[2] Baker C A *et al.* 2006 *Phys.Rev.Lett.* **97** 131801
[3] Kharzeev D, Pisarski R and Tytgat M H 1998 *Phys.Rev.Lett.* **81** 512–515
[4] Kharzeev D 2006 *Phys.Lett.* B **633** 260–264
[5] Voloshin S A 2004 *Phys.Rev.* C **70** 057901
[6] Abelev B *et al.* (STAR Collaboration) 2009 *Phys.Rev.Lett.* **103** 251601
[7] Abelev B *et al.* (STAR Collaboration) 2010 *Phys.Rev.* C **81** 054908
[8] Abelev B *et al.* (ALICE Collaboration) 2013 *Phys. Rev. Lett. 110,* **012301**
[9] Adamczyk L *et al.* (STAR Collaboration) 2013 *Phys.Rev.* **C88** 064911
[10] Pratt S, Schlichting S and Gavin S 2011 *Phys.Rev.* C **84** 024909
[11] Bzdak A, Koch V and Liao J 2010 *Phys.Rev.* C **81** 031901
[12] Bzdak A, Koch V and Liao J 2011 *Phys.Rev.* C **83** 014905
[13] Hori Y (ALICE Collaboration) 2013 *Nucl.Phys.* A **904-905** 475c–478c
[14] Abelev B B *et al.* (ALICE Collaboration) 2014 Elliptic flow of identified hadrons in Pb-Pb collisions at $\sqrt{s_{NN}}$ = 2.76 TeV (*Preprint* 1405.4632)
[15] Aamodt K *et al.* (ALICE Collaboration) 2008 *JINST* **3** S08002
[16] Poskanzer A M and Voloshin S 1998 *Phys.Rev.* C **58** 1671–1678
[17] Alver B *et al.* 2008 *Phys.Rev.* C **77** 014906
[18] Ollitrault J Y, Poskanzer A M and Voloshin S A 2009 *Phys.Rev.* C **80** 014904
[19] Onderwaater J (for the ALICE Collaboration) 2014 QM2014 poster Estimates of heavy-ion collision symmetry planes with the alice detector https://indico.cern.ch/event/219436/session/2/contribution/312

Hot Quarks 2014 IOP Publishing
Journal of Physics: Conference Series **612** (2015) 012045 doi:10.1088/1742-6596/612/1/012045

Pion-kaon femtoscopy in Au+Au collisions at STAR

Katarzyna Poniatowska for the STAR collaboration
Warsaw University of Technology, Koszykowa 75, 00-662 Warsaw, Poland
E-mail: poniatowska@if.pw.edu.pl

Abstract. In non-identical particle correlations, e.g. pion-kaon femtoscopy, one can obtain information about source size and asymmetry in emission processes of pions and kaons. Such asymmetry give us knowledge of which type of particles is emitted first/second and/or from which region of the source. The studies of non-identical particle femtoscopy for Beam Energy Scan energies give us the opportunity to study how the source size and asymmetry in particle emission depend on the initial conditions of the collision. It also allows one to examine these parameters in the vicinity of the theoretical critical point. In these proceedings, we present STAR results of pion-kaon femtoscopy at mid-rapidity in Au+Au collisions at $\sqrt{s_{NN}}$ = 7.7, 19.6 and 39 GeV.

1. Introduction

Femtoscopy, the momentum correlations of two or more particles with small relative momenta in their center-of-mass are used to measure source sizes of the order 10^{-15} m. With this technique, the size of the particle-emission source formed during heavy-ion collisions can be measured. Correlations of non-identical particles provide information about the space-time asymmetry between two sources of particle emission such as the relative position of the sources and the average times of particle emissions [1]. The spacial shift is coming from the transverse flow of particles with different masses [2]. By analyzing the size of the emission source and the asymmetry of the space-time at the Beam Energy Scan (BES) program energies, we will examine the characteristics of the source as a function of the initial conditions after collision. As part of the BES program, the STAR (Solenoidal Tracker at RHIC) experiment collected Au+Au data at $\sqrt{s_{NN}} = 7.7, 11.5, 14.5, 19.6, 27, 39$ GeV. These proceedings show an analysis at center-of-mass energies 7.7, 19.6 and 39 GeV. The results for $\sqrt{s_{NN}} = 19.6$ GeV may be particularly interesting because they contain information from the region possibly near the critical point, as hinted at by the proton kurtosis measurement [3][4]. Near a critical point, fluctuations may occur of strangeness [5], which may manifest as a smaller number of pairs and therefore cause abnormal correlation functions.

2. Methodology of non-identical particle femtoscopy

Non-identical particle correlations result from Final State Interactions: Coulomb and strong force. The dominant influence on the correlation of pions and kaons is the Coulomb force [6].

To calculate the correlation function $C(k^*)$ one should calculate the ratio of k^* distribution of pairs taken from the same event (correlated pairs) and k^* distribution constructed with pairs taken from different events (uncorrelated pairs). The k^* is the magnitude of the three-momentum of either particle in the pair rest frame.

The correlation functions can be calculated in two groups of pairs, which are divided depending on the space-time separation between pion and kaon emission points. The first group represents the case where the pions catch up with the kaons, whereas the second case is when the pions move away from the kaons. From each group, a different correlation function is constructed, $C_+(k^*)$ and $C\_(k^*)$, where the sign index reflects the sign of $\vec{v} \cdot \vec{k}^*_\pi$, with $\vec{v}$ the pair velocity and $\vec{k}^*_\pi$ the pion momentum vector in the PRF (pair rest frame) [7]. When both correlation functions are identical, the average space-time emission points of pions and kaons coincide. If the correlation function C_+ shows a larger deviation from unity than $C\_$ it can be concluded that the pions are emitted closer to the center of the source than kaons. In this case pions with a larger velocity will tend to catch up with the kaons, and the Coulomb correlation strength will be enhanced compared with the case where the pions are slower than the kaons [6].

The longitudinally comoving system (LCMS) is used in this analysis. In the LCMS, the *out* axis is parallel to the pair velocity, the *long* axis is the beam axis and the *side* axis is perpendicular to the other two. Correlation functions are prepared for three projections of $\vec{k}^*_\pi$: k_{side}, k_{long} and k_{out}. The corresponding projections of the three-vector $\vec{r}^*$, the relative distance between the particle freeze-out points in the PRF, are r^*_{out}, r^*_{side}, and r^*_{long}. Due to azimuthal symmetry and symmetry about $mid-rapidity$: $\langle r_{side}\rangle = \langle r_{long}\rangle = 0$ [8]. Thus $C_+/C\_$ ("double ratio"), defined with respect to the signs of k^*_{side} and k^*_{long}, should not deviate from unity.

3. Results

The cuts used for selecting pions and kaons measured in Au+Au collisions at $\sqrt{s_{NN}} = 7.7$, 19.6 and 39 GeV are shown in Table 1. In this analysis, we use information from the Time Projection Chamber (TPC) and the time-of-flight (TOF) detector [9][10]. If the particle had a momentum greater than 0.55 GeV/c it had to be registered by both detectors. In identification of particles registered in TPC detector, the particles that were within 3 standard deviations (N σ) of the Bichsel function were selected. Tracks with the distance of closest approach (DCA) to the primary vertex less than 3 cm were selected. Only statistical uncertainties are represented.

Table 1. Selection criteria used in the analysis at $\sqrt{s_{NN}} = 7.7$, 19.6 and 39 GeV. The TPC particle identification cuts for $\sqrt{s_{NN}} = 7.7$ and 19.6 GeV energies are different from the 39 GeV one.

	7.7 GeV / 19.6 GeV	39 GeV
analysed	3.4 M / 17.5 M events	91.2 M events
π momentum	[0.1, 1.2] GeV/c	[0.15, 0.7] GeV/c
K momentum	[0.1, 1.2] GeV/c	[0.3, 0.8] GeV/c
N σ for π and K	< 3.0	< 3.0
π m^2	[0.0176, 0.022] GeV^2/c^4	[0.017, 0.026] GeV^2/c^4
K m^2	[0.23, 0.26] GeV^2/c^4	[0.22, 0.27] GeV^2/c^4
$\mid\eta\mid$	< 0.5	< 0.5
DCA	< 3 cm	< 3 cm

3.1. The comparison of correlation functions for Au + Au collisions at 7.7, 19.6 and 39 GeV

The magnitude and the shape of the correlation functions for Au+Au collisions at $\sqrt{s_{NN}} = 7.7$, 19.6 and 39 GeV are the same (Fig. 1). The similarity of these functions (taking into account the statistical errors) means, that the source created in collisions at these three energies have the same size. The work on the estimation of the source is still in progress.

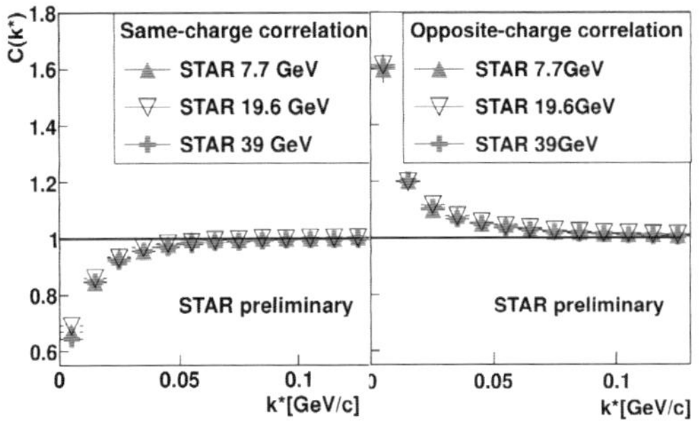

Figure 1. The comparison of pion-kaon correlation functions for $Au + Au$ collisions at $\sqrt{s_{NN}}$ = 7.7, 19.6 and 39 GeV for the minimum-bias collisions (0 – 80%). Results are presented for two types of correlations: same-charge correlation ($\pi^+K^+ + \pi^-K^-$), opposite-charge correlation ($\pi^+K^- + \pi^-K^+$).

3.2. The asymmetry in $Au + Au$ collisions at $\sqrt{s_{NN}} = 7.7$, 19.6 and 39 GeV

Within the larger uncertainties at $\sqrt{s_{NN}}$ = 7.7 GeV, the "double ratio" functions at $\sqrt{s_{NN}}$ = 7.7, 19.6 and 39 GeV agree well and show similar trends (Fig. 2). This indicates that the asymmetry in the emission process is observed for all the presented energies and is consistent. The same trend in asymmetry was also observed for higher energy, e.g., $\sqrt{s_{NN}}$ = 130 GeV [11]. Sizeable transverse flow is observed even for the lowest BES energy. Work on defining the size of the asymmetry is in progress.

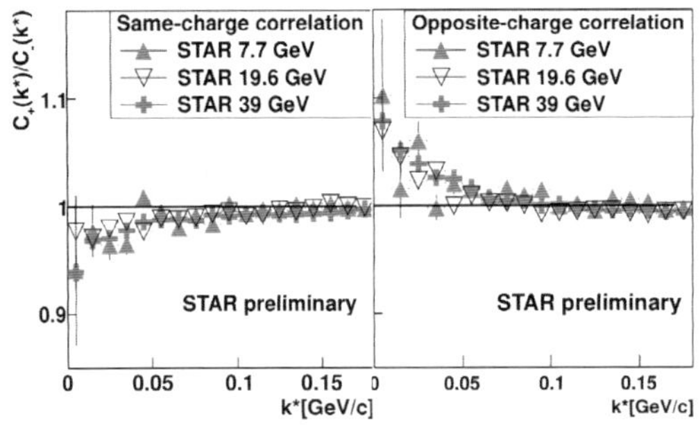

Figure 2. Comparison of C_+/C_- functions for pions and kaons for $Au + Au$ minimum-bias (0 – 80%) at $\sqrt{s_{NN}}$ = 7.7, 19.6 and 39 GeV for the "out" direction. Results are presented for both types of correlations: same-charge correlation ($\pi^+K^+ + \pi^-K^-$), opposite-charge correlation ($\pi^+K^- + \pi^-K^+$).

4. Summary

In these proceedings, we present an analysis of femtoscopy correlation functions and asymmetry calculations for Au+Au collisions at $\sqrt{s_{NN}}$ = 7.7, 19.6 and 39 GeV. It is shown that the source sizes of these energies look similar and that such behaviour is observed for same-charge and opposite-charge correlations. Further calculations have to be made to get exact source sizes. The asymmetry in the emission is observed for all investigated energies here. The shape of the C_+/C_- function indicates that most of the pions are emitted closer to the system's center or/and later than most of the kaons for these collisions. This indicates that transverse flow is sizeable even for low energy collisions such as $\sqrt{s_{NN}} = 7.7$ GeV. At this stage, we do not observe abnormal correlation functions and asymmetry functions at $\sqrt{s_{NN}} = 19.6$ GeV.

5. Acknowledgements

This work was supported by the Grants: UMO-2012/07/D/ST2/02123 and supported by National Science Centre, Poland.

References

[1] R. Lednicky, Nucl. Phys. A774 **189-198**, 2006
[2] STAR Collaboration, Nucl.Phys. A715 **591-594**, 2003
[3] STAR Collaboration, Phys. Rev. Lett. **112**, 032302, 2014
[4] STAR Collaboration, "Studying the Phase Diagram of QCD Mattter at RHIC", SN0598, 2014
[5] J. Rafelski, B. Muller 1982 Phys. Rev. Lett. **48** 1066
[6] R. Lednicky, V.L. Lyuboshitz, B. Erazmus, D. Nouais, Physics Letters B **373**, 30-34, 1996
[7] A. Kisiel, *Studies of non-identical meson-meson correlation at low relative velocities in relativistic heavy-ion collisions registered in the STAR experiment*, PhD Thesis, Warsaw Univeristy of Technology, Dec 2004
[8] R. Lednicky, Proc. 8th Int. Workshop on Multiparticle production, Correlations and Fluctuations '98, p. 148-155, 1998
[9] M. Anderson et al.,[STAR Collaboration], Nucl. Instrum. Meth. A **499**, 659 (2003)
[10] STAR Time-of-Flight Proposal: http://www.star.bnl.gov/STAR/tof/publications/TOF 20040524.pdf.
[11] STAR Collaboration, Phys. Rev. Lett. **91**, 262302, 2003

Hot Quarks 2014 IOP Publishing
Journal of Physics: Conference Series **612** (2015) 012046 doi:10.1088/1742-6596/612/1/012046

Higher order fluctuations of conserved charges in heavy ion collisions

Miki Sakaida, Masayuki Asakawa, Masakiyo Kitazawa
Department of Physics, Osaka University, Toyonaka, Osaka 560-0043, Japan
E-mail: sakaida@kern.phys.sci.osaka-u.ac.jp

Abstract.
We investigate the effect of the global charge conservation on the cumulants of conserved charges in relativistic heavy ion collisions in a finite rapidity window by studying the time evolution of cumulants in the hadronic medium. It is argued that the global charge conservation does not affect the experimental result of the net-electric charge fluctuation observed by ALICE, because of the finite diffusion distance of charged particles in the hadronic stage.

1. Introduction

Bulk fluctuations of conserved charges measured by event-by-event analyse s in relativistic heavy ion collisions, especially higher order cumulants, are expected to be promising observables to characterize properties of the hot medium created by collision events and to find the QCD critical point [1, 2, 3]. Recently, experimental investigations of these observables in heavy ion collisions have been actively performed at the RHIC [4] and LHC [5].

One of the important properties of these fluctuations in heavy ion collisions is that experimentally-observed fluctuations had experienced non-equilibrium evolution, because of the dynamical nature of the hot medium created in heavy ion collisions [3, 6]. In fact, the recent experimental result on the rapidity window dependence of net electric charge fluctuation by the ALICE Collaboration at the LHC [5] shows that the value of the second order cumulant of the net electric charge is significantly suppressed compared with the one in the equilibrated hadronic medium. This result is reasonably understood if one interprets the suppression as a sign of the survival of the small fluctuation created in the primordial deconfined medium [2, 3, 6].

However, there exists another mechanism called the global charge conservation (GCC) to cause the suppression of fluctuations compared with the equilibrated value. If one counts a conserved charge in the total system generated by experiments, there are no event-by-event fluctuations because of the charge conservation. This fact is referred to as the GCC [1]. In Ref. [7], on the assumption that the equilibration is established in the final state of the heavy ion collision, the magnitude of this effect is estimated. It, however, should be noted that the suppression of charge fluctuation observed at ALICE [5] is stronger than the one estimated in Ref. [7]. This result shows that there exists another contribution for the suppression in addition to the GCC, such as the nonequilibrium effect as originally addressed in Ref. [2]. The effect of the GCC at LHC energy, therefore, has to be discussed again with the nonthermal effects incorporated.

In the present study, we investigate the effect of the GCC on cumulants of conserved charges under such nonequilibrium circumstances, by describing the time evolution of fluctuations in a finite volume system [8]. We also discuss the effects of the GCC on higher-order cumulants of conserved charges for the first time. By comparing our result with the experimental one at ALICE in Ref. [5], we find that the effect of GCC on the net electric charge fluctuation in the rapidity window observed by this experiment is almost negligible.

2. Stochastic formalism to describe diffusion of hadrons

In heavy ion collisions with sufficiently large $\sqrt{s_{\rm NN}}$, the hot medium created at the mid-rapidity has an approximate boost invariance. We denote the net number of a conserved charge per unit space-time rapidity η as $n(\eta,\tau)$ with the proper time τ, and set $\tau=\tau_0$ at hadronization. Because of the local charge conservation, the probability distribution of $n(\eta,\tau_0)$ at hadronization inherits from the one that existed in the deconfined medium [1]. After hadronization, particles diffuse and rescatter, and the distribution of $n(\eta,\tau)$ continues to approach the one of the equilibrated hadronic medium until kinetic freezeout at $\tau=\tau_{\rm fo}$.

In this study, to describe the time evolution of fluctuation of the particle number at mid-rapidity $Q(\Delta\eta,\ \tau)=\int_{-\Delta\eta/2}^{\Delta\eta/2}d\eta\ n(\eta,\tau)$ in hadronic medium, with the size of the rapidity window to count the charge number $\Delta\eta$, until $\tau=\tau_{\rm fo}$ we adopt the diffusion master equation [3, 8]. In this model, we divide the system with the total rapidity length $\eta_{\rm tot}$ into M discrete cells with an equal finite length $a=\eta_{\rm tot}/M$. We then consider a single species of particles for the moment, and denote the particle number existing in the mth cell as n_m and the probability distribution that each cell contains n_m particles as $P(\boldsymbol{n},\tau)$ with $\boldsymbol{n}=(n_0,n_1,\cdots,n_m,\cdots,n_{M-2},n_{M-1})$. Finally, we assume that each particle moves to the adjacent cells with a probability $\gamma(\tau)$ per unit proper time, as a result of microscopic interactions. The probability $P(\boldsymbol{n},\tau)$ then obeys the diffusion master equation,

$$\partial_\tau P(\boldsymbol{n},\tau)=\gamma(\tau)\sum_{m=0}^{M-1}[(n_m+1)\{P(\boldsymbol{n}+\boldsymbol{e}_m-\boldsymbol{e}_{m+1},\tau)+P(\boldsymbol{n}+\boldsymbol{e}_m-\boldsymbol{e}_{m-1},\tau)\}-2n_mP(\boldsymbol{n},\tau)]],\tag{1}$$

where $\boldsymbol{e}_m$ is a unit vector whose all components are zero except for mth one, which takes unity.

The average and the Gaussian fluctuation of $n(\eta,\tau)$ in Eq. (1) in the continuum limit, $a\to 0$, agree with the ones of the stochastic diffusion equation, [6]

$$\partial_\tau n(\eta,\tau)=D(\tau)\partial_\eta^2 n(\eta,\tau)+\partial_\eta\xi(\eta,\tau),\tag{2}$$

when one sets the diffusion coefficient as $D(\tau)=\gamma(\tau)a^2$ [3]. Here, $\xi(\eta,\tau)$ is the temporarily-local stochastic force. It is known that the fluctuation of Eq. (2) in equilibrium becomes of Gaussian [3, 8]. On the other hand, the diffusion master equation Eq. (1) can give rise to nonzero higher order cumulants in equilibrium [3]. This is the reason why we employ the diffusion master equation instead of Eq. (2).

3. Solution of diffusion master equation

Now, we determine the time evolution of cumulants by solving Eq. (1) and taking the continuum limit $a\to 0$. In order to take account of the finiteness of the size of the hot medium, we set reflecting boundaries at $\eta=\pm\eta_{\rm tot}/2$. After some algebra [8], the cumulants of $Q(\tau)$ with the fixed initial condition $n(\eta,\tau_0)=N(\eta)$ are calculated to be

$$\langle(Q(\Delta\eta,\ \tau))^n\rangle_{\rm c}=\int_{-\eta_{\rm tot}/2}^{\eta_{\rm tot}/2}d\eta\ N(\eta)H_n(\eta),\tag{3}$$

where $H_2(\eta) = I(\eta) - I(\eta)^2$, $H_4(\eta) = I(\eta) - 7I(\eta)^2 + 12I(\eta)^3 - 6I(\eta)^4$ with

$$I(\eta) = \frac{\Delta\eta}{\eta_{\rm tot}} \sum_{k=-\infty}^{\infty} \cos\left(\frac{\pi k \eta}{\eta_{\rm tot}}\right) \sin\left(\frac{\pi k \Delta\eta}{2\eta_{\rm tot}}\right) \cos\left(\frac{\pi k}{2}\right) \exp\left[-\frac{1}{2}\left(\frac{\pi k d(\tau)}{\eta_{\rm tot}}\right)^2\right] / \left(\frac{\pi k \Delta\eta}{2\eta_{\rm tot}}\right). \quad (4)$$

Here, $d(\tau) = \sqrt{2\int_{\tau_0}^{\tau} d\tau' D(\tau')}$ is the average diffusion length of each Brownian particle.

Next, we extend the above results to the cases with general initial conditions with non-vanishing initial fluctuations. In the following, we also extend the formula to treat cumulants of the difference of densities of two particle species, $n_1(\eta,\tau)$ and $n_2(\eta,\tau)$, $Q_{(\rm net)}(\Delta\eta,\ \tau) = \int_{-\Delta\eta/2}^{\Delta\eta/2} d\eta\ (n_1(\eta,\tau) - n_2(\eta,\tau))$, in order to consider cumulants of conserved charges in QCD.

Next, let us constrain the initial condition. Because we consider a finite system, the initial condition at $\tau = \tau_0$ should be determined in accordance with the GCC. In order to obtain the initial conditions which conform the GCC, here we model the initial configuration by an equilibrated free classical gas in a finite volume. In this system, one can obtain the correlation functions by taking the continuum limit of the multinomial distribution [8]. On the other hand, we assume that the total charge number at $\tau = \tau_0$ in the total system obeys the one given by the grand canonical ensemble [3]. Using these initial conditions and Eq. (3), one obtains the higher order cumulants of $Q_{(\rm net)}(\Delta\eta,\ \tau)$.

4. Effects of the GCC

Next, let us study how the GCC affects the rapidity window dependence of the cumulants of conserved charges. To describe the time evolution of cumulants, we use the dimensionless parameters $T \equiv d(\tau)/\eta_{\rm tot}$. To make the argument simple, we consider the $\Delta\eta$ dependence for the fixed initial condition, i.e., all fluctuations vanish at $\tau = \tau_0$. In Fig. 1, we show $\langle (Q_{(\rm net)}(\Delta\eta))^n \rangle_{\rm c} / \langle Q_{(\rm tot)}(\Delta\eta) \rangle_{\rm c}$ for $n = 2$ (left) and 4 (right), as a function of $\Delta\eta/\eta_{\rm tot}$ with the total length $\eta_{\rm tot}$ for five values of T including infinity. For comparison, we also show the result with an infinite volume by the dashed and dotted lines. The figure shows that $\langle (Q_{(\rm net)}(\Delta\eta))^2 \rangle_{\rm c}$ vanishes at $\Delta\eta/\eta_{\rm tot} = 1$ for each T, namely $\Delta\eta = \eta_{\rm tot}$, which is a trivial consequence of the GCC. On the other hand, as $\Delta\eta$ becomes smaller from this value with fixed T, the result approaches the one with infinite volume. The effect of the GCC vanishes almost completely except for the range,

$$\frac{\eta_{\rm tot} - \Delta\eta}{2} \lesssim d(\tau). \quad (5)$$

In this expression, the left-hand side is the distance between the left (right) boundary and the left (right) edge of the rapidity window, while the right-hand side is the diffusion length of each Brownian particle. When Eq. (5) is satisfied, particles which are reflected by one of the boundaries at least once can enter the rapidity window. Therefore, the existence of the boundaries can affect the fluctuations of conserved charges in the rapidity window. On the other hand, when the condition Eq. (5) is not satisfied, particles inside the rapidity window do not know the existence of the boundaries; in other words, the fact that the system is finite. In the latter case, therefore, the fluctuations in the rapidity window are free from the effect of the GCC.

With fixed $\eta_{\rm tot}$, larger T corresponds to larger τ. Figure 1(left) shows that as T becomes larger the fluctuation increases and approaches a linear function representing the equilibration, which is consistent with the result obtained in Ref. [7].

From Fig. 1 (right), which shows the fourth order cumulants for several values of T, one obtains completely the same conclusion on the effect of the GCC: The effect of the GCC is visible only for large values of $\Delta\eta/\eta_{\rm tot}$.

In this analysis, it is assumed that the particle current vanishes on the two boundaries. This

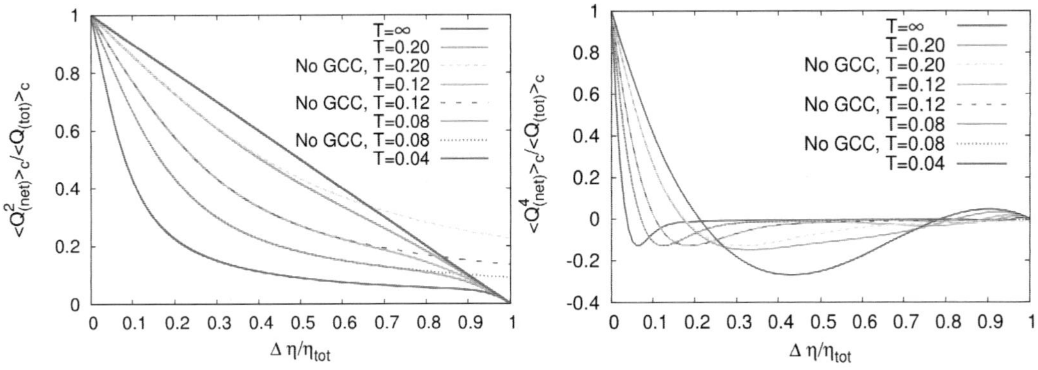

Figure 1. Second (left panel) and forth (right panel) order cumulants of conserved charges without initial fluctuation as a function of $\Delta\eta/\eta_{\rm tot}$ with five values of the parameter T. The corresponding results in an infinite volume are also plotted for several values of T.

assumption would not be suitable to describe the hot medium created by heavy ion collisions, since the hot medium does not have such hard boundaries but only baryon rich regions. From the above discussion, however, it is obvious that the effect of boundaries does not affect the fluctuations in the rapidity window unless the condition Eq. (5) is satisfied irrespective of the types of the boundaries.

We finally inspect the $\Delta\eta$ dependence of the net electric charge fluctuations observed at ALICE [5] in more detail. In order to estimate the effect of the GCC, we must determine the magnitude of $\eta_{\rm tot}$. From the pseudo-rapidity dependence of charged-particle yield at LHC energy, $\sqrt{s_{\rm NN}} = 2.76$ TeV, in Ref. [9], we take the value $\eta_{\rm tot} = 8$ in the following. With this $\eta_{\rm tot}$, and the maximum rapidity coverage of the 2π detector, TPC, of ALICE, $\Delta\eta = 1.6$ [5], one obtains $\Delta\eta/\eta_{\rm tot} = 0.2$. From Fig. 1, one can immediately conclude that the suppression of $\langle (Q_{(\rm net)}(\Delta\eta))^2\rangle_c$ in this experiment cannot be explained solely by the naive formula of the GCC [7]. From the discussion in the previous subsections, it is also concluded that the effect of the GCC on the diffusion in the hadronic stage is negligible in this experimental result.

References

[1] Koch V 2008 Hadronic Fluctuations and Correlations *Preprint* arXiv:0810.2520 [nucl-th]

[2] Asakawa M, Heinz U W, and Müller B 2000 Fluctuation probes of quark deconfinement *Phys. Rev. Lett.* **85** 2072

[3] Kitazawa M, Asakawa M, and Ono H 2014 Non-equilibrium time evolution of higher order cumulants of conserved charges and event-by-event analysis *Phys. Lett. B* **728** 386

[4] Adamczyk L, *et al.* [STAR Collaboration] 2014 Beam energy dependence of moments of the net-charge multiplicity distributions in Au+Au collisions at RHIC *Phys. Rev. Lett.* **113** 092301

[5] Abelev B, *et al.* [ALICE Collaboration] 2013 Net-Charge Fluctuations in Pb-Pb collisions at $\sqrt{s_{\rm NN}} = 2.76$ TeV *Phys. Rev. Lett.* **110** 152301

[6] Shuryak E V and Stephanov M A 2001 When can long range charge fluctuations serve as a QGP signal? *Phys. Rev. C* **63** 064903

[7] Bleicher M, Jeon S, and Koch V 2000 Event-by-event fluctuations of the charged particle ratio from nonequilibrium transport theory *Phys. Rev. C* **62** 061902

[8] Sakaida M, Asakawa M and Kitazawa M 2014 Effects of global charge conservation on time evolution of cumulants of conserved charges in relativistic heavy ion collisions *Preprint* arXiv:1409.6866 [nucl-th]

[9] Abbas E, *et al.* [ALICE Collaboration] 2013 Centrality dependence of the pseudorapidity density distribution for charged particles in Pb-Pb collisions at $\sqrt{s_{\rm NN}} = 2.76$ TeV *Phys. Lett. B* **726** 610

Hot Quarks 2014 IOP Publishing
Journal of Physics: Conference Series **612** (2015) 012047 doi:10.1088/1742-6596/612/1/012047

Recent results on event-by-event fluctuations in ALICE at the LHC

Satyajit Jena, on behalf of the ALICE Collaboration
University of Houston, Texas, USA
E-mail: sjena@cern.ch

Abstract. Non-statistical event-by-event fluctuations in relativistic heavy-ion collisions have been proposed as a probe of phase instabilities near the QCD phase transition. In a thermodynamical picture of the strongly interacting system formed in heavy-ion collisions, the fluctuations of the mean transverse momentum, mean multiplicity fluctuations, the balance function, higher moments of net-particle multiplicity distributions, etc., are related to the fundamental properties of the system, hence they may reveal information about the QCD phase transition. We present and discuss recent results on event-by-event measurements at LHC energies.

1. Introduction

Quantum chromodynamics (QCD) [1] predicts that at a sufficiently high energy density, on the order of 0.5 GeV/fm^3, a deconfined state of quarks and gluons is produced, the so-called quark gluon plasma (QGP). Heavy-ion collisions at ultra-relativistic energies can produce this state of matter, which is characterized by high temperatures and energy densities, where the degrees of freedom are no longer determined by hadrons, but by their constituents, the quarks and gluons [2]. Many different observables are used to verify the existence of and subsequently study the properties of the QGP. In this article, we will discuss on the net-charge fluctuations, the balance functions and the mean $\langle p_{\rm T} \rangle$ fluctuation results obtained with the ALICE [3] detector at the Large Hadron Collider (LHC) at CERN. The central detectors at ALICE cover a large pseudorapidity region with good momentum resolution as well as good particle identification capabilities. This gives us an excellent opportunity to study the fluctuations and correlations of physical observables on an event-by-event basis. In these analyses, the Time Projection Chamber (TPC) [3] is used for selecting tracks, the Inner Tracking System (ITS) [3] is used for vertexing and triggering and the VZERO [3] scintillator hodoscopes are used for estimating centrality as well as for triggering.

2. Net-charge fluctuations

The net–charge fluctuations are sensitive to the number of charges in the system, thus the fluctuations in the QGP, with fractionally charged partons, are significantly different from those of a hadron gas with unit charged particles. The net–charge fluctuations can be expressed by the quantity D [4], defined as:

$$D = 4\frac{\langle \delta Q^2 \rangle}{N_{\rm ch}} \approx \nu_{(+-,dyn)} \times \langle N_{\rm ch} \rangle + 4 \quad (1)$$

where $\langle \delta Q^2 \rangle$ is the variance of the net–charge Q with $Q = N_+ - N_-$ and $N_{\rm ch} = N_+ + N_-$. Here N_+ and N_- are the numbers of positive and negative particles. In the experiment, the net–charge fluctuations are best studied by calculating the quantity $\nu_{+-,dyn}$:

$$\nu_{(+-,dyn.)} = \frac{\langle N_+(N_+ - 1)\rangle}{\langle N_+\rangle^2} + \frac{\langle N_-(N_- - 1)\rangle}{\langle N_-\rangle^2} - 2\frac{\langle N_- N_+\rangle}{\langle N_-\rangle\langle N_+\rangle}. \tag{2}$$

Taking into account global charge conservation and finite acceptance, one obtains the corrected value of $\nu^{\rm corr}_{+-,\rm dyn} = \nu_{+-,\rm dyn} + \frac{4}{\langle N_{\rm total}\rangle}$ where $\langle N_{\rm total}\rangle$ is the average total number of charged particles produced over the full phase space. In the figure 1 (a), the variation of $\nu_{(+-,dyn.)} \times N_{ch}$ with

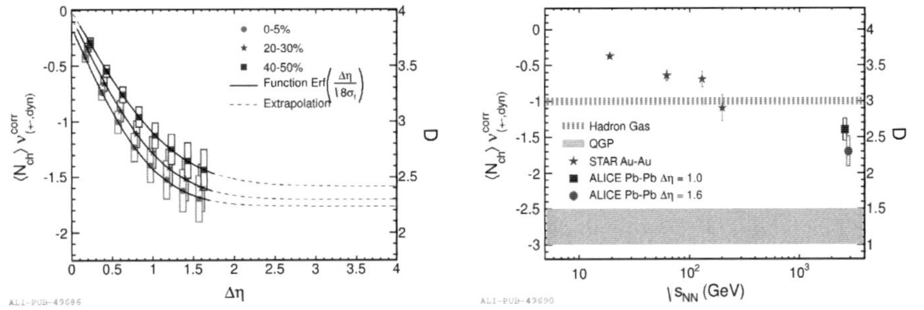

(a) (b)

Figure 1. (a): $\Delta\eta$ dependence of net–charge fluctuation, expressed in term $\langle N_{\rm ch}\rangle\nu^{\rm corr}_{+-,\rm dyn}$ (left axis) and D (right axis) for three different centrality percentiles. (b): The energy dependence of the net–charge fluctuations, expressed in $\langle N_{\rm ch}\rangle\nu^{\rm corr}_{+-,\rm dyn}$ (left axis) and D (right axis) of ALICE at $\Delta\eta = 1$ and $\Delta\eta = 1.6$ along with RHIC [5] results. The expectations for a hadron resonance gas and the QGP are shown as shaded bands [4]. Figures are taken from [6].

respect to $\Delta\eta$ is studied. The relative value of $\nu_{(+-,dyn.)} \times N_{ch}$ reduces smoothly with increasing $\Delta\eta$ window. This behavior was predicted earlier in Ref. [7] and was attributed to the dissipation of the primordial signal arising from hadronic diffusion during the evolution from quark gluon plasma stage to the freeze-out stage. The data points are fitted with an error function of the form $\mathrm{erf}(\Delta\eta/\sqrt{8}\sigma_f)$, representing the diffusion process. In figure 1(b), the energy dependence of the net–charge fluctuations is shown in terms of $\nu_{(+-,dyn.)} \times N_{ch}$ and D (left– and right–axis, respectively). By confronting the measured value with the theoretically predicted fluctuations [4, 7], it is observed that the ALICE results are within the limits of the QGP and the HG scenarios.

3. Balance Function

The system that is produced in a heavy-ion collision undergoes an expansion, during which it exhibits collective behavior and can be described in terms of hydrodynamics. It was proposed to measure the creation time of particles via the correlations between positively and negatively charged pairs as a function of rapidity by the balance function[8]. The definition of the balance function for the pseudorapidity difference $\Delta\eta$ (similar equation for $B(\Delta\varphi)$) reads:

$$B_{+-}(\Delta\eta) = \frac{1}{2}\Big(C_{+-}(\Delta\eta) + C_{-+}(\Delta\eta) - C_{--}(\Delta\eta) - C_{++}(\Delta\eta)\Big). \tag{3}$$

Each term of Eq. 3, is corrected for detector and tracking inefficiencies as well as for acceptance effects and can be written as $C_{ab} = (N_{ab}/N_b)/f_{ab}$. The factors f_{ab} (where in the case of charged particles, a and b correspond to the charge i.e. f_{+-}, f_{-+}, f_{++} and f_{--}) represent the probability given that a particle a is reconstructed, a second particle emitted at a relative pseudorapidity or azimuthal angle ($\Delta\eta$ or $\Delta\varphi$, respectively), would also be detected.

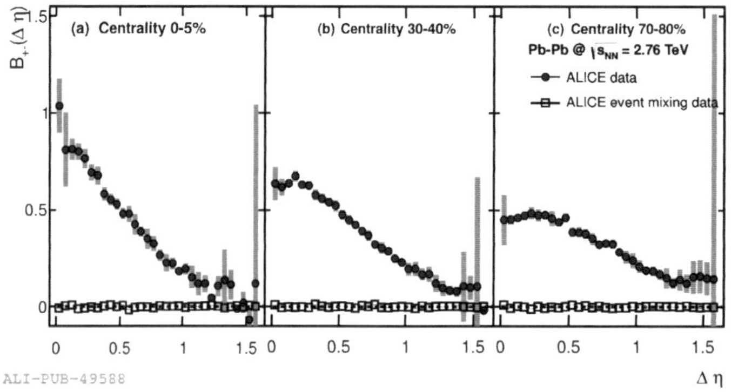

Figure 2. Balance function as a function of $\Delta\eta$ for different centrality classes: 0-5% (a), 30-40% (b) and 70-80% (c). Mixed events results, not corrected for the detector effects, are shown by open squares. See text for details. Figures are taken from [9].

First balance functions from the ALICE were obtained from Pb–Pb collisions at $\sqrt{s_{NN}}$ = 2.76 TeV and can be found at [9]. Figure 2 presents the balance function as a function of the relative pseudorapidity $\Delta\eta$ for three different centrality classes: the (a) 0-5% (most central), the (b) 30-40% (midcentral) and the (c) 70-80% (most peripheral) centrality bins. It is seen that the width of the balance function gets smaller for more central collisions.

4. Mean p_{T} fluctuations

Event-by-event fluctuations of the mean transverse momentum, i.e. $\langle p_{\mathrm{T}}\rangle$, contain information about the dynamics of and correlations in heavy-ion collisions. It has been suggested that the fluctuations in $\langle p_{\mathrm{T}}\rangle$ may be related to the occurrence of thermalization and collectivity [10]. In general, there may be a variation in the fluctuations (increase or decrease) in heavy-ion collisions with respect to those that occur in pp collisions, which serve as the baseline for comparison. The two-particle correlator $C_m = \langle \Delta p_{\mathrm{T},i}, \Delta p_{\mathrm{T},j}\rangle$ [11] is a measure of the dynamical component of the variance of $\langle p_{\mathrm{T}}\rangle$ and is defined by

$$C_m = \frac{1}{\sum_{k=1}^{n_{\mathrm{ev}}} N_k^{\mathrm{pairs}}} \cdot \sum_{k=1}^{n_{\mathrm{ev}}} \sum_{i=1}^{N_k} \sum_{j=i+1}^{N_k} (p_{\mathrm{T},i} - \langle p_{\mathrm{T}}\rangle_m) \cdot (p_{\mathrm{T},j} - \langle p_{\mathrm{T}}\rangle_m), \quad (4)$$

where n_{ev} is the number of events in a given multiplicity class m, N_k^{pairs} is the number of pairs constructed out of N_k number of particles in an event and equal to $0.5 \cdot N_k \cdot (N_k - 1)$ and $\langle p_{\mathrm{T}}\rangle$ is the average p_{T} of all tracks of all events in class m. C_m vanishes in the presence of only statistical fluctuations.

In figure 3 (a), the relative dynamical fluctuation $\sqrt{C_m}/M(p_{\mathrm{T}})_m$ ($M(p_{\mathrm{T}})$ is the average p_{T} of all tracks in all events of class m) is shown as a function of the average charged-particle multiplicity $\langle \mathrm{d}N_{\mathrm{ch}}/\mathrm{d}\eta\rangle$ in pp collisions at $\sqrt{s}$ = 0.9, 2.76 and 7 TeV. The non-zero values of $\sqrt{C_m}/M(p_{\mathrm{T}})_m$ indicate significant dynamical event-by-event $M(p_{\mathrm{T}})$ fluctuations which decrease with the increase of multiplicity. No beam energy dependence is observed in the relative fluctuation $\sqrt{C_m}/M(p_{\mathrm{T}})_m$. Figure 3 compares the ALICE results for $\sqrt{C_m}/M(p_{\mathrm{T}})_m$ to measurements in Au–Au collisions at $\sqrt{s_{NN}}$ = 200 GeV by the STAR experiment at RHIC [12] as a function of (b)$\langle \mathrm{d}N_{\mathrm{ch}}/\mathrm{d}\eta \rangle$ and (c)$\langle N_{\mathrm{part}} \rangle$. The agreement between LHC and the RHIC energies, albeit within the large experimental uncertainties, points to a relation between the observed fluctuation patterns and the collision geometry.

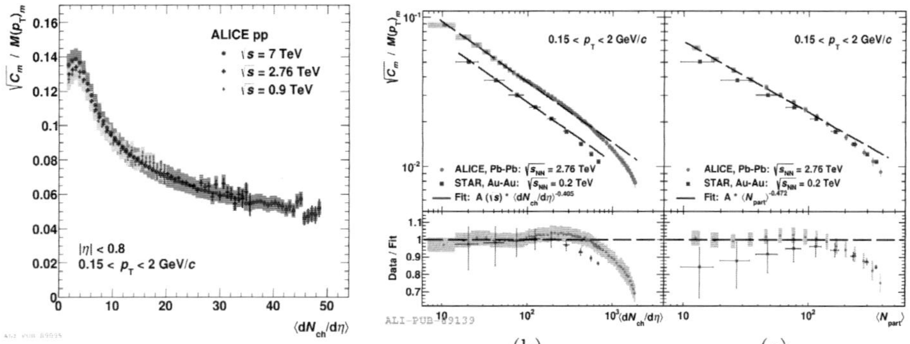

(a) (b) (c)

Figure 3. (a): Relative fluctuations $\sqrt{C_m}/M(p_{\mathrm{T}})_m$ as a function of $\langle \mathrm{d}N_{\mathrm{ch}}/\mathrm{d}\eta \rangle$, for pp collisions at $\sqrt{s} = 0.9$, 2.76 and 7 TeV. The relative dynamic fluctuations in Pb–Pb collisions measured by ALICE and Au–Au collisions measured by STAR as a function of $\langle \mathrm{d}N_{\mathrm{ch}}/\mathrm{d}\eta \rangle$ (b) and $\langle N_{\mathrm{part}} \rangle$(c) along with power-law fits to the data as shown in dashed lines. The lower panels of (b) & (c) show ratios to power law fits. For the STAR data points, statistical and systematic uncertainties are combined. Figures are taken from [13].

5. Summary and outlook

In summary, dynamical net-charge fluctuations were presented as a function of centrality and pseudorapidity. Their value is below the expectation for a hadron resonance gas and above that of the QGP. The widths of the balance functions are found to decrease when moving from peripheral to central collisions. The fluctuations of mean transverse momentum get diluted for higher multiplicities. An additional reduction of the fluctuations is observed for central Pb–Pb collisions. Future studies will extend the analysis of net–charges to higher moments net–particle multiplicity distributions and will add moment analyses of identified quark flavors. In addition, the balance functions and mean transverse momentum fluctuations will be studied in pp and p–Pb collisions as well as for identified particles with respect to their transverse momentum and event–plane dependence.

References

[1] H. Satz, Rep. Prog. Phys. **63**, (2000) 1511; S.A. Bass, M. Gyulassy, H. Stöcker, W. Greiner, J. Phys. **G25**, (1999) R1; J. Cleymans, R.V. Gavai, E. Suhonen, Phys. Rep. **130**, (1986) 217.
[2] H. Satz, Nucl. Phys. A **862**, (2011) 4.
[3] K. Aamodt *et al.* ALICE Collaboration, JINST, **3**, (2008), S08002.
[4] S. Jeon, V. Koch, Phys. Rev. Lett. **83**, (1999) 5435; S. Jeon, V. Koch, Phys. Rev. Lett. **85**, (2000) 2076.
[5] B. I. Abelev *et al.* STAR Collaboration, Phys. Rev. C **79**, 024906 (2009).
[6] B. Abelev *et al.* ALICE Collaboration, Phys. Rev. Lett. **110**, (2013), 152301.
[7] E. V. Shuryak *et al.* Phys. Rev. C **63**, (2001) 064903; M. A. Aziz *et al.* Phys. Rev. C **70**, (2004) 034905.
[8] S. Bass, P. Danielewicz, and S. Pratt, Phys. Rev. Lett. **85**, (2000) 2689.
[9] B. Abelev *et al.*ALICE Collaboration, Phys. Lett. B **723** (2013) 267279.
[10] S. Gavin, Phys. Rev. Lett. **92**, (2004) 162301; S. Gavin, J. Phys. G **30**, (2004) S1385.
[11] S.A. Voloshin, V. Koch and H.G. Ritter, Phys. Rev. C, **60**, (024901), 1999.
[12] J. Adams *et al.* STAR Collaboration, Phys. Rev. C **72**, (2005) 044902.
[13] B. Abelev *et al.* ALICE Collaboration, Eur. Phys. J. C (2014) 74-3077

A 'soft+hard' model for Pion, Kaon, and Proton Spectra and v_2 measured in PbPb Collisions at $\sqrt{s} = 2.76A$TeV

Gergely Gábor Barnaföldi [1], **Károly Ürmössy** [1], **Gábor Bíró** [1,2]

[1] Wigner Research Center for Physics of the HAS,
29–33 Konkoly–Thege Miklós Str. H-1121 Budapest, Hungary
[2] Loránd Eötvös University, 1/A Pázmány Péter Sétány, H-1117 Budapest, Hungary

E-mail: barnafoldi.gergely@wigner.mta.hu

Abstract. Hadron spectra measured in high-energy collisions present distributions which can be derived from the non-extensive statistical and thermodynamical phenomena. Based on earlier theoretical developments, it seems, the methods are very applicable for jets hadronization processes in electron-positron, proton-proton, and even in heavy-ion collisions.

Here, we present what can was learnt from the recent theoretical and phenomenological developments: transverse momentum spectra and azimuthal anisotropy (v_2) of charge averaged pions, kaons and protons stemming from central Pb+Pb collisions at $\sqrt{s}$ = 2.76 ATeV are described *analytically* in a 'soft + hard' model.

In this model, we propose that hadron yields produced in heavy-ion collisions are simply the sum of yields stemming from jets (hard yields) in addition to the yields originating from the Quark-Gluon Plasma (soft yields). The hadron spectra in both types of yields are approximated by the Tsallis – Pareto like distribution.

1. Introduction

Perturbative quantum-chromodynamics improved parton model (pQCD) calculations describe the spectra of hadrons stemming from hard processes in proton-proton (pp) collisions [1, 2]. Keeping track of parton model calculations at leading order, it can be argued that the Tsallis – Pareto-like (or cut power-law) distribution is a good approximation for the resulting hadron spectra [3, 4]. This observation has been tested by many groups [5]–[13].

In central heavy-ion (AA) collisions, however, the power of the hadron spectra changes dramatically around $p_T \approx 6$ GeV/c, thus, it cannot be fitted by one single cut power-law distribution. As the $p_T \lessapprox 6$ GeV/c part of hadron spectra cannot be reproduced by pQCD methods, thermal models based on the Tsallis statistics have been developed for this purpose [14]–[21]. Beyond these models, lies the conjecture that the Quark-Gluon Plasma (QGP) created in high-energy AA collisions has non-conventional thermodynamical properties. Among others, a possible scenario is that negative-binomial (NBD) hadron multiplicity distributions are responsible for the emergence of the non-Boltzmannian (Tsallis – Pareto like) distribution in high energy reactions [9, 22, 23, 24].

In this paper, we persue the conjecture presented in Refs. [25, 26, 27] and make out the hadron spectra measured in AA collisions as the sum of yields stemming from the QGP (we refer to as 'soft yields') and yields coming from jets (we refer to as 'hard yields'). In these papers, transverse spectra and azimuthal anisotropy (v_2) of charged hadrons obtained from various centrality Pb+Pb collisions at $\sqrt{s} = 2.76A$TeV have been described. In this paper, we turn our attention to the spectra and v_2 of identified pions, kaons, and protons created from central Pb+Pb collisions at the same LHC energy.

We note that transverse spectra and v_2 of various identified hadrons measured at RHIC energy have been described by a similar model [15, 16]. In that model, spectra measured in pp collisions have been used as hard yields, and it has been conjectured that hard yields are suppressed at low p_T.

2. Fits to transverse spectra and v_2

We made fits on recent data measured in Pb+Pb collisions at $\sqrt{s} = 2.76A$TeV by the ALICE experiment. We used the identified hadron spectra from Ref. [28] as a basis of our 'soft+hard' fits and then these parameters were converted to obtain asymuthal anisotropy data. Our v_2 results were compared to the measured values based on Ref. [29].

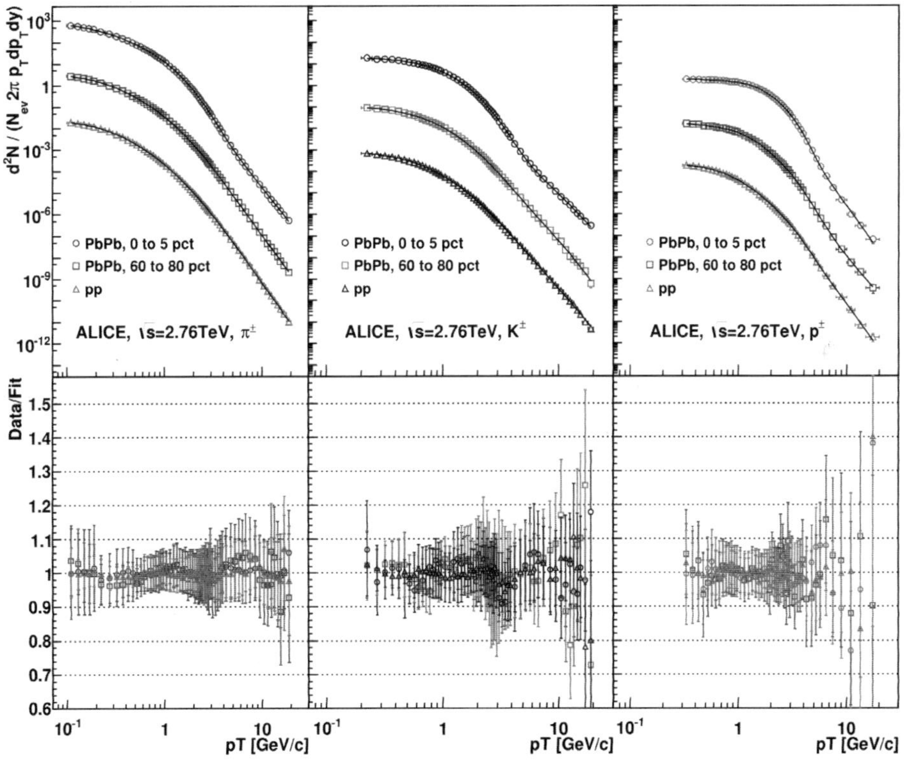

Figure 1. *Top panels*: Transverse momentum spectra of charge averaged pions, kaons and protons stemming from central (0−5%) and periferal (60%−80%) Pb+Pb as well as pp collisions at $\sqrt{s} = 2.76A$TeV [28]. Curves are fits of Eq. (1). *Bottom panels:* Data over fit plots.

We conjecture that hadrons stemming both 'soft' and 'hard' yields are distributed according to differently parametrized Tsallis distribution:

$$\left.\frac{dN}{2\pi p_T dp_T dy}\right|_{y=0} = \sum_i f_i = \sum_i A_i \left[1 + \frac{(q_i - 1)}{T_i}[\gamma_i(m_T - \nu_{0,i}p_T) - m]\right]^{-1/(q_i-1)} , \quad (1)$$

with i = 'soft' or 'hard', $m_T = \sqrt{p_T^2 + m^2}$, $\nu_{0,soft}$ being the transverse flow velocity of the QGP on the freeze-out hyper-surface as given in Ref. [18], and $\nu_{0,hard}$ being an average jet velocity (see e.g. in Ref. [23]), finally, $\gamma_i = 1/\sqrt{1 - \nu_{0,i}^2}$. These yields have maxima at $p_{T,i}^{max} = \gamma_i\, m\, \nu_{0,i}$. Due to the pion mass, these maxima are below the measurement range in the case of pions and charged averaged hadrons, and thus $\nu_{0,i}$ cannot be determined accurately. The argument in Eq. (1) may be approximated by $[\gamma_i(m_T - \nu_{0,i}p_T) - m]/T_i \approx p_T/T_i^{Dopp}$ with the Doppler-shifted parameters $T_i^{Dopp} = T_i\sqrt{(1 + \nu_{0,i})/(1 - \nu_{0,i})}$, which be compared to experimantal data.

Fits were plotted on Fig. 1: transverse momentum spectra of charge averaged pions, kaons and protons stemming from central $(0 - 5\%)$ and peripheral $(60\% - 80\%)$ Pb+Pb as well as pp collisions at $\sqrt{s} = 2.76A$TeV [28]. Curves are fits of Eq. (1). Data/fit ratios were also presented at the lower row, which agreement supports the validity of the theory within 10%. As

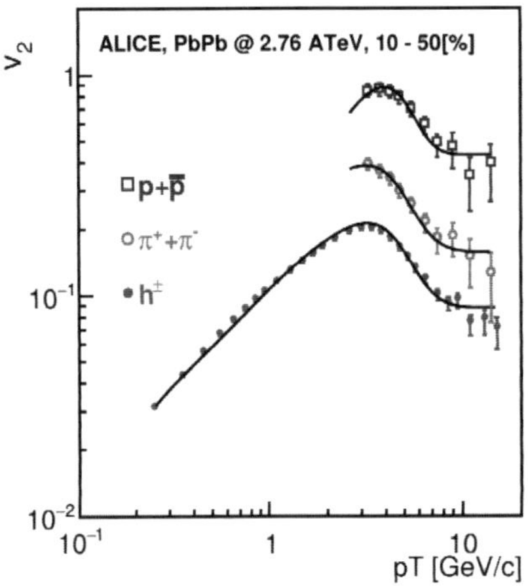

Figure 2. The v_2 of charge averaged pions, protons and charged hadrons stemming from $10\% - 50\%$ centrality Pb+Pb collisions at $\sqrt{s} = 2.76$ ATeV [29]. Curves are fits of Eq. (2).

discussed in Refs. [26, 27], Tsallis-like hadron distributions and small modular fluctuations of the transverse flow velocities as a function of the azimuthal angle $\nu_i(\varphi) = \nu_{0,i} + \sum_{n=1}^{\infty} \delta\nu_{n,i}\cos(n\varphi)$, result in an azimuthal anisotropy of the measured one for $10 - 50\%$ central Pb+Pb on Fig. 2:

$$v_2 = \frac{w_{hard}\, f_{hard} + w_{soft}\, f_{soft}}{f_{hard} + f_{soft}} , \quad (2)$$

where the coefficient functions are

$$w_i = \frac{\delta\nu_{2,i}\,\gamma_i^3}{2T_i} \frac{p_T - \nu_{0,i}\, m_T}{1 + \dfrac{q_i - 1}{T_i}\left[\gamma_i(m_T - \nu_{0,i}\, p_T) - m\right]} . \quad (3)$$

Fits of Eqs. (1) and (2) are shown in Figs. 1 and 2 with their parameters are shown in Tab. 1. These are in agreement, what was obtained for the 'soft' component: $\nu_{0,soft} = 0.693 \pm 0.051$ and $\delta\nu_{2,soft} = 0.053 \pm 0.017$, as well as, we got for charge averaged hadrons in Refs. [26, 27].

Type	Hadron	q_i	T_i [GeV]	A_i	$\nu_{0,i}$
	$\pi^{\pm}$	1.046 ± 0.016	0.260 ± 0.025	$1252. \pm 324.$	—
Soft	$p/\bar{p}$	1.006 ± 0.057	0.153 ± 0.024	4.453 ± 0.83	0.71 ± 0.0337
	$\pi^{\pm}$	1.172 ± 0.010	0.091 ± 0.012	$2947. \pm 513.$	—
Hard	$p/\bar{p}$	1.133 ± 0.031	0.238 ± 0.0712	9.18 ± 1.82	—

Table 1. Parameters for the charge averaged indentified pion and proton spectra.

3. Conclusions

The 'soft+hard' model were presented and applied on ALICE data in pp and Pb+Pb collisons.

Acknowledgments

This work was supported by Hungarian OTKA grants K104260, NK106119, and NIH TET 12 CN-1-2012-0016. Author GGB also thanks the János Bolyai Research Scholarship of the H.A.S.

References

[1] Y. Zhang et.al., *Phys. Rev. C*, **65**, 034903 (2002).
[2] G. G. Barnaföldi, G. Kalmár, K. Urmossy et.al, *Gribov 80 - Memorial Volume*, (ISBN: 978-981-4350-18-1), p. 357., World Scientific, Singapore, (2011)
[3] Cheuk-Yin Wong, Grzegorz Wilk, Conference: C13-05-30, (2013)
[4] Cheuk-Yin Wong, Grzegorz Wilk, *Phys. Rev. D*, **87**, 114007 (2013)
[5] The PHENIX Collab., Phys. Rev. D, 83, 052004 (2011)
[6] Cheuk-Yin Wong, G. Wilk, *Acta Physica Polonica B*, **42**, 2047, (2012)
[7] J. Cleymans, D. Worku, *J. Phys. G: Nucl. Part. Phys.*, **39**, 025006, (2012)
[8] G. G. Barnaföldi et al., *J. Phys. Conf. Ser.*, **270**, 012008, (2011)
[9] K. Urmossy, arXiv:1212.0260
[10] T. Wibig, I. Kurp, *JHEP*, **0312**, 039, (2003);
[11] T. Wibig, *Eur. Phys. J. C*, **74**, (2014), 2966
[12] Khandai, *Int. J. Mod. Phys. A*, **28** (2013) 1350066,
[13] M. Praszalowicz, *Physics Letters B*, **727**, (2013) 461
[14] G. Wilk, Z. Wlodarczyk, *Eur. Phys. J. A*, **48**, (2012) 161
[15] Z. Tang, L. Yi, L. Ruan, et al., *Chin. Phys. Lett.*, **30** (2013) 031201
[16] Z. Tang, Y. Xu, L. Ruan et.al., *Phys. Rev. C:* **79**, 051901 (R), (2009)
[17] D. D. Chinellato, J. Takahashi, I. Bediaga, *J. Phys. G:* **37**, 094042, (2010)
[18] K. Urmossy, T. S. Biro, *Phys. Lett. B:* **689**, 14-17, (2010)
[19] B. De, G. Sau, S. K. Biswas et.al., *J. Mod. Phys. A:* **25**, 1239-1251, (2010)
[20] M. Biyajima, T. Mizoguchi, N. Nakajima et.al., *Eur. Phys. J. C:* **48**, 597-603, (2006)
[21] T. S. Biro, A. Jakovac, *Phys. Rev. Lett.*, **94** (2005) 132302
[22] T.S. Biró, G.G. Barnaföldi, P. Ván, K. Urmossy, arXiv:1404.1256 (2014)
[23] K. Urmossy, G. G. Barnaföldi, T. S. Biró, *Phys. Lett. B*, **718**, 125-129, (2012),
[24] K. Urmossy, G. G. Barnaföldi, T. S. Biró, *Phys. Lett. B*, **701**, 111-116, (2011),
[25] K. Urmossy, G. G. Barnaföldi, Sz. Harangozó, Z. Xu, Proceedings of the *10th International Workshop on High-pT Physics in the RHIC/LHC era*, arXiv:1501.02352, (2014)
[26] K. Urmossy, T. S. Biró, G. G. Barnaföldi, Z. Xu, Proceedings of the *Workshop on Particle Correlation and Femtoscopy 2014*, arXiv:1501.05959, (2014)
[27] K. Urmossy, G. G. Barnaföldi, T. S. Biró, Z. Xu, arXiv:1405.3963, (2014)
[28] B. Abelev etal. (ALICE Collab.), *Phys. Lett. B*, **736** (2014) 196-207
[29] B. Abelev etal. (ALICE Collab.), *Phys. Lett. B*, **719** (2013) 18-28

Identified light-flavour particle production measured with ALICE at the LHC as a probe of soft QCD and hot hadronic matter

Barbara Guerzoni for the ALICE Collaboration

Dipartimento di Fisica dell'Università di Bologna and INFN sezione di Bologna, via Irnerio 46, Bologna, Italy

E-mail: barbara.guerzoni@bo.infn.it

Abstract. ALICE is a general-purpose heavy-ion experiment able to identify particles over a wide momentum range thanks to excellent vertexing and tracking performance, low material budget and different Particle IDentification (PID) techniques. In this paper the measurement of the production of identified light flavour particles in pp, p–Pb and Pb–Pb collisions is reported. It is of fundamental importance to study the particle production mechanisms playing a role in the different momentum ranges and probe the hot medium formed in heavy-ion collisions. Information on the effects of the medium produced in Pb–Pb collisions on particle production can be obtained comparing the particle ratios in pp and Pb–Pb events and studying the nuclear modification factor (R_{AA}). The transverse momentum (p_T) integrated yields and ratios are then discussed in terms of thermal models to extract the properties of the medium produced in Pb–Pb collisions at the chemical freeze-out.

1. Introduction

The ALICE Collaboration has shown that for peripheral Pb–Pb collisions the shapes of the invariant yields are similar to those observed in pp collisions [1]. Moving to central Pb–Pb events the spectra get flatter at low p_T (the heavier is the particle mass, the flatter is the spectrum) and are strongly suppressed at high p_T when compared to pp events scaled by the number of binary collisions. The spectra of identified particles provide hence a tool to study the dynamics of the quark-gluon plasma created in Pb–Pb collisions and the way it affects the particle production mechanisms in different momentum regions. Studying the low p_T spectra ($p_T \lesssim 2$ GeV/c), information about the production of soft particles from the thermalized medium can be obtained. The high p_T region (p_T approximately > 8 GeV/c) is important to study the hard production in perturbative regime and jet quenching phenomena. The intermediate p_T range provides possible explanation of the so-called baryon anomaly phenomena first observed at RHIC, that is the enhancement of the baryon over meson ratio as a function of p_T with increasing centrality of the collision. In the following, results on particle ratios and R_{AA} will be shown. Further information can be obtained comparing the integrated yields of identified particles produced in pp, p–Pb, Pb–Pb collisions. Finally, the comparison of the integrated yields of identified particles with predictions from thermal models, fundamental to characterize the medium at the chemical freeze-out, will be discussed.

2. Results

In Fig. 1 the p/π and K/π ratios (left panels) are shown as a function of $p_{\rm T}$ for pp and central Pb–Pb collisions at $\sqrt{s_{\rm NN}} = 2.76$ TeV. It is remarkable the increase of the enhancement of the p/π ratio at $p_{\rm T} \sim 3$ GeV/c from pp to Pb–Pb collisions, in line with a similar observation made at RHIC in central Au–Au collisions [2, 3]. At LHC energies a slight enhancement of the K/π ratio, not seen at RHIC, can also be observed. These observations can be understood in terms of a hydrodynamic expansion of the medium or parton recombination.

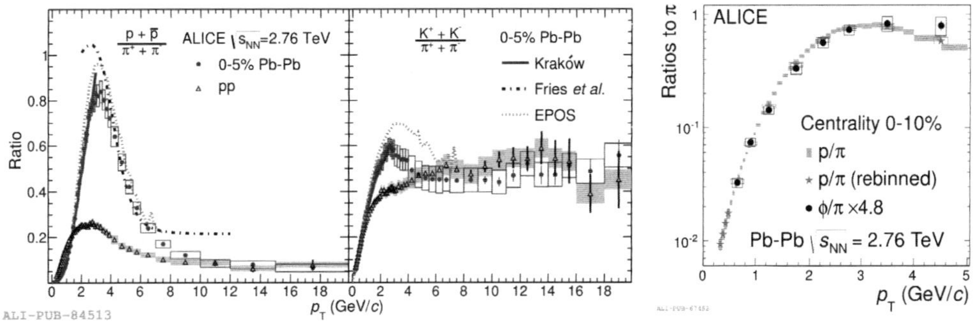

Figure 1. (left) p/π and K/π ratios as a function of $p_{\rm T}$ for pp and central Pb–Pb collisions at $\sqrt{s_{\rm NN}} = 2.76$ TeV [1]. Kraków, Fries and EPOS predictions are superimposed to the data. (right) p/π and ϕ/π ratios as a function of $p_{\rm T}$ measured in central Pb–Pb collisions at $\sqrt{s_{\rm NN}} = 2.76$ TeV [4].

The model proposed by Fries et al. [5], that introduces the recombination for soft thermal radially flowing partons, is able to describe qualitatively the data. The rise of the ratios up to $p_{\rm T} \sim 3$ GeV/c can be also described by the predictions of the Kraków hydrodynamic model [6]. The EPOS generator [7], that combines a hydrodynamic description of the medium with high $p_{\rm T}$ physics and jet quenching, can also describe the data although it overestimates the peak. In the right panel of Fig. 1 it is evident that, in central Pb–Pb collisions, the shape of the ϕ/π and p/π ratios as a function of $p_{\rm T}$ [4] are consistent. This could suggest that the peak is the results of the hydrodynamic radial flow whose effect depends on the particle mass, in contrast to other hadronization processes that would only increase the baryon production. For $p_{\rm T} > 10$ GeV/c, the p/π and K/π ratios obtained in pp and Pb–Pb collisions are in agreement suggesting that the parton fragmentation (jet chemistry) is not modified by the medium produced in Pb–Pb collisions. This hypothesis is supported by Fig. 2 where the nuclear modification factor $R_{\rm AA}$ for pions, kaons, protons and all charged particles in central (left) and peripheral (right) Pb–Pb collisions are reported [1]. The $R_{\rm AA}$ is defined as the ratio between the Pb–Pb spectra normalized to the number of binary collisions and the pp spectra. While at low momentum ($p_{\rm T} < 8$ GeV/c) there is a mass hierarchy in particle suppression, namely protons are less suppressed than kaons and pions, at high momenta ($p_{\rm T} > 10$ GeV/c) the suppression is the same for all the particle species. This is an indication that particle composition and ratios at high momenta are the same in medium and in vacuum disfavouring models where large energy loss is associated with mass ordering or large fragmentation differences between baryons and mesons.

In the left panel of Fig. 3, momentum-integrated particle ratios measured in pp, high multiplicity p–Pb and central Pb–Pb collisions are compared to study the system-size dependence of particle production. Looking at the K/π, Ξ/π and Ω/π ratios, one observes that the production of strange hadrons increases with the system size even if the Ξ/π ratio seems to decrease moving from p–Pb to Pb–Pb collisions. This behavior still needs to be

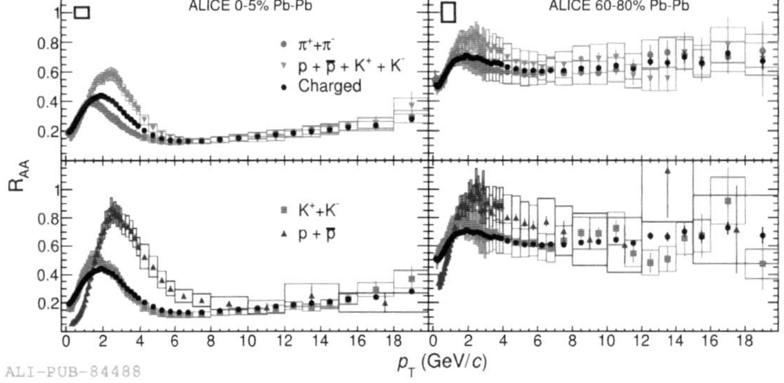

Figure 2. Nuclear modification factor $R_{\rm AA}$ as a function of $p_{\rm T}$ in central (left) and peripheral (right) Pb–Pb collisions at $\sqrt{s_{\rm NN}} = 2.76$ TeV for all charged (black), pions (red), kaons (green) protons (blue) and protons + kaons (purple) [1].

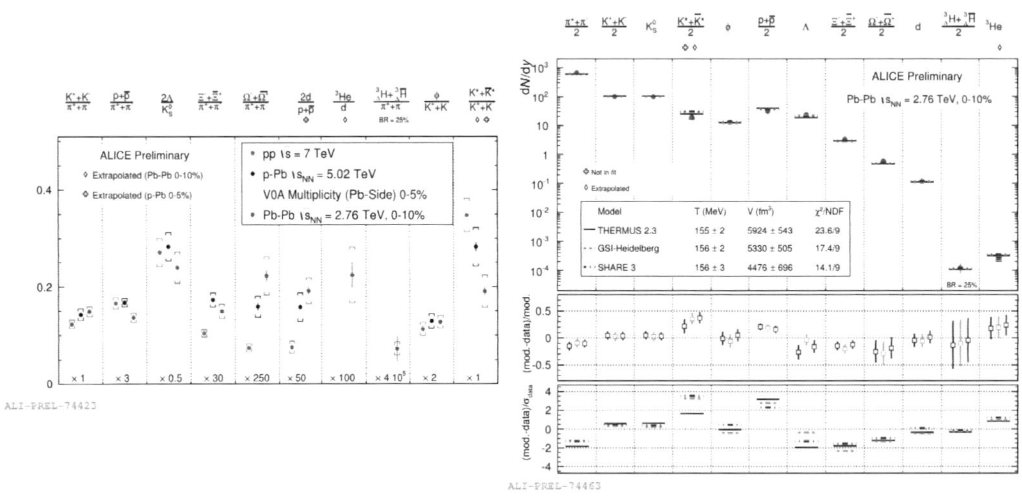

Figure 3. (left) Particle ratio in pp, high multiplicity p–Pb and central Pb–Pb collisions at $\sqrt{s_{\rm NN}} = 7, 5.02$ and 2.76 TeV respectively. (right) Particle yields measured in central Pb–Pb collisions compared to results of fit with THERMUS, GSI and SHARE thermal models.

clarified. The K*/K ratio decreases with increasing the system size, a phenomenon that can be understood as due to the rescattering of the K* decay products in the hadronic medium. The d/p ratio increases with the system size, providing further constraints on the proposed deuteron production mechanisms, based on thermal and coalescence models. Looking at the p/π and $\Lambda/{\rm K}^0_s$ ratios, it is also possible to notice a hint of momentum-integrated baryon suppression moving from p–Pb to Pb–Pb collisions.

In the right panel of Fig. 3, the integrated yields (dN/dy) of identified particles measured in central Pb–Pb collisions are reported. They are interpreted in terms of thermal models that describe the properties of the system at chemical freeze-out. These models are based on six main parameters: the chemical freeze-out temperature $T_{\rm ch}$, the baryochemical potential $\mu_{\rm B}$ that at the LHC energies is fixed to zero, the volume V and the parameters that take into account deviation from equilibrium production for strange, charm or light quarks, $\gamma_{s,c,q}$. Equilibrium thermal models ($\gamma_s = \gamma_c = \gamma_q = 1$) were able to reproduce the yields measured at RHIC but, once their parameters were extrapolated to LHC energies, their were found to overestimate the proton production by a factor 1.5. In Fig. 3 the results of the fit to the data of three different models implementing full equilibrium, Thermus 2.3 [8], GSI-Heidelberg [9] and SHARE [10], are reported. The disagreement of the protons with respect to the equilibrium model expectations persists. Note that the K* resonances are not included in the fit. It can be also noticed that the value of the temperature parameter obtained from the fit by the three models are compatible ($\sim$ 156 MeV) and lower than the value predicted from the extrapolation of RHIC data ($\sim$ 164 MeV).

To explain the deviations from the thermal models many hypotheses have been formulated. At lower energies particle abundances were described by thermal models assuming thermal and chemical equilibrium, assuming the modification of particle ratios in the hadronic phase negligible. These deviations from thermal predictions can be explained assuming that final state interactions in the hadronic phase are not negligible and affecting in different way baryons and mesons [11, 12]. One process that might be important is baryon annihilation, in line with the observation of the lower value of the protons compared to the model predictions. A non-equilibrium configuration and a quark-gluon plasma undergoing a sudden hadronization without rescattering can be rapresented by the SHARE model [10] with the γ_q parameter not fixed to 1. This gives to the hadrons the possibility to be produced out of equilibrium. In such a configuration it is possible to fit all hadrons except nuclei with $\chi^2/ndf \approx 1$. Another possible explanation invokes a flavour-dependent transition temperature [13] hence a single freeze-out temperature could not be able to describe the data.

3. Conclusions

Light flavour measurements performed by the ALICE Collaboration allow to study the effect of the medium produced in Pb–Pb collisions on particle production. p/π, K/π and ϕ/π ratios provide information to understand the baryon anomaly and suggest that spectra are characterised by hadron masses rather than their (valence) quark contentent. The R_{AA} at high p_{T} suggests that the parton fragmentation is not modified by the medium. Finally, equilibrium thermal models succesfully describe the integrated yields of the measured particle species, though overestimating proton and K* production in central Pb–Pb collisions.

References

[1] Abelev B et al. (ALICE Collaboration) 2014 *Phys. Lett.* B **736** 196.
[2] Abelev B et al. (STAR Collaboration), 2006 *Phys. Rev. Lett.* **97** 152301.
[3] Adare A et al. (PHENIX Collaboration), 2013 *Phys. Rev.* C **88** 024906.
[4] Abelev B et al. (ALICE Collaboration), arXiv:1404.0495.
[5] Fries RJ, Muller B, Nonaka C and Bass SA 2003 *Phys. Rev.* C **68** 044902.
[6] Bozek P and Wyskiel-Piekarska 2012 *Phys. Rev.* C **85** 064915.
[7] Werner K et al. 2012 *Phys. Rev.* C **85** 064907.
[8] Wheaton S et al., 2009 *Comp. Phys. Commun.* **180** 84.
[9] Andronic A et al., 2009 *Phys. Lett* B **673** 142.
[10] Petran M et al., arXiv:1310.5108v2.
[11] Becattini F et al., arXiv:1405.0710.
[12] Becattini F et al., 2013 *Phys. Rev. Lett* **111** 082302.
[13] Bellwied R et al., 2013 *Phys. Rev. Lett* **111** 202302.

Transport properties of the hot and dense sQGP

H. Berrehrah[1], E. Bratkovskaya[1], W. Cassing[2] and R. Marty[1,3]

[1] Frankfurt Institute for Advanced Studies and Institute for Theoretical Physics, Johann Wolfgang Goethe Universität, Ruth-Moufang-Strasse 1, 60438 Frankfurt am Main, Germany
[2] Institut für Theoretische Physik, Universität Giessen, 35392 Giessen, Germany
[3] Institut de physique nucléaire de Lyon (IPNL), Université Claude Bernard Lyon 1, Villeurbanne, France

E-mail: berrehrah@fias.uni-frankfurt.de

Abstract. The transport properties of the quark gluon plasma (QGP) are studied in a QCD medium at finite temperature and chemical potential. We calculate the shear viscosity $\eta(T, \mu_q)$ and the electric conductivity $\sigma_e(T, \mu_q)$ for a system of interacting massive and broad quasi-particles as described by the dynamical quasi-particle model "DQPM" at finite temperature T and quark chemical potential μ_q within the relaxation time approximation. Our results are in a good agreement with lattice QCD at finite temperature and show clearly the increase of the transport coefficients with increasing T and μ_q. Our results provide the basic ingredients for the study of the hot and dense matter in the Beam Energy Scan (BES) at RHIC and CBM at FAIR.

1. Introduction

The exploration of the phase diagram of strongly interacting matter is a major field of modern high-energy physics. The transition from a hadronic to a partonic medium at small net-baryon densities is known to be a crossover. At high baryon densities one expects new phases of the strongly interacting matter. Besides the understanding of the thermodynamic properties of such systems, the transport properties at high temperature and density are of interest for many purposes. They are the key ingredients for hydrodynamic calculations (η/s, ξ/s) and transport simulations which compare the in and out-off equilibrium systems. They are also useful to confront different transport models, which study the expansion of the plasma created in relativistic heavy-ion collisions.

It is the purpose of this study to evaluate the transport coefficients for a system of interacting massive and broad quasi-particles as described by the dynamical quasi-particle model "DQPM" at finite temperature T and quark chemical potential μ_q within the relaxation time approximation. The relaxation times for these particles are calculated using the transition rates corresponding to qq, $q\bar{q}$, qg, gg elastic and $gg \leftrightarrow q\bar{q}$ inelastic scattering processes, where q ($\bar{q}$) denotes the light (anti-)quark and g the gluons of the sQGP.

2. The DQPM at finite (T, μ_q)

To evaluate the transition rates and cross sections of the partonic scattering processes in the sQGP using the leading order Born Feynman diagrams one needs to specify- in such a finite T and μ_q medium- the values of the fundamental ingredients which are the coupling α_s, the infrared regulator "IR" and the masses of external q,$\bar{q}$ and g legs. These ingredients are provided by the DQPM in our calculations.

The DQPM coupling constant and gluon mass (which is considered also as IR) are shown in figure 1 (a) and (b) as a function of T and μ_q. One sees that α_s is much larger than one near $T_c(\mu_q)$; the non-perturbative effects then become most apparent at these temperatures. The increase of μ_q leads to a decrease of α_s and $m_{q,\bar{q},g}$. The DQPM provides also the particle width γ and the Breit-Wigner spectral

Journal of Physics: Conference Series **612** (2015) 012050 doi:10.1088/1742-6596/612/1/012050

function ρ^{BW} since the constituents of the sQGP are considered as strongly interacting massive effective quasiparticles [1].

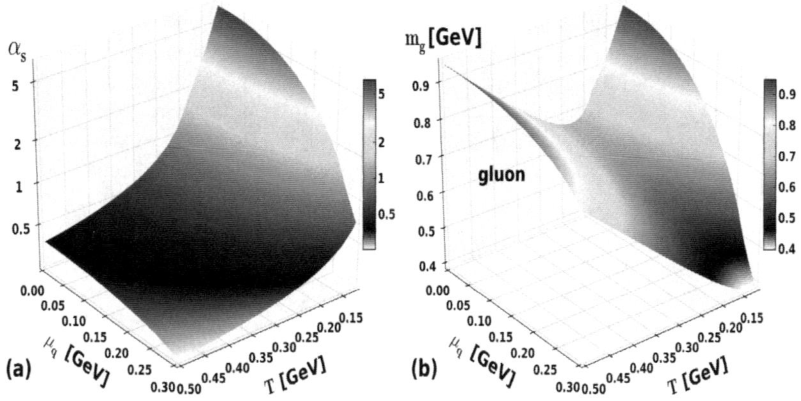

Figure 1. *(Color online) The DQPM coupling constant α_s (a) and gluon mass (b) as a function of the temperature T and quark chemical potential μ_q.*

3. g, q and $\bar{q}$ scattering at finite (T, μ_q)

By dressing the quark and gluon lines with the effective propagators at finite temperature and chemical potential, we derive the on- and off-shell cross sections and transition rates for the dominant partonic reactions in the sQGP. The partons are dressed by effective DQPM pole masses in the on-shell case and are dressed by the DQPM spectral functions with a finite width in the off-shell case. We refer to the on-shell study by DpQCD approach (Dressed pQCD) and to the off-shell case by IEHTL approach (Infrared Enhanced Hard Thermal Loop) [2, 3].

The effect of the spectral function in the off-shell approach is demonstrated to be small because of the small DQPM width [2, 3]. However the off-shell IEHTL calculation shows a reduction of the kinematic threshold as compared to the DpQCD on-shell non-perturbative approach. Such an effect leads only to a small shift in the partonic transition rates. Therefore, we show in Figs.2 (a) and (b) the DpQCD $u\,u \to u\,u$ cross section σ as a function of T and $\sqrt{s}-\sqrt{s_0}$ (s_0 is the energy threshold) and transition rate ω as a function of T and μ_q, respectively.

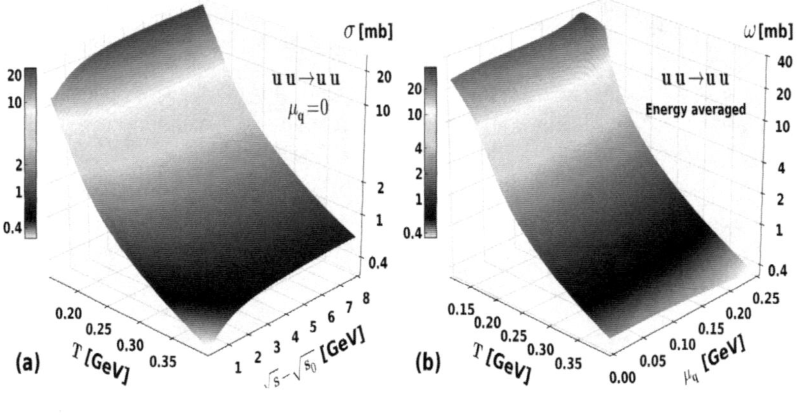

Figure 2. *(Color online) $u\ u \to u\ u$ cross section as a function of $\sqrt{s}-\sqrt{s_0}$, with s_0 denoting the threshold (a) and thermal transition rate as of function of (T, μ_q) (b).*

Figs.2 (a) and (b) show a large enhancement of σ and ω for T close to $T_c(\mu_q)$ due to the infrared enhancement of $\alpha_s(T, \mu_q)$ as seen in Fig.1. Despite the lower value of the IR regulator at higher values of μ_q the smaller value of α_s at finite μ_q explains the decrease of σ and ω at finite and large μ_q.

The thermal transition rate ω can be parametrized by a power law in T for each value of μ_q. Having almost the same power laws at finite μ_q as compared to $\mu_q = 0$ for temperatures larger than $T_c(\mu_q = 0)$, we observe some scaling effects in the transport coefficients at these temperatures.

4. sQGP transport properties at finite (T, μ_q)

We compute transport coefficients using the relaxation time approximation. In the dilute gas approximation the relaxation time τ is obtained for on-shell particles $(\tau_c^{-1})_{\rm DpQCD}$ and for off-shell quasi-particles $(\tau_c^{-1})_{\rm IEHTL}$,$(\tau_c^{-1})_{\rm DQPM}$ by Eq.(1) [4], where n_i is the $q/\bar{q}$ or g density. For the DQPM we do not need the explicit cross sections since the inherent quasi-particle width $\gamma_i(T)$ directly provides the total interaction rate [1].

$$(\tau_i^{-1})_{\rm DpQCD}=\sum_{j\in q,\bar{q},g} n_j^{\rm on}(T,\mu_j)\,\omega_{\rm ji}^{DpQCD}(T,\mu_j),\;\; n_j^{\rm on}(T,m_j,\mu_j)=\int\frac{d^3p}{(2\pi)^3}f_j(p,T,m_j,\mu_j),\;\; (\tau_i^{-1})_{DQPM}=\frac{\hbar c}{\gamma_i(T,\mu_i)}$$

$$(\tau_i^{-1})_{\rm IEHTL}=\sum_{j\in q,\bar{q},g} n_j^{\rm off}(T,\mu_j)\,\omega_{\rm ji}^{IEHTL}(T,\mu_j),\quad n_j^{\rm off}(T,\mu_j)=\iint\frac{d^3p}{(2\pi)^3}\rho_j^{BW}(m_j)dm_j\, f_j(p,T,m_j,\mu_j). \tag{1}$$

Figs. 3-(a) and (b) show the u-quark and gluon relaxation times for on- and off-shell partons at finite T and finite (T,μ_q), respectively. Figs. 3-(a) shows that τ decreases with temperature since the quark and gluon densities are increasing functions of temperature. We can evaluate -in terms of powers of T- the behavior of $\tau_{u,g}$ for the different approaches. The higher power coefficients in the relaxation time for $T < 1.2T_c$ in DpQCD/IEHTL approaches can be traced back to the infrared enhancement of the effective coupling. A finite μ_q increases smoothly τ_u especially close to $T_c(\mu_q)$, as shown in Fig. 3-(b)

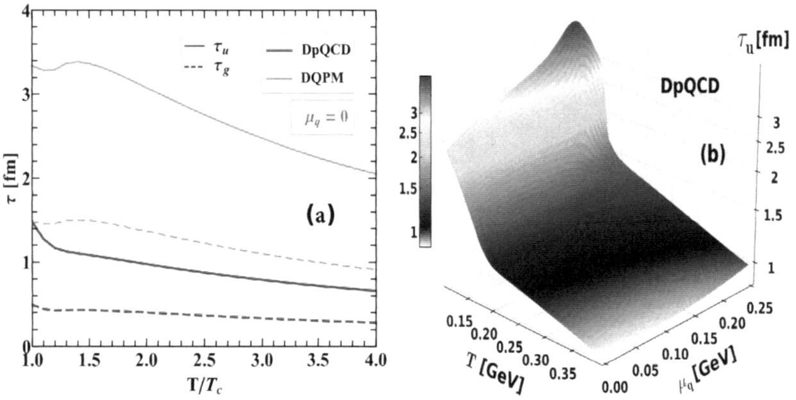

Figure 3. *(Color online) u-quark and gluon relaxation times $\tau_{u,g}$ following the on- and off-shell models as a function of temperature T for $\mu_q = 0$ (a) and τ_u given by DpQCD as a function of (T,μ_q) (b).*

Using the relaxation times τ shown above we compute the different transport coefficients of the sQGP. As an example we show the shear viscosity η and electric conductivity σ_e as a function of (T,μ_q) following our models where η and σ_e are given (for the case of on-shell partons) by

$$\eta^{\rm on}(T,\mu_q)=\frac{1}{15T}g_g\int\frac{d^3p}{(2\pi)^3}\tau_g^{\rm on}f_g\frac{\mathbf{p}^4}{E_g^2}+\frac{1}{15T}\frac{g_q}{6}\int\frac{d^3p}{(2\pi)^3}\left[\sum_q^{u,d,s}\tau_q^{\rm on}f_q+\sum_{\bar{q}}^{\bar{u},\bar{d},\bar{s}}\tau_{\bar{q}}^{\rm on}f_{\bar{q}}\right]\frac{\mathbf{p}^4}{E_q^2},$$

$$\sigma_e^{\rm on}(T,\mu_q)=\sum_{f,\bar{f}}^{u,d,s}\frac{e_f^2}{m_f(T,\mu_q)}n_f^{\rm on}(T,\mu)\tau_f^{\rm on}(T,\mu_q), \tag{2}$$

with e_q the electric charge of quarks. For off-shell partons Eqs.(2) can be generalized by taking into account the parton spectral functions, off-shell relaxation times and parton densities. Figs.4-(a) and (b) show that η/s, where s is the DQPM entropy density, and σ_e/T given by DpQCD/IEHTL and DQPM models are in the range of the lQCD data. On the other side –going from pQCD to non-pertrubative based models– leads to a reduction of η/s and σ_e/T. As a function of temperature η/s shows a minimum around T_c and then increases slowly for higher temperatures. σ_e/T increases slowly until it reaches at high temperatures the pQCD regime where $\sigma_e/T \sim const.$

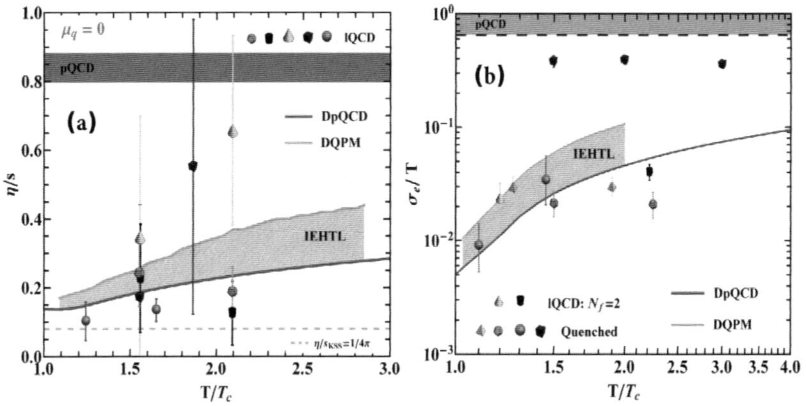

Figure 4. *(Color online) η/s (a) and σ_e/T (b) following the on- and off-shell models as a function of temperature T for $\mu_q = 0$. The lattice QCD data are given from [5].*

Finally the (T, μ_q) dependencies of η/s and σ_e/T are shown in Figs.5-(a) and (b) respectively. One sees that η increases like T^3 for large temperature for all chemical potentials in such a way that η/s is almost constant. η/s and σ_e/T show a smooth increase as a function of (T, μ_q) for finite but small μ_q, since our calculations are not applicable for large μ_q ($\mu_q \geq 0.22$ GeV).

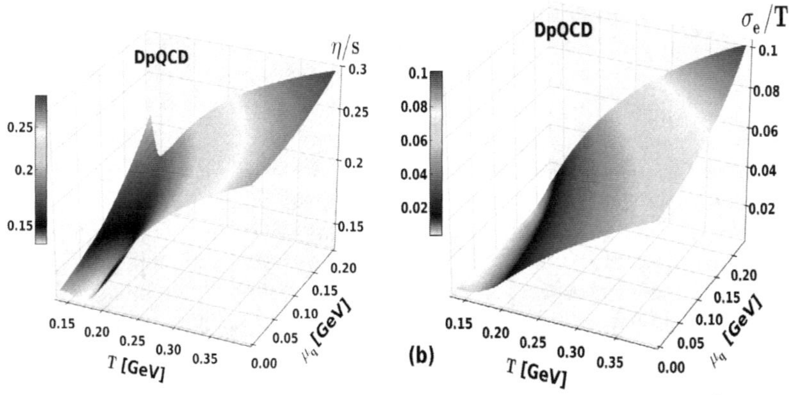

Figure 5. *(Color online) η/s (a) and σ_e/T (b) following the on-shell DpQCD model as a function of (T, μ_q).*

5. Summary

We have presented a microscopic computation of sQGP transport properties at finite temperature and chemical potential using the relaxation times of the constituents of the system. The partons are considered as strongly interacting massive quasi-particles described by the DQPM [1–3] in which quarks and gluons have a finite mass and width that vary with T and μ_q. We have demonstrated that our η/s and σ_e/T are in the range of lQCD data. The transport coefficients in our models show a smooth dependence on (T, μ_q). This is consistent with a crossover transition at small μ_q in line with lattice calculations.

References

[1] Cassing W 2009 *The European Physical Journal Special Topics* **168**(1) 3–87
[2] Berrehrah H *et al*, Phys.Rev. **C89** (2014) 054901; J.Phys.Conf.Ser. **509** (2014) 012076; arXiv:1405.3243.
[3] Berrehrah H, Gossiaux P, Aichelin J, Cassing W, Torres-Rincon J *et al.* 2014 *Phys.Rev.* **C90** 051901
[4] Marty R, Bratkovskaya E, Cassing W, Aichelin J and Berrehrah H 2013 *Phys.Rev.* **C88** 045204
[5] Meyer H.B., Phys.Rev. **D76**, 101701 (2007); Nakamura A and Sakai S, Phys.Rev.Lett. **94**, 072305 (2005); Sakai S and Nakamura A, PoS LAT2007, 221.; Ding H.-T *et al.*, PoS **185** (2012).; Aarts G *et al*, Phys.Rev.Lett. **99**, 022002 (2007).; Gupta S, Phys.Lett. **B597**, 57 (2004).; Brandt B.B *et al*, PoS **186** (2012).

A new shock-capturing numerical scheme for ideal hydrodynamics

Z. Fecková[1,2] and B. Tomášik[2,3]

[1] University of Pavol Jozef Šafárik, Šrobárova 2, 04001 Košice, Slovakia
[2] Matej Bel University, Tajovského 40, 97401 Banská Bystrica, Slovakia
[3] FNSPE, Czech Technical University in Prague, Břehová 7, 11519 Prague 1, Czech Republic

E-mail: zuzana.feckova1@student.upjs.sk, boris.tomasik@umb.sk

Abstract. We present a new algorithm for solving ideal relativistic hydrodynamics based on Godunov method with an exact solution of Riemann problem for an arbitrary equation of state. Standard numerical tests are executed, such as the sound wave propagation and the shock tube problem. Low numerical viscosity and high precision are attained with proper discretization.

1. Introduction

Ultrarelativstic heavy ion collisions are nowadays extensively studied both in theory and experiment. The collision leads to the creation of hot and dense matter, quark gluon plasma, that soon thermalizes and expands. This expansion has been successfully described by relativistic hydrodynamics [1, 2]. Since the initial conditions used in hydrodynamic modeling are often complicated, possibly leading to shock waves, sophisticated numerical methods must be used. A class of high-resolution shock-capturing methods particularly suited to solve this problem are Godunov methods. We present a new numerical scheme for ideal relativistic hydrodynamics using the exact solution of Riemann problem for an arbitrary equation of state (EoS).

2. Hydrodynamical modeling

The ideal relativistic hydrodynamic equations have the following form:

$$\begin{aligned} \partial_\mu n^\mu &= 0, \\ \partial_\mu T^{\mu\nu}_{(0)} &= 0, \end{aligned} \tag{1}$$

where n^μ is the flow of a conserved charge and $T^{\mu\nu}_{(0)}$ is the energy and momentum tensor in the non-dissipative case. In nuclear collisions the relevant charge is the baryon number, thus $n^\mu = n_B u^\mu$, where $u^\mu = \gamma(1, \vec{v})$ is the flow velocity, γ being the Lorentz factor. Our model covers nuclear collisions at highest energies, where the net baryon density is practically zero. Therefore we do not consider the first equation at all and solve only the second equation that expresses the conservation of energy and momentum. The energy and momentum tensor has, in the ideal case, this explicit form:

$$T^{\mu\nu}_{(0)} = (\epsilon + p)u^\mu u^\nu - pg^{\mu\nu}, \tag{2}$$

where ϵ is the energy density, p is the pressure and $g^{\mu\nu} = \text{diag}(1, -1, -1, -1)$ is the Minkowski metric. The second equation can be rewritten in a different form, useful for numerical implementation, where we employ a vector of conserved variables U and the spatial flow of the conserved variables is $F(U)$:

$$\partial_t U + \partial_x F(U) = 0, \tag{3}$$

where

$$U = \big((\epsilon + p)\gamma^2 - p, (\epsilon + p)\gamma^2 v^1, \tag{4}$$
$$(\epsilon + p)\gamma^2 v^2, (\epsilon + p)\gamma^2 v^3\big)^T,$$
$$F^i = \big((\epsilon + p)\gamma^2 v^i, (\epsilon + p)\gamma^2 v^i v^1 + \delta^{i1} p, \tag{5}$$
$$(\epsilon + p)\gamma^2 v^i v^2 + \delta^{i2} p, (\epsilon + p)\gamma^2 v^i v^3 + \delta^{i3} p\big)^T.$$

The presented equations are solved numerically in the latter form using Godunov method with an exact Riemann solution at the interface. In this method we consider a piecewise constant distribution of variables inside the cells of a numerical grid. To obtain the value of conserved variables at the next time-step, a local Riemann problem is solved at each interface of two neighbouring cells. Its solution allows us to compute fluxes of conserved variables $F(U)$ at the interface [3, 4]. For a given cell, we then obtain the values of conserved variables at the next time-step by averaging the fluxes on the left and right boundary of the cell:

$$U_i^{t+\Delta t} = U_i^t + \frac{\Delta x}{\Delta t}(F_{i+1/2} - F_{i-1/2}), \tag{6}$$

where U_i^t is the value of a variable in the ith cell at time t and $F_{i-1/2}(F_{i+1/2})$ is the flux at its left (right) boundary.

3. Numerical tests

The sound wave propagation test consists of simulating a sound wave of one wavelength in the numerical grid. We impose the following initial conditions:

$$p_{init}(x) = p_0 + \delta p \sin\frac{2\pi x}{\lambda}, \tag{7}$$
$$v_{init}(x) = \frac{\delta p}{c_s(\epsilon_0 + p_0)} \sin\frac{2\pi x}{\lambda},$$

with parameters $p_0 = 10^3$ fm^{-4}, $\delta p = 10^{-1}$ fm^{-4}. Since the variation of pressure is sufficiently small $\delta p \ll p_0$, we can consider the linearized analytic solution, a sound wave of velocity c_s, and compare this solution to the values obtained by numerical computation to evaluate the precision of our numerical scheme. The precision is studied with $L1$ norm evaluated after one time-step, that corresponds to the period of the wave, with different numbers of cells in the numerical grid N_{cell} [5]:

$$L(p(N_{cell}), p_s) = \sum_{i=1}^{N_{cell}} |p(x_i; N_{cell}) - p_s(x_i)| \frac{\lambda}{N_{cell}}, \tag{8}$$

The dependence of the $L1$ norm on the number of cells is shown in the left panel of Fig.1. The precision improves with finer discretization, as expected, and is very good for grids larger than $N_{cell} \gtrsim 500$. Despite the fact that we are using ideal hydrodynamics, the numerical computation introduces dissipation into our solution. Since quark gluon plasma is expected to have a very

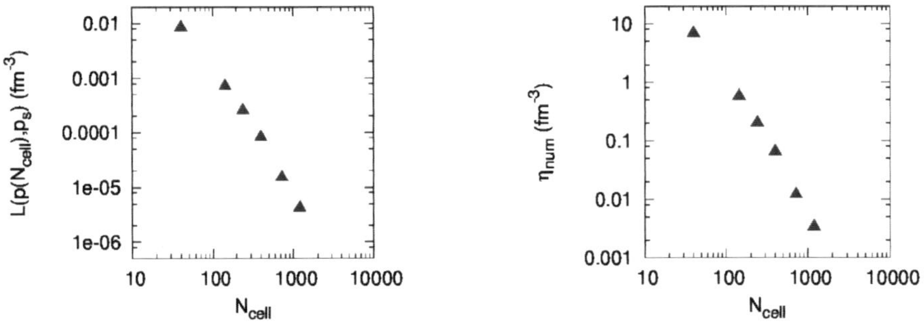

Figure 1. Dependence of the $L1$ norm (left) and the numerical viscosity η_{num} (right) on the number of cells in the numerical grid.

low viscosity, the artificial dissipation of the numerical scheme must be kept very low too. We have evaluated the numerical viscosity of the scheme η_{num}, using $L1$ norm:

$$\eta_{num} = -\frac{3\lambda}{8\pi^2} c_s(\epsilon_0 + p_0) \ln\left[1 - \frac{\pi}{2\lambda\delta p} L(p(N_{cell}), p_s)\right]. \tag{9}$$

We present the dependence of numerical viscosity η_{num} on the number of cells in the grid in the right panel of Fig.1. Similarly to the dependence of $L1$ norm, it decreases with the number of cells and its values are adequately small. We also evaluated a more suitable parameter - the ratio of numerical viscosity and entropy density in our scheme η_{num}/s. We present its values in Fig.2 together with values of η/s for pion gas [6] and the limiting value of η/s for quark-gluon plasma from AdS/CFT calculations $1/4\pi$ [7]. From this comparison we can see that the ratio η_{num}/s is sufficiently small for modelling of quark-gluon plasma and will not influence the results due to numeric effects more than the physical viscosity. For simulations in 1D this test shows that suitable grid sizes are $N_{cell} \gtrsim 400$. The 3D case remains to be checked.

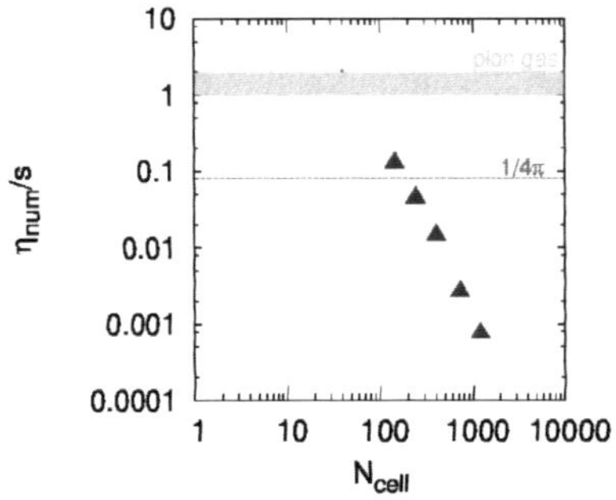

Figure 2. Dependence of the numerical viscosity to entropy density ratio η_{num}/s (blue points) compared to the limiting value $\eta/s = 1/4\pi$ (red) and η/s of pion gas (green band).

We continue with the shock tube problem. It consists of imposing special initial conditions with a discontinuity in energy density in the middle of the spatial domain between two constant states. The initial conditions in energy density are $\epsilon_L = 16$ GeV ($\epsilon_R = 1$ GeV) in the left(right) half of the numerical grid. The initial velocity is zero over the whole grid. With time, we expect to see the dissolution of the discontinuity into a rarefaction wave propagating to the left, to the region of higher energy density, and a shock wave propagating to the right, where energy density is lower. The shock tube problem has an analytic solution, which allows a good comparison between the numerical and exact solution. In the left panel of Fig.3 we show the

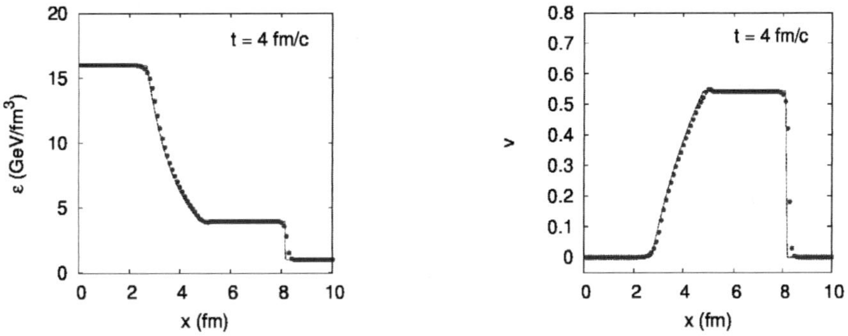

Figure 3. Profile of the energy density ϵ (left) and the velocity v (right) in the numerical grid after 100 time-steps (our scheme in blue, analytic solution in black).

profile of energy density in the grid after 100 time-steps. We see that the numerical solution (blue) is comparable to the analytic solution (black). The scheme is able to handle the initial discontinuity very well. The profile of velocity after 100 time-steps, which we show in the right panel of Fig.3, replicates the analytic solution similarly well.

4. Outlook

We have built and tested an ideal relativistic hydrodynamic scheme based on the exact solution of Riemann problem. The presented tests in one spatial dimension show a good resolution and ability to capture shock and rarefaction very well. We will extend this scheme to three dimensions and then apply it in description of the flow in ultrarelativistic nuclear collisions.

5. Acknowledgments

This work has been partially supported by grants APVV-0050-11, VEGA 1/0457/12 (Slovakia). ZF also acknowledges support from VVGS-PF-2014-442 (Slovakia). BT also acknowledges support from MŠMT grant LG13031 (Czech Republic).

6. References

[1] Romatschke P 2010 *Int. J. Mod. Phys. E* **19** 1
[2] Gale C, Jeon S and Schenke B 2013 *Int. J. Mod. Phys. A* **28** 1340011
[3] Godunov S K 1959 *Mat. Sb.* **47** 357
[4] Pons J A 2000 *J. Fluid Mech.* **422** 152
[5] Akamatsu Y, Inutsuka S, Nonaka C and Takamoto M 2014 *J. Comput. Phys.* **256** 34
[6] Mitra S, Sarkar S 2013 *Phys. Rev. D* **87** 094026
[7] Policastro G, Son D T and Starinets A O 2001 *Phys. Rev. Lett.* **87** 081601

3+1 dimensional viscous hydrodynamics at high baryon densities

Iu. Karpenko[1,2], M. Bleicher[1,3], P. Huovinen[1,3] and H. Petersen[1,3]

[1] Frankfurt Institute for Advanced Studies, Ruth-Moufang-Straße 1, 60438 Frankfurt am Main, Germany

[2] Bogolyubov Institute for Theoretical Physics, 14-b, Metrolohichna str., 03680 Kiev, Ukraine

[3] Institute for Theoretical Physics, Johann Wolfgang Goethe Universität, Max-von-Laue-Str. 1, 60438 Frankfurt am Main, Germany

E-mail: karpenko@fias.uni-frankfurt.de

Abstract. A 3+1 dimensional event-by-event viscous hydrodynamic + cascade model is applied for the simulation of heavy ion collision reactions at $\sqrt{s_{NN}} = 6.3\ldots 200$ GeV. UrQMD cascade is used for the pre-thermal (pre-hydro) and final (post-hydro) stages of the reaction. The baryon, as well as electric charge densities are consistently taken into account in the model. For this aim the equation of state based on a Chiral model coupled to the Polyakov loop is used in hydrodynamic phase of evolution.

As a result of the model adjustment to the experimental data, the effective values of the shear viscosity over entropy density η/s are obtained for different collision energies in the BES region. A decrease of the effective values of η/s from 0.2 to 0.08 is observed as collision energy increases from $\sqrt{s} \approx 7$ to 39 GeV.

1. Introduction

During the last years an extensive analysis of the experimental data from Au-Au collisions with maximal energy at BNL RHIC collider and Pb-Pb collisions at CERN LHC has been done in the various models based on the viscous hydrodynamic approach, with the aim to extract the values of transport coefficients of the medium created, e.g. shear viscosity to entropy density ratio η/s. There are still uncertainties coming from the different models of the pre-thermal state used, however a combined analysis of top RHIC/LHC data within the same model points to the temperature-dependent η/s of QGP medium which has to increase with temperature [1].

The aim of the present study is to conduct a similar analysis for heavy ion collisions at lower collision energies corresponding to RHIC Beam Energy Scan program and highest collision energies achievable at the future FAIR/NICA experiments, and estimate the effective value of η/s of hydrodynamic medium created. For this aim, a state-of-the-art event-by-event viscous hydrodynamic + cascade model is used.

2. Model description

The first key ingredient for the applications of the hydro-based model to the BES collision energy range is a full 3-dimensional hydrodynamic evolution. This requires non-trivial space-time rapidity dependence of initial conditions for the hydrodynamic phase. This feature is often missing in the present initial state models, such as (Monte Carlo) Glauber approach,

CGC-based models, IP-Glasma. We use UrQMD cascade [2, 3] for the pre-thermal phase of the evolution, which includes and employs Pythia for the calculation of initial hard scatterings. The interactions during the pre-thermal phase are calculated until a hypersurface of a constant Bjorken proper time $\tau = \sqrt{t^2 - z^2} = \tau_0$, the latter being a parameter of the model.

On the $\tau = \tau_0$ hypersurface the system is fluidized in a following way. The energy, momenta, baryon and electric charges of hadrons crossing the hypersurface are distributed to the corresponding quantities of the (geometrically) neighbouring fluid cells using Gaussian profiles:

$$\left\{\Delta P^{\mu}_{ijk}, \Delta N^{0}_{ijk}\right\} = \left\{P^{\mu}, N^{0}\right\} \cdot C \cdot \exp\left(-\frac{\Delta x_i^2 + \Delta y_j^2}{R_\perp^2} - \frac{\Delta \eta_k^2}{R_\eta^2}\gamma_\eta^2 \tau_0^2\right) \quad (1)$$

where Δx_i, Δy_j, $\Delta \eta_k$ denote the particle's position with respect to the center of the hydrodynamic cell $\{i, j, k\}$, and $\gamma_\eta = \cosh(y_p - \eta)$ is the longitudinal Lorentz factor of a particle moving with respect to the reference frame with rapidity η. The Gaussians are normalized so that the total energy, momentum and charge deposition to the fulid from each particle equals to its energy, momentum and charge.

The second key ingredient of the model is finite baryon density, which is consistently taken into account. It is naturally present in the pre-thermal phase via UrQMD dynamics, and initial baryon and electric charge densities for the hydrodynamic expansion are calculated as mentioned above, whereas initial strangeness density is taken to be zero for all hydrodynamic system. Next, in hydrodynamic phase the additional charge conservation equations are solved, which provides densities distributions at the end of the hydro phase. The equation of state for hydrodynamics depends on the baryon density as well, provided by Chiral model [5].

The Israel-Stewart framework [4] is employed for the causal relativistic viscous hydrodynamic equations. In particular, the energy-momentum conservation equations and additionally the evolutionary equations for the shear stress tensor:

$$\langle u^{\gamma} \partial_{;\gamma} \pi^{\mu\nu} \rangle = -\frac{\pi^{\mu\nu} - \pi^{\mu\nu}_{\text{NS}}}{\tau_\pi} - \frac{4}{3}\pi^{\mu\nu}\partial_{;\gamma}u^{\gamma} \quad (2)$$

are solved, see [11] for the details. For the purpose of the current study neither bulk viscosity nor baryon diffusion are taken into account. The numerical 3+1D viscous hydrodynamic solution is performed with `vHLLE` code, which is extensively described in [6] and is tested through.

The particlization, i.e. transition from hydrodynamic medium to a hadron gas, is fixed to happen at the hypersurface of constant energy density in the rest frame of the fluid, $\epsilon = \epsilon_{\text{crit}} = 0.5$ GeV/fm^3. A Cooper-Frye prescription [7] is used to produce hadron distribution at the hypersurface. Numerically, the particlization hypersurface is assembled from many finite flat elements $\Delta\sigma_\mu$, which are found via Cornelius subroutine [8]. This results in a following distribution for the hadron species i:

$$p^0 \frac{d^3 N_i}{d^3 p} = \sum_k \Delta\sigma_{k,\mu} p^{\mu} f(p \cdot u_k, T_k, \mu_{k,i}), \quad (3)$$

where the distribution function f is taken as the one for local equilibrium with the non-equilibrium corrections according to shear the shear stress tensor $\pi^{\mu\nu}$ [11].

The produced particles are inserted as an input to the UrQMD cascade, which calculates its interactions and decays short-lived and long-lived resonances, serving as an afterburner component of the model.

3. Results

The model parameters: Gaussian radii $R_\perp$, R_η, shear viscosity over entropy rati η/s and starting time for the hydrodynamic evolution τ_0 have to be fixed from the fit to the experimental data.

$\sqrt{s}$ [GeV]	τ_0 [fm/c]	$R_\perp$ [fm]	R_η [fm]	η/s
7.7/8.8(SPS)	3.2/2.83	1.4	0.5	0.2
11.5	2.1	1.4	0.5	0.2
19.6/17.3(SPS)	1.22/1.42	1.4	0.5	0.15
27	1.0	1.2	0.5	0.12
39	0.9	1.0	0.7	0.08
62.4	0.7	1.0	0.7	0.08
200	0.4	1.0	1.0	0.08

Table 1. Collision energy dependence of the model parameters, which is used to adjust the model to the experimental data in BES region and higher RHIC energies.

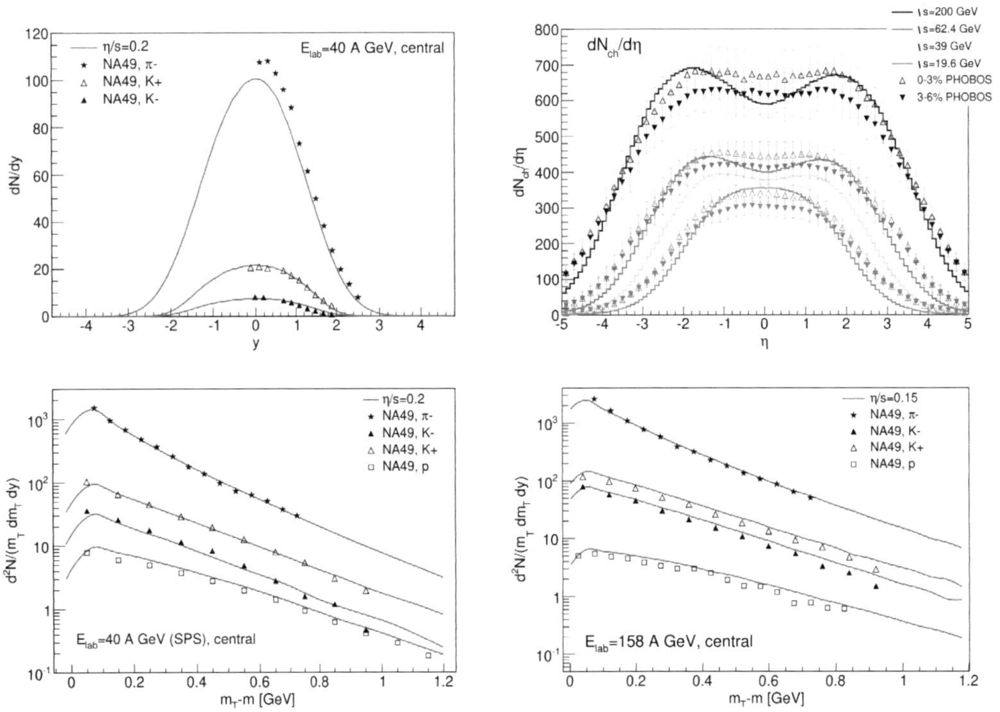

Figure 1. Rapidity distributions of pions and kaons at $E_{\rm lab} = 40$ A GeV Pb-Pb collisions (top left), pseudorapidity distributions of all charged hadrons at $\sqrt{s} = 19.6, 62.4$ and 200 GeV Au-Au collisions (top right), calculated in the model and compared to the experimental data from NA49 [9] and PHOBOS [10]

However, doing a χ^2 fit is hardly feasible since evaluating every point in the parameter space, i.e. running an event-by-event hydrodynamic+cascade simulation typically requires $\propto 10^3$ CPU hours. Thus, the parameters are manually adjusted instead, basing on a visual reproduction of the experimental observables: rapidity spectra, transverse momentum spectra and p_T integrated elliptic flow, at every particular collision energy. The resulting values of the parameters are listed in Table 1. Note that at $\sqrt{s} < 20$ GeV collision energy the value of the starting time τ_0 is taken to be equal to the time for two nuclei to completely pass through each other on average, $\tau_0 = 2R/(\gamma v_z)$, where R is a radius of the nucleus and v_z is its velocity in the laboratory frame. However, at $\sqrt{s} > 40$ GeV the latter formula leads to the values of the starting time $\tau_0 < 0.5$ fm/c, which is too early time for the hydrodynamic picture to be applicable. Thus for higher collision energies τ_0 is adjusted as well as another free parameter of the model.

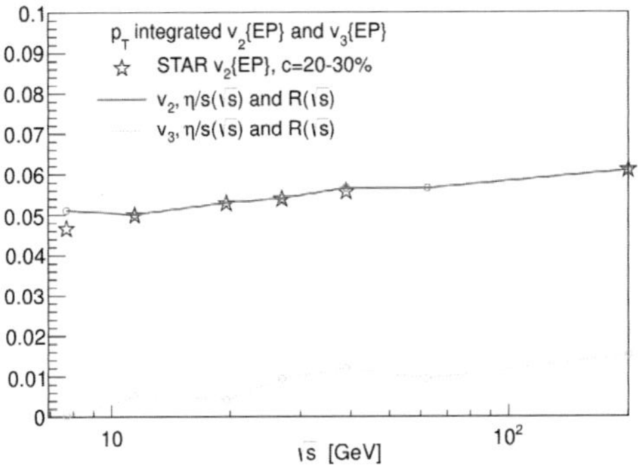

Figure 2. p_T integrated elliptic flow of all charged hadrons for 20-30% central Au-Au collisions at $\sqrt{s_{NN}} = 7.7-200$ GeV. The data points are from STAR collaboration [12].

The resulting (pseudo)rapidity and transverse momentum distributions are shown in Fig. 1. Note that, whereas the parameters are chosen so that only the overall height of the $dN/d\eta$ approaches the experiment, the trends in the shape change as a function of collision energy (going from baryon compression to baryon transparency) are generally reproduced as well. However, for higher collision energies there is a shallow dip in the pseudorapidity distribution (see Fig. 1, top right), which is not much visible in the data. The shapes of p_T spectra for pions, kaons and protons are reproduced as well, which indicates a correct interplay between the radial flow and thermal motion of the produced hadrons in the model.

Finally we turn to the results for the flow coefficients in Fig. 2. Whereas the elliptic flow v_2 is used to adjust the parameters, and is therefore following the experimental data points, the triangular flow v_3 can be interpreted as a model prediction.

In summary, event-by-event analysis in the 3+1D viscous hydrodynamic+cascade model is performed for heavy ion collisions at $\sqrt{s} = 7.7 - 200$ GeV. As a result of the model adjustment to the experimental data the effective values of η/s for the hydrodynamic phase are found for the set of BES collision energy points. A decrease of the effective values of η/s from 0.2 to 0.08 is observed as collision energy increases from $\sqrt{s} \approx 7$ to 39 GeV. This hints that in a more realistic hydrodynamic setup, where η/s would be taken temperature- and chemical potential dependent, it should increase with increasing baryon chemical potential.

Acknowledgements

The simulations have been performed at the Center for Scientific Computing (CSC) at the Goethe-University Frankfurt. The authors acknowledge the financial support by the Helmholtz International Center for FAIR and Hessian LOEWE initiative. HP acknowledges funding by the Helmholtz Young Investigator Group VH-NG-822. The work of P.H. was supported by BMBF under contract no. 06FY9092.

References

[1] H. Niemi, G. S. Denicol, P. Huovinen, E. Molnar and D. H. Rischke, Phys. Rev. C **86** (2012) 014909
[2] S. A. Bass *et al.*, Prog. Part. Nucl. Phys. **41** (1998) 255 [nucl-th/9803035].
[3] M. Bleicher *et al.*, J. Phys. G **25** (1999) 1859 [hep-ph/9909407].
[4] W. Israel and J. M. Stewart, Annals Phys. **118** (1979) 341.
[5] J. Steinheimer, S. Schramm and H. Stocker, J. Phys. G **38** (2011) 035001 [arXiv:1009.5239 [hep-ph]].
[6] I. Karpenko, P. Huovinen and M. Bleicher, Comput. Phys. Commun. **185** (2014) 3016
[7] F. Cooper and G. Frye, Phys. Rev. D **10** (1974) 186.
[8] P. Huovinen and H. Petersen, Eur. Phys. J. A **48** (2012) 171 [arXiv:1206.3371 [nucl-th]].
[9] S. V. Afanasiev *et al.* [NA49 Collaboration], Phys. Rev. C **66** (2002) 054902 [nucl-ex/0205002].
[10] B. Alver *et al.* [PHOBOS Collaboration], Phys. Rev. C **83** (2011) 024913 [arXiv:1011.1940 [nucl-ex]].
[11] I. A. Karpenko, M. Bleicher, P. Huovinen and H. Petersen, J. Phys. Conf. Ser. **509** (2014) 012067
[12] L. Adamczyk *et al.* [STAR Collaboration], Phys. Rev. C **86** (2012) 054908 [arXiv:1206.5528 [nucl-ex]].

Journal of Physics: Conference Series **612** (2015) 012053 doi:10.1088/1742-6596/612/1/012053

Identified charged hadron production in Pb-Pb collisions with event shape engineering

Marco Antonio Tangaro for the ALICE collaboration
Dipartimento Interateneo di Fisica "M. Merlin" and Sezione INFN, Via Orabona 4, 70126 Bari, Italy

E-mail: marco-antonio.tangaro@cern.ch

Abstract. The Event Shape Engineering technique allows the selection of different event shapes for a definite centrality and colliding system. The event selection is based on the azimuthal distribution of produced particles, using the so-called flow vector. For such shape selected events, the elliptic flow coefficient (v_2) is significantly different with respect to the unbiased events. Moreover, recent Monte-Carlo studies show a strong correlation between the (final state) event shape selection and the (initial state) eccentricity of the collision. This opens the opportunity to characterize events according to the initial geometry. An approach to select the eccentricity of the event with the Event Shape Engineering is presented. Then the effect of this selection on identified particle spectra, mean transverse momentum and v_2 of charged particles in heavy-ion collisions at $\sqrt{s_{\rm NN}} = 2.76$ TeV center-of-mass energy is discussed.

1. Introduction

At present, the overlap geometry in nucleus-nucleus collisions is usually studied by fixing the collision centrality or by using colliding nuclei of different size and shape. The Event Shape Engineering (ESE) technique, recently proposed in [1], allows to bias the event sample by selecting events characterized by a well defined initial geometry, by means of the flow vector evaluated from the azimuthal distribution of produced particles. Monte Carlo (MC) simulations [1] show that anisotropic flow coefficients and flow vector are strongly correlated. Moreover a strong positive correlation between the flow vector and the eccentricity of the collision has been observed [2]. These results suggest that the shape of the initial geometry can be constrained using the flow vector in the final state.

2. Event Shape Engineering

In this analysis data from Pb–Pb collisions at $\sqrt{s_{\rm NN}} = 2.76$ TeV collected by ALICE [3, 4] in 2010, are used.

The main tracking detectors used are the Inner Tracking System (ITS) [5] and the Time Projection Chamber (TPC) [6]. The ALICE detector is capable of performing Particle IDentification (PID) over a broad range of transverse momentum using different PID techniques. In this analysis a combined selection of the TPC-TOF (Time Of Flight [7]) signal has been used to cleanly identify pions, kaons and protons over a wide momentum range, without loss of efficiency. The centrality of the Pb–Pb collisions has been evaluated using the information of the event multiplicity measured in the V0 detectors [8], which consist of two rings of plastic scintillators named V0A ($2.8 < \eta < 5.1$) and V0C ($-3.7 < \eta < -1.7$).

The event selection is based on the magnitude of the reduced flow-vector q_2 [9, 10], defined as:

$$q_2 = \frac{|Q_2|}{\sqrt{M}} \tag{1}$$

where

$$Q_2 = Q_{2,x} + iQ_{2,y} \tag{2}$$

and

$$Q_{2,x} = \sum_i w_i cos(n\phi_i), \quad Q_{2,y} = \sum_i w_i sin(n\phi_i) \tag{3}$$

here the sum goes over all V0 detector channels, ϕ_i and w_i are the azimuthal angle and the multiplicity of the channel i, respectively, and M is the event multiplicity.

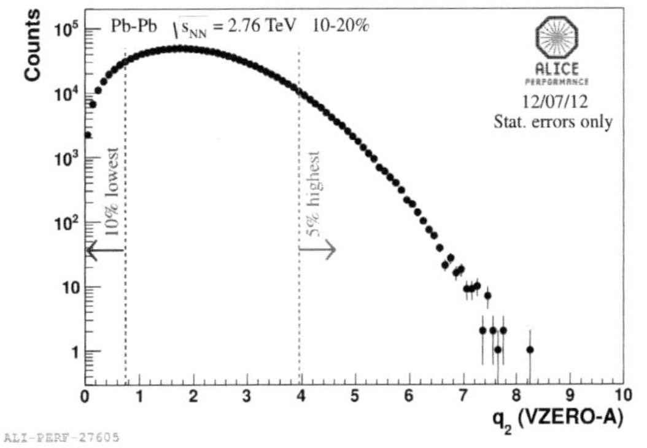

Figure 1. q_2 distribution from V0A in 10-20% centrality class.

In Fig. 1 the q_2 distribution from the V0A is reported. From the flow vector distribution events with 5% highest and 10% lowest q_2 can be selected, called, in the following, large-q_2 and small-q_2 respectively.

To avoid trivial bias from non-flow processes (e.g. resonance decay, jet fragmentation and Bose-Einstein correlations) two sub-events, with large η-gap, are considered. The flow vector is evaluated from V0A ($2.8 < \eta < 5.1$), while the experimental observables are measured in the TPC. This ensures a large $|\Delta\eta|$ separation between the detectors involved in the analysis, leading to non-flow suppression.

3. Results

The v_2 and hadron spectra are reported for primary particles, defined as particles produced directly in the collision or in the decay of short-lived resonances and not particles from the weak decays of strange hadrons or from interactions with material.

In the analysis of v_2 only tracks reconstructed in the TPC with $0.2 < p_T < 20$ GeV/c are used, since the detector provides the smallest azimuthal non-uniformities. The v_2 is evaluated with the event plane method [9], correlating tracks in the TPC ($-0.8 < \eta < 0.8$) with the event plane from V0C ($-3.7 < \eta < -1.7$).

The effect of the event-shape selection on the distribution of $v_2(p_T)$ of unidentified charged hadron is shown in Fig. 2 (left) for the 30-40% centrality class. The ratio between large(small)-q_2 and unbiased sample is reported on the right. The event shape selection results in a change in the value of the v_2 of 20% for large-q_2 and 10% for the small-q_2. Ratios are constant up to

p_T= 6 GeV/c, while for p_T > 6 GeV/c the ratios seem to approach unity, even though large statistical uncertainties do not allow a strong conclusion.

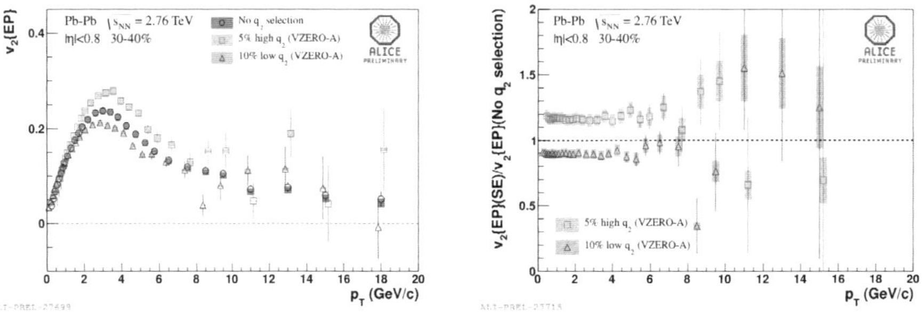

Figure 2. (left) $v_2(p_T)$ distribution for the unbiased (black), 5% large-q_2(red) and 10% small-q_2(blue) sample, in 30-40% centrality class. (right) Ratio between large(small)-q_2 and unbiased sample in red(blue).

We characterize the selected sample with a measurement of transverse momentum distributions. This allows to study the correlation between anisotropic and radial flow, by measuring p_T distributions of primary particles in strongly or weakly anisotropic environments, at fixed impact parameter.

Figure 3 shows the ratio between p_T-spectra in 10% large(small)-q_2 and the unbiased sample, in 30-40% centrality class. In the ratio the spectra corrections cancels out, allowing a precise measurement of the effect of the event shape selection on p_T spectra.

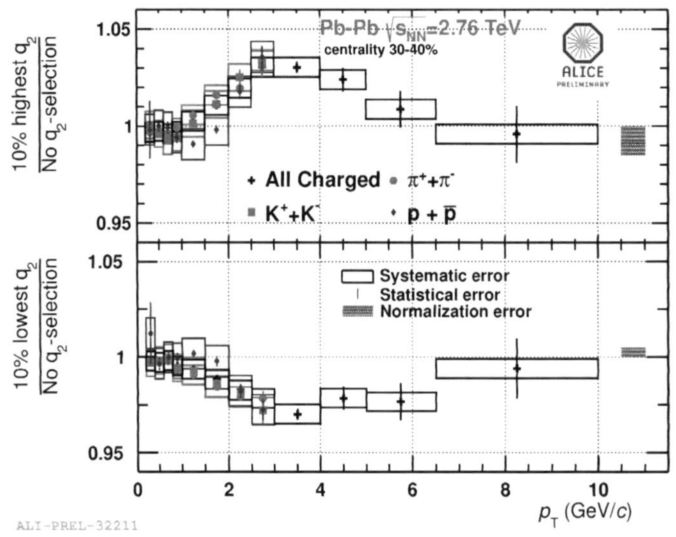

Figure 3. Ratio between the transverse momentum spectra in the sample with 10% highest (lowest) q_2 and the unbiased sample.

A modification of the p_T-spectrum is observed for $1 \leq p_T \leq 5$ GeV/c: in the large-q_2 sample the p_T distribution is harder than the unbiased, while the opposite happens in small-q_2 sample, where softer p_T spectra are observed. The modification seems to vanish at high p_T, at the limit of applicability of the hydrodynamical picture. The selection does not introduce any trivial bias related to the multiplicity shift or jet contribution, which would have influenced the p_T spectra shape even at high p_T.

The trend is the same for pions, kaons and protons. A hint of mass ordering is observable in the transverse momentum range $p_T \leq 3$ GeV/c suggesting a stronger(weaker) radial flow in the large(small)-q_2 event sample.

4. Summary

The effect of the event shape engineering technique on v_2 and spectra has been reported. The event shape selection allows to select events with larger or smaller elliptic flow. The $v_2(p_T)$ is independent on p_T. This suggests that the q_2 selection allows to select event based on the initial geometry of the collision.

Particle spectra in event shape selected events has been also presented: large-q_2 values lead to harder spectra, while softer spectra have been found in small-q_2 events. The effect vanishes at high p_T, showing a behaviour similar to that observed for the elliptic flow in shape engineered events. This, with the mass ordering observed at low p_T, suggests that the origin of the effect is related with the hydrodynamic response of the system to the initial geometry rather than with hard processes.

References

[1] Schukraft J *et al* 2013 *Phys. Lett.* B **719** 394–98
[2] Huo P *et al* 2014 *Phys. Rev.* C **90** 024910
[3] ALICE Collaboration (2004) *J. Phys. G: Nucl. Part. Phys.* **30** 1517
[4] ALICE Collaboration (2006) *J. Phys.* G **32** 1295–2040
[5] ALICE Collaboration (2010) *J. Inst.* **5** 3003
[6] Alme J *et al* (2010) *Nuclear Instruments and Methods in Physics Research* A **622** 316
[7] ALICE Collaboration (2014) *Int. J. Mod. Phys.* A **29** 1430044
[8] ALICE Collaboration (2013) *JINST* P **8** P10016
[9] Voloshin S A, Poskanzer A M and Snellings R 2010 *Landolt-Boernstein, Relativistic Heavy Ion Physics* vol 1/23 (Springer-Verlag) p 5–54
[10] Poskanzer A M and Voloshin S A 1998 *Phys. Rev.* C **58** 1671

Exact solutions of the (0+1)-dimensional kinetic equation in the relaxation time approximation

Ewa Maksymiuk
Institute od Physics, Jan Kochanowski University, PL-25460 Kielce, Poland
E-mail: `MaksymiukEwa@gmail.com`

Abstract. We present exact solutions of the (0+1)-dimensional kinetic equation for a massive gas in the relaxation time approximation. At first, we analyse the case of classical statistics and argue that the traditional second-order hydrodynamics misses the shear-bulk coupling. In the next step, we include Bose-Einstein and Fermi-Dirac statistics in the calculations and show that they are important for the description of the effects connected with bulk viscosity.

1. Introduction

In this paper we present exact solutions of the (0+1)-dimensional kinetic (Boltzmann) equation for a massive gas in the relaxation time approximation [1]. Our results describe a purely longitudinal boost-invariant expansion and may be useful for the description of very early stages of relativistic heavy-ion collisions. At first, we analyse the case of classical statistics and show that the traditional second-order hydrodynamics has problems to correctly reproduce the kinetic result. This discrepancy is due to the missing shear-bulk coupling in the standard second-order hydrodynamics. We further take into account the effects of quantum statistics [2]. They turn out to be important for the bulk viscous pressure.

The results presented here generalise several earlier results obtained for massless particles [3, 4]. We note that the exact solutions of the kinetic equation help us to select the right form of the kinetic coefficients [3, 4] and the correct structure of the hydrodynamic equations, as has been demonstrated recently in Refs. [3, 4, 5, 6, 7, 8, 9].

2. The boost-invariant Boltzmann equation in the relaxation time approximation

Our approach uses a simple form of the kinetic equation, namely

$$p^\mu \partial_\mu f = C[f], \qquad C[f] = -\frac{p_\alpha u^\alpha}{\tau_{\rm eq}}(f - f_{\rm eq}), \tag{1}$$

where f is the one-particle phase-space distribution function depending on the parton space-time coordinates x and momentum p, and C is the collision term written in the relaxation time approximation [10, 11, 12]. The parameter $\tau_{\rm eq}$ is the relaxation time. In our present calculations we use the value $\tau_{\rm eq} = 0.25$ fm/c. The boost-invariance implies that the kinetic equation (1) can be rewritten in the form

$$\frac{\partial f}{\partial \tau} = \frac{f_{\rm eq} - f}{\tau_{\rm eq}}, \tag{2}$$

where $\tau = \sqrt{t^2 - z^2}$ is the proper time. In addition, the function f may depend only on the three variables: τ, w and p_T, where $w = tp_L - zE$. The background equilibrium distribution function may be written as

$$f_{\rm eq}(\tau, w, p_T) = \frac{2}{(2\pi)^3}\left\{\exp\left[\frac{\sqrt{w^2 + (m^2 + p_T^2)\,\tau^2}}{T(\tau)\tau}\right] - \epsilon\right\}^{-1}. \tag{3}$$

In Eq. (3) the parameter ϵ specifies the appropriate quantum statistics. With $\epsilon = +1, 0, -1$ we include Bose-Einstein, Boltzmann or Fermi-Dirac statistics, respectively.

The first moment of the left-hand side of the kinetic equation describes the divergence of the energy-momentum tensor that must be conserved, namely

$$T^{\mu\nu}(\tau) = g_0 \int dP p^\mu p^\nu f(\tau, w, p_T), \qquad \partial_\mu T^{\mu\nu} = 0. \tag{4}$$

Here g_0 is the number of internal degrees of freedom and dP is the momentum integration measure. Equation (4) is fulfilled if the energy densities calculated with the distribution functions f and $f_{\rm eq}$ are equal. This leads us to our main equation

$$\begin{aligned} & T^4(\tau)\tilde{\mathcal{H}}_2\left[1, \frac{m}{T(\tau)}\right] \\ &= D(\tau,\tau_0)\Lambda_0^4\tilde{\mathcal{H}}_2\left[\frac{\tau_0}{\tau\sqrt{1+\xi_0}}, \frac{m}{\Lambda_0}\right] + \int\limits_{\tau_0}^{\tau}\frac{d\tau'}{\tau_{\rm eq}}D(\tau,\tau')T^4(\tau')\tilde{\mathcal{H}}_2\left[\frac{\tau'}{\tau}, \frac{m}{T(\tau')}\right]. \end{aligned} \tag{5}$$

This is an integral equation for the effective temperature $T(\tau)$ that can be solved using the iterative method [13]. The function $\tilde{\mathcal{H}}_2(y,z)$ is defined in Ref. [2]. Here we used the initial condition given by the Romatsche-Strickland form

$$f_0(w, p_T) \;=\; \frac{1}{4\pi^3}\left\{\exp\left[\frac{\sqrt{(1+\xi_0)w^2 + (m^2 + p_T^2)\tau_0^2}}{\Lambda_0\tau_0}\right] - \epsilon\right\}^{-1}. \tag{6}$$

We note that the form of Eq. (5) is the same for the three different statistics. The differences are hidden in the implicit dependence of the functions $f_{\rm eq}$, f_0, and $\tilde{\mathcal{H}}_2$ on the quantum statistics parameter ϵ.

3. Shear and bulk viscosities of a relativistic massive quantum gas

To find the shear viscosity η for Bose–Einstein and Fermi–Dirac gases, we use the formula [14]

$$\eta(T) = \frac{2g_0\tau_{\rm eq}}{15T}\int\frac{d^3p}{(2\pi)^3}\frac{p^4}{E^2}f_{\rm eq}(1 + \epsilon f_{\rm eq}). \tag{7}$$

On the other hand, we determine the effective shear viscosity using the exact solution of the kinetic equation

$$\eta_{\rm eff}(\tau) = \frac{1}{2}\tau\left[\mathcal{P}_T(\tau) - \mathcal{P}_L(\tau)\right]. \tag{8}$$

We treat the bulk viscosity ζ in the similar way as the shear viscosity. For a massive quantum gas, the formula for the bulk viscosity is the following [14]

$$\zeta(T) = \frac{2g_0\tau_{\rm eq}}{3T}\int\frac{d^3p}{(2\pi)^3}\frac{m^2}{E^2}f_{\rm eq}(1+\epsilon f_{\rm eq})\left(c_s^2E - \frac{p^2}{3E}\right). \tag{9}$$

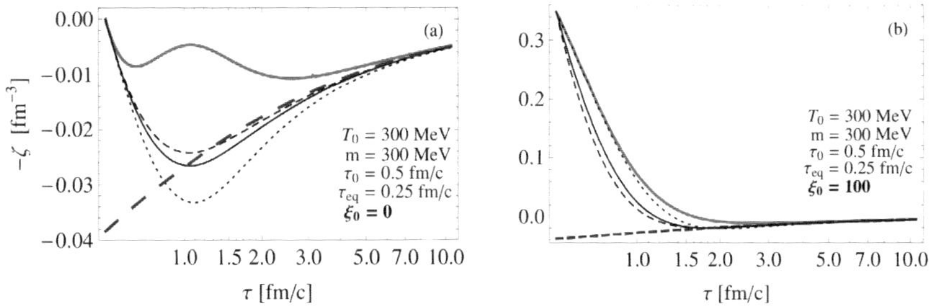

Figure 1. Comparison of the bulk viscosity for Boltzmann statistics. The mass used in the calculations is $m = 300$ MeV, which is an effective mass of quarks in a proton. The panel (a) describes a system which is initially isotropic, while the panel (b) describes a system which is initially highly oblate. Red solid lines represent the effective bulk viscosity obtained from the kinetic theory, blue-dashed lines describe the bulk viscosity given by Eq. (9), finally, black lines show our results obtained for three versions of second–order hydrodynamics [1].

To find the effective bulk viscosity, which is obtained from the exact solution we use the formula

$$\zeta_{\rm eff}(\tau) = -\frac{1}{3}\tau \left[\mathcal{P}_L(\tau) + 2\mathcal{P}_T(\tau) - 3\mathcal{P}_{\rm eq}(\tau)\right]. \tag{10}$$

The sound velocity appearing in Eq. (9) is obtained from the formula $c_s^2(T) = \partial\mathcal{P}_{\rm eq}(T)/\partial\mathcal{E}_{\rm eq}(T)$. In the classical limit, $\epsilon \to 0$, the integrals (7) and (9) become analytic. The appropriate formulas can be found in Ref. [1].

4. Results

Results shown in Fig. 1 represent the time evolution of the bulk viscosity calculated directly from kinetic theory and compared with three formulations of second-order hydrodynamics [1]. In the case of an initially isotropic system we can see that the simplest formulation of second-order hydrodynamics (dotted line) gives the worst agreement with the kinetic theory. On the other hand, this formulation gives the best agreement for an initially highly oblate system. The problems illustrated in Fig. 1 were useful to identify the importance of the shear-bulk couplings [15, 16, 17] in the hydrodynamic approach. To have a good agreement it is necessary to use equations where the bulk and shear viscosities are correlated [16, 17]. The proper description of the bulk pressure is important as it may affect different physical observables studied in relativistic heavy-ion collisions [18, 19, 20].

To show the effects of quantum statistics on the evolution of matter we calculated the shear viscosity using Eqs. (7) and (8) and the bulk viscosity using Eqs. (9) and (10). In Fig. 2 we present our results. In the case of the shear viscosity, we find only small differences between the results obtained for Bose-Einstein, Boltzmann, and Fermi-Dirac statistics. This is in contrast to the bulk-viscosity case, where we find important differences, especially, for the initially isotropic systems.

5. Conclusions

We have constructed the exact solutions of the Boltzmann equation using analytical and numerical methods. This allowed us to find the effective bulk and shear viscosities of the (0+1)–dimensional system and to compare them with the analytic formulas appearing in the literature. We have shown that standard equations of second-order hydrodynamics do not work

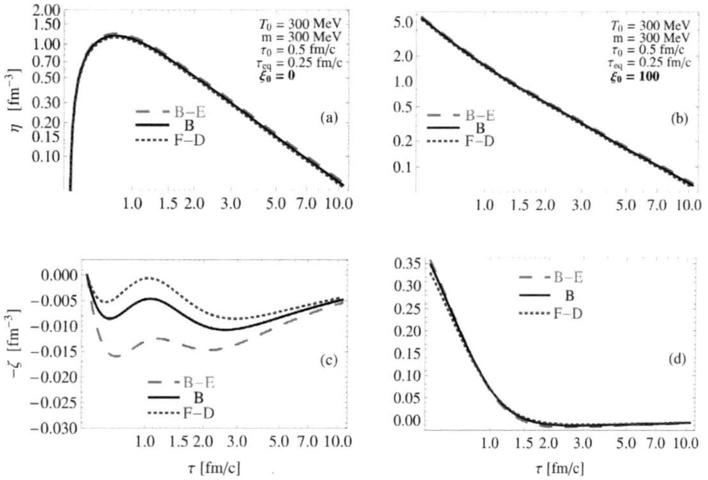

Figure 2. Comparison of exact solutions for different quantum statistics. The values of the parameters are displayed in the panels. Panels (a) and (c) differ from (b) and (d) by the value of the initial anisotropy parameter.

properly — although the quantum statistics effects are not essential for the shear viscosity they become quite important for the correct description of the bulk viscous effects.

Acknowledgments

I would like to thank my supervisor Wojciech Florkowski for valuable advices and interesting discussions. This work was supported by Polish National Science Center grant No. DEC-2012/06/A/ST2/00390.

References

[1] Florkowski W, Maksymiuk E, Ryblewski R and Strickland M 2014 *Phys.Rev.* **C89** 054908
[2] Florkowski W, Maksymiuk E 2014 (*Preprint* 1411.3666)
[3] Florkowski W, Ryblewski R and Strickland M 2013 *Nucl.Phys.* **A916** 249–259
[4] Florkowski W, Ryblewski R and Strickland M 2013 *Phys.Rev.* **C88** 024903
[5] Florkowski W, Ryblewski R, Strickland M and Tinti L 2014 *Phys.Rev.* **C89** 054909
[6] Nopoush M, Ryblewski R and Strickland M 2014 *Phys.Rev.* **C90** 014908
[7] Denicol G S, Heinz U W, Martinez M, Noronha J and Strickland M 2014 (*Preprint* 1408.5646)
[8] Denicol G S, Heinz U W, Martinez M, Noronha J and Strickland M 2014 (*Preprint* 1408.7048)
[9] Nopoush M, Ryblewski R and Strickland M 2014 (*Preprint* 1410.6790)
[10] Bhatnagar P L, Gross E P and Krook M 1974 *Physical Review* **94** 511-525
[11] Anderson J and Witting H 1974 *Physica* **74** 466-488
[12] Czyż W and Florkowski W 1986 *Acta Physica Polonica* **B17** 819-837
[13] Banerjee B, Bhalerao R S and Ravishankar V 1989 *Physics Letters* **B224** 16
[14] Sasaki C and Redlich K 2009 *Physical Review* **C79** 055207
[15] Denicol G, Jeon S and Gale C 2014 (*Preprint* 1403.0962)
[16] Denicol G S, Florkowski W, Ryblewski R and Strickland M 2014 *Phys.Rev.* **C90** 044905
[17] Jaiswal A, Ryblewski R and Strickland M 2014 *Phys.Rev.* **C90** 044908
[18] Noronha-Hostler J, Denicol G S, Noronha J, Andrade R P G and Grassi F 2013 *Physical Review* **C88** 044916
[19] Noronha-Hostler J, Denicol G S, Noronha J, Andrade R P G and Grassi F 2013 *Journal of Physics: Conference Series* **458** 012018
[20] Rose J B, Paquet J F, Denicol G S, Luzum M, Schenke B *et al.* 2014 (*Preprint* 1408.0024)

Directed flow in relativistic heavy-ion collisions within the PHSD transport approach and 3FD hydrodynamical model

V P Konchakovski[1], **W Cassing**[1], **Y B Ivanov**[2,3] **and V D Toneev**[4]

[1] Institute for Theoretical Physics, University of Giessen, 35392 Giessen, Germany
[2] Kurchatov Institute, 123098 Moscow, Russia
[3] National Research Nuclear University "MEPhI", 115409 Moscow, Russia
[4] Joint Institute for Nuclear Research, 141980 Dubna, Russia

Abstract. We analyze recent STAR data for the directed flow of protons, antiprotons and charged pions obtained within the beam energy scan program within the Parton-Hadron-String-Dynamics (PHSD) transport model and the 3-Fluid hydroDynamics (3FD) approach. We clarify the role of partonic degrees of freedom in the kinetic PHSD approach. The PHSD results, simulating a partonic phase and its coexistence with a hadronic one, are roughly consistent with data. The hydrodynamic results are obtained for two EoS, a pure hadronic EoS and an EoS with a crossover type transition. The latter case is favored by the STAR experimental data. Special attention is paid to the description of antiproton directed flow based on the balance of $p\bar{p}$ annihilation and the inverse processes for $N\bar{p}$ pair creation from multi-meson interactions. Generally, a semi-qualitative agreement between the measured data and model results supports the idea of a crossover type quark-hadron transition which softens the nuclear EoS.

1. Introduction

The study of the particle azimuthal angular distribution in momentum space is an important tool to probe the hot, dense matter created in heavy-ion collisions [1, 2]. The directed flow refers to a collective sidewards deflection of particles and is characterized by the first-order harmonic v_1 of the Fourier expansion of the particle azimuthal angular distribution with respect to the reaction plane [3]. It is generally assumed that the directed flow is generated during the early nuclear passage time [4, 5]. The directed transverse flow therefore probes the onset of bulk collective dynamics during thermalization, thus providing valuable information on the pre-equilibrium stage [6, 7, 8, 9]. Apart from first measurements in the early nineties and till recent times, the directed flow was studied mainly theoretically [10, 11, 12] although some experimental information from the Schwerionen-Synchrotron (SIS), Alternating Gradient Synchrotron (AGS) and Super-Proton-Synchrotron (SPS) is available [13]. The interest in the directed flow $v_1(y)$ has recently been enhanced considerably due to new STAR data obtained in the framework of the beam energy scan (BES) program [14]. The directed flow of identified hadrons – protons, antiprotons, positive and negative pions – has been measured with high precision for semi-central Au+Au collisions in the energy range $\sqrt{s_{NN}}$ = (7.7-200) GeV. We analyse these data in the framework of the Parton-Hadron-String-Dynamics (PHSD) transport model and a 3-Fluid hydroDynamics (3FD) approach and refer the reader to Ref. [15] for the details.

Journal of Physics: Conference Series **612** (2015) 012055 doi:10.1088/1742-6596/612/1/012055

2. Reminder of PHSD transport approach

The PHSD model is a covariant dynamical approach for strongly interacting systems formulated on the basis of Kadanoff-Baym equations [16, 17] or off-shell transport equations in phase-space representation, respectively. In the Kadanoff-Baym theory the field quanta are described in terms of dressed propagators with complex selfenergies. Whereas the real part of the selfenergies can be related to mean-field potentials of Lorentz scalar, vector or tensor type, the imaginary parts provide information about the lifetime and/or reaction rates of time-like particles [18]. Once the proper complex selfenergies of the degrees of freedom are known, the time evolution of the system is fully governed by off-shell transport equations for quarks and hadrons (as described in Refs. [16, 18]). The PHSD model includes the creation of massive quarks via hadronic string decay - above the critical energy density ~ 0.5 GeV/fm^3 - and quark fusion forming a hadron in the hadronization process. With some caution, the latter process can be considered as a simulation of a crossover transition since the underlying EoS in PHSD is a crossover [18]. At energy densities close to the critical energy density the PHSD describes a coexistence of the quark-hadron mixture. This approach allows for a simple and transparent interpretation of lattice QCD results for thermodynamic quantities as well as correlators and leads to effective strongly interacting partonic quasiparticles with broad spectral functions. For a review on off-shell transport theory we refer the reader to Ref. [18]; PHSD model results and their comparison with experimental observables for heavy-ion collisions from the lower SPS to RHIC energies can be found in Refs. [18, 19, 20, 21]. In the hadronic phase, i.e. for energies densities below the critical energy density, the PHSD approach is identical to the Hadron-String-Dynamics (HSD) model [22, 23, 24].

3. Reminder of 3FD hydrodynamic approach

The 3FD model [25] is a straightforward extension of the two-fluid model with a radiation of direct pions [26, 27, 28] and the (2+1)-fluid model [29, 30]. These models have been extended to treat the baryon-free fluid on an equal footing with the baryon-rich ones. A certain formation time, τ, is allowed for the fireball fluid, during which the matter of the fluid propagates without interactions. The formation time τ is associated with the finite time of string formation and decay and is incorporated also in the kinetic transport models such as PHSD and HSD. The 3FD model [25] treats a nuclear collision from the very beginning, i.e., from the stage of the incident cold nuclei to the final freeze-out stage. Contrary to the conventional hydrodynamics, where a local instantaneous stopping of projectile and target matter is assumed, the specific feature of the 3FD is a finite stopping power resulting in a counter streaming regime of leading baryon-rich matter. The basic idea of a 3FD approximation to heavy-ion collisions [31, 32] is that at each space-time point a generally nonequilibrium distribution of baryon-rich matter can be represented as a sum of two distinct contributions initially associated with constituent nucleons of the projectile and target nuclei. In addition, newly produced particles, populating predominantly the midrapidity region, are associated with a fireball fluid. Therefore, the 3FD approximation is a minimal way to simulate the finite stopping power at high incident energies.

4. Directed flow results

The characteristic slope of the $v_1(y)$ distributions at midrapidity, $\frac{dv_1}{dy}|_{y=0} = F$, is presented in Fig. 1 for protons, antiprotons and pions. We compare (as in Ref. [15]) the results of transport approaches (l.h.s) and hydrodynamical models (r.h.s) with the experimental data from the STAR collaboration [14] as well as the results of prior experiments [34, 35].

For protons in Fig.1 (l.h.s) there is a qualitative agreement of the HSD/PHSD results with the experimental measurements. The standard UrQMD model results, as cited in the experimental paper [14] and in the more recent theoretical work [33], are displayed in Fig. 1 (l.h.s.) by the wide and narrow shaded areas, respectively. These results for protons are close to those from the HSD

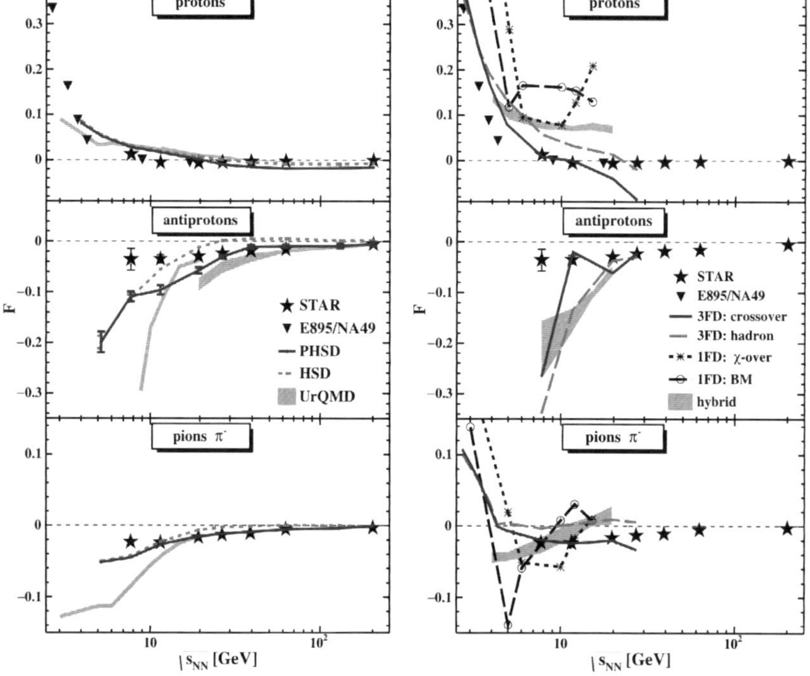

Figure 1. The beam energy dependence of the directed flow slope F at midrapidity for protons, antiproton and charged pions from semicentral Au+Au collisions for transport approaches (l.h.s.: PHSD/HSD, UrQMD [33, 14]) and hydrodynamical models (r.h.s.: 3FD, 1FD, hybrid [33]) in comparison to STAR experimental data [14] and results of prior experiments [34, 35].

and essentially overestimate the slope for energies below ~ 30 GeV. The predictions for the pure hadronic version of the transport model HSD (dotted lines in Fig. 1 (l.h.s.) slightly differ from the PHSD results. For the antiproton slopes we again observe an almost quantitative agreement with the BES experiment [14]. The differences between the calculations and experimental data become apparent for the charged pion slopes at $\sqrt{s_{NN}} \leq 11$ GeV: the negative minimum of the charged pion slope is deeper than the measured one. The HSD and PHSD results practically coincide at low energy (due to a minor impact of partonic degrees of freedom) but dramatically differ from those of the UrQMD model for $\sqrt{s_{NN}} \leq 20$ GeV. This difference might be attributed to a neglect of the inverse processes for antiproton annihilation in UrQMD since in PHSD/HSD the antiprotons pick up some flow from the fusing mesons that is lower in magnitude that the direct flow for primary $\bar{p}$.

The excitation functions for the slopes of the v_1 distributions at midrapidity for 3FD model with different EoS are presented in Fig. 1 (r.h.s.). The discrepancies between experiment and the 3FD model predictions are larger for the purely hadronic EoS (dashed line) then for the crossover EoS (solid line) though it is far from being perfect. In Ref. [33] an essential part of the STAR data (for $\sqrt{s_{NN}} \leq 20$ GeV) is analyzed within collective approaches: the one-fluid (1F) hydrodynamical model with a first-order phase transition simulated by the bag model (BM) and a crossover chiral transition (χ-over), as well as within a modern hybrid model combining hydrodynamics with a kinetic model. The results of this work are also displayed in the Fig. 1 (r.h.s.) for comparison (open circles and stars).

5. Conclusions

The microscopic PHSD transport approach reproduces the general trend in the differential $v_1(y)$ excitation function and leads to an almost quantitative agreement for protons, antiprotons, and pions especially at higher energies. It is noteworthy that 3FD demonstrates high sensitivity to the nuclear EoS and provides the best results with a crossover for the quark-hadron phase transition being in a reasonable agreement with the STAR results in the considered energy range $\sqrt{s_{NN}} < 30$ GeV. Note also that a crossover transition is implemented by default in PHSD. Our flow analysis shows no indication of a first-order transition.

Acknowledgments

This work in part was supported by the LOEWE Center HIC for FAIR as well as BMBF.

References

[1] S. A. Voloshin, A. M. Poskanzer and R. Snellings, in Landolt-Boernstein New Series, I/23, p. 5-54, edited by R. Stock (Springer-Verlag, 2010).
[2] P. Sorensen, In Quark-Gluon Plasma 4, ed. by R. Hwa and X.N. Wang, World Scientific (2010).
[3] A.M. Poskanzer and S.A. Voloshin, Phys. Rev. C **58**, 1671 (1998).
[4] H. Sorge, Phys. Rev. Lett. **78**, 2309 (1997).
[5] N. Herrmann, J. P. Wessels, and T. Wienold, Ann. Rev. Nucl. Part. Sci. **49**, 581 (1999).
[6] E. Schnedermann and U. Heinz, Phys. Rev. Lett. **69**, 2908 (1992).
[7] D. E. Kahana, D. Keane, Y. Pang, T. Schlagel and S. Wang, Phys. Rev. Lett. **74**, 4404 (1995).
[8] J. Barrette *et al.* (E877 Collaboration), Phys. Rev. Lett. **73**, 2532 (1994).
[9] I. G. Bearden *et al.* (NA44 Collaboration), Phys. Rev. Lett. **78**, 2080 (1997).
[10] P. Danielewicz, Nucl. Phys. A **673**, 375 (2000).
[11] P. K. Sahu and W. Cassing, Nucl. Phys. A **712**, 357 (2002).
[12] M. Isse, A. Ohnishi *et al.*, Phys. Rev. C **72**, 064908 (2005).
[13] P. Senger *et al.*, Lect. Notes Phys. **814**, 681 (2011).
[14] STAR Collaboration: L. Adamczyk, *et al.*, Phys. Rev. Lett. **112**, 162301 (2014).
[15] V. P. Konchakovski, W. Cassing, Y. B. Ivanov and V. D. Toneev, Phys. Rev. C **90**, 014903 (2014); W. Cassing, V. P. Konchakovski, A. Palmese, V. D. Toneev and E. L. Bratkovskaya, arXiv:1408.4313.
[16] S. Juchem, W. Cassing, and C. Greiner, Phys. Rev. D **69**, 025006 (2004); Nucl. Phys. A **743**, 92 (2004).
[17] W. Cassing, E. L. Bratkovskaya, Nucl. Phys. A **831**, 215 (2009); Phys. Rev. C **78**, 034919 (2008); W. Cassing, Nucl. Phys. A **791**, 365 (2007).
[18] W. Cassing, E. Phys. J. ST **168**, 3 (2009).
[19] V. D. Toneev *et al.*, Phys. Rev. C **85**, 034910 (2012).
[20] V. P. Konchakovski *et al.*, Phys. Rev. C **85**, 011902 (2012).
[21] O. Linnyk *et al.*, Phys. Rev. C **84** (2011) 054917; Phys. Rev. C **85** (2012) 024910; Phys. Rev. C **87** (2013) 014905.
[22] W. Ehehalt and W. Cassing, Nucl. Phys. A **602**, 449 (1996).
[23] W. Cassing and E. L. Bratkovskaya, Phys. Rep. **308**, 65 (1999).
[24] W. Cassing, E. L. Bratkovskaya, S. Juchem, Nucl. Phys. A **674**, 249 (2000).
[25] Yu. B. Ivanov, V. N. Russkikh, and V. D. Toneev, Phys. Rev. C **73**, 044904 (2006).
[26] I. N. Mishustin, V. N. Russkikh, and L. M. Satarov, Yad. Fiz. **48**, 711 (1988).
[27] V. N. Russkikh, Yu. B. Ivanov, Yu. E. Pokrovsky, and P. A. Henning, Nucl. Phys. **A572**, 749 (1994).
[28] I. N. Mishustin, V. N. Russkikh, and L. M. Satarov, Yad. Fiz. **54**, 429 (1991).
[29] U. Katscher, D. H. Rischke, J. A. Maruhn, W. Greiner, I. N. Mishustin, and L. M. Satarov, Z. Phys. **A346**, 209 (1993).
[30] J. Brachmann, A. Dumitru, J. A. Maruhn, H. Stöcker, W. Greiner, and D. H. Rischke, Nucl. Phys. **A619**, 391 (1997).
[31] Y. B. Ivanov, Yad. Fiz. **46**, 100 (1987).
[32] Yu. B. Ivanov, Nucl. Phys. A474, 669 (1987).
[33] J. Steinheimer, J. Auvinen, H. Petersen, M. Bleicher, and H. Stöcker, Phys. Rev. C **89**, 054913 (2014).
[34] C. Alt et al.(NA49 Collaboration), Phys. Rev. C **68**, 034903 (2003).
[35] H. Liu et al. (E895 Collaboration), Phys. Rev. Lett. **84**, 5488 (2000).

Electric conductivity of the quark-gluon plasma investigated using Boltzmann transport theory

Moritz Greif, Ioannis Bouras, Zhe Xu and Carsten Greiner
Institut für Theoretische Physik, Johann Wolfgang Goethe-Universität, Max-von-Laue-Strae 1, D-60438 Frankfurt am Main, Germany
E-mail: greif@th.physik.uni-frankfurt.de

Abstract. The electric conductivity of a hot quark-gluon plasma is obtained numerically by solving the relativistic Boltzmann equation. We use a relativistic, semi-classical partonic cascade including screened binary and inelastic, radiative $2 \leftrightarrow 3$ perturbative QCD scattering. We employ the Green-Kubo formula and, independently, evaluate the static electric current established by the influence of an external electric field, in order to extract the conductivity. Both numerical methods give the same result. By using only constant, isotropic cross sections, we are able to compare the numerical results and analytic formulas with excellent agreement. Using pQCD scattering with running coupling in the transport code allows us to contrast our results with recent lattice QCD data.

1. Introduction

The quark-gluon plasma (QGP), generated temporarily in ultrarelativisitic collisions of heavy nuclei, is a poorly understood state of matter, governed by quantum chromodynamics (QCD). Relativistic transport coefficients, although not directly measurable, provide a meaningful way by which the QGP can be characterized.

The precise calculation of transport coefficients in a realistic medium requires numerically solvable theories, such as the relativistic Boltzmann transport theory. In the past, the shear viscosity over entropy density ratio η/s has been evaluated from the numerical solution of the Boltzmann equation [1–3], as well as the heat conductivity coefficient κ [4].

The longitudinal static electric conductivity σ_{el} relates the response of the charged particle diffusion current density $\vec{j}$ to an externally applied static electric field $\vec{E}$, $\vec{j} = \sigma_{\mathrm{el}}\vec{E}$. Recently, several scientific groups focused on the electric conductivity [5–17]. It is related to the soft dilepton production rate [18] and the diffusion of magnetic fields in the medium [19–21]. Results for the conductivity in different theoretical frameworks provide us furthermore with the possibility to compare effective cross sections of charged particles.

Here we present a systematic study of the electric conductivity coefficient for a hot QCD plasma [22], applying the microscopic relativistic transport code *Boltzmann Approach to Multi-Parton Scatterings* (BAMPS) [23].

2. Numerical methods

We solve the relativistic 3+1-dimensional Boltzmann equation,

$$\left(\frac{\partial}{\partial t} + \frac{\vec{p}}{E} \cdot \vec{\nabla}\right) f_k(x,t) = \mathcal{C}_k^{2\to 2}[f] + \mathcal{C}_k^{2\leftrightarrow 3}[f], \tag{1}$$

using the BAMPS framework, developed and previously employed in Refs. [1, 2, 4, 23–32]. On-shell particles (3 quark flavors and gluons) propagate freely but collide within the stochastic interpretation of transition rates, including all leading-order $2 \to 2$ and $2 \leftrightarrow 3$ pQCD processes, or alternatively, using a fixed, isotropic cross section. The inelastic cross sections are obtained through the Gunion-Bertsch matrix element [33], which was further improved [30] and applied within BAMPS computations in Ref. [34]. Within BAMPS, the running of the strong coupling α_s is evaluated explicitly at the microscopic scale of the momentum transfer of the respective channel [28, 29, 34]. The Landau-Pomeranchuk-Migdal (LPM) is effectively modeled, see Ref. [34]. To extract the electric conductivity, we employ the zero-frequency Green-Kubo [35, 36] formula for an equilibrated system with volume V and temperature $T = \beta^{-1}$,

$$\sigma_{\mathrm{el}} = \beta V \int_0^\infty \langle j^x(0) j^x(t)\rangle \,\mathrm{d}t, \tag{2}$$

with the total electric current density

$$j^\mu = \sum_{k=1}^{M} q_k V_k^\mu, \quad N_k^\mu = n_k u^\mu + V_k^\mu, \tag{3}$$

where the sum runs over all particle species k with density n_k and charge q_k. The particle flow $N_k^\mu(x)$ can be decomposed with respect to the timelike total four velocity u^μ and the spacelike part V_k^μ. The electric current is extracted from BAMPS in equilibrium box setups, for different temperatures. Separately, we apply an external static electric field in x direction, $\vec{E} = E_x \vec{e}_x$. Because of momentum transfer due to collisions a finite electric current will establish in the simulation after a sufficiently large time. The average magnitude of this static current is proportional to the electric field, and defines the zero-frequency electric conductivity σ_{el} directly, $\langle j^x(t)\rangle_{\mathrm{static}} = \sigma_{\mathrm{el}} E^x$. It can be seen that the results of both methods are equal within the numerical uncertainties [22]. For constant, isotropic cross sections, results from BAMPS could be precisely reproduced analytically by solving the linearized Boltzmann equation within kinetic theory. Using the relaxation time approximation of the collision term, the result for the conductivity $\sigma_{\mathrm{el}} = 1/(3T)\sum_{k=1}^{M} q_k^2 n_k \tau$ deviates by about 10% from BAMPS, using the transport relaxation time $\tau^{-1} = n_{\mathrm{total}}\sigma_{\mathrm{transport\ cross\ section}}$.

3. pQCD-based cross sections

The methods to obtain the electric conductivity presented in Sec. 2 can be readily applied to more realistic scenarios, where leading-order pQCD-based cross sections [23,30,34] are employed within BAMPS. Figure 1 depicts our results for the electric conductivity using pQCD cross sections.

The filled red squares show results for a scenario using only elastic $2 \leftrightarrow 2$ pQCD interactions when the strong coupling constant is fixed to $\alpha_s = 0.3$. The ratio σ_{el}/T is constant within the uncertainties. This is expected by dimensional analysis, as $\sigma_{\mathrm{el}}/T \sim T^{-2}\sigma_{pQCD}^{-1}$ and leading-order pQCD cross sections $\sigma_{\mathrm{pQCD}} \sim T^{-2}$. In Ref. [5], the authors predict the same temperature dependence.

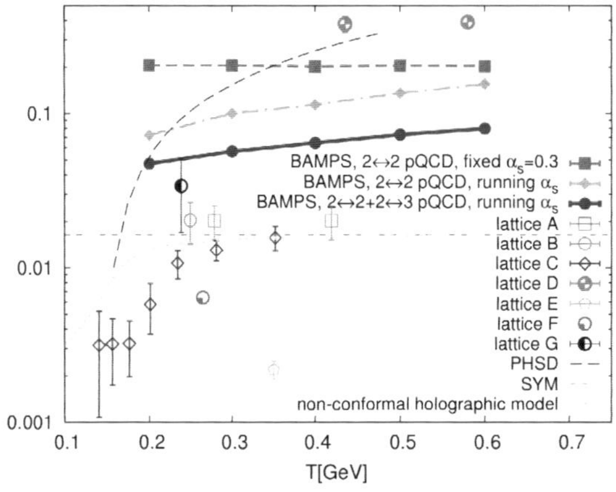

Figure 1. Numerical results for the electric conductivity (filled symbols) compared to recent results from literature [22]. The open symbols represent results from lattice QCD. PHSD: [14], SYM: [37], nonconformal holographic model: [17], lattice A: [8], lattice B: [12], lattice C: [13], lattice D: [7], lattice E: [9], lattice F: [11], lattice G: [10]. The electric charge is explicitly multiplied out, $e^2 = 4\pi/137$.

The filled yellow diamonds depict the electric conductivity over temperature for a running coupling constant α_s [28]. The QCD running coupling is evaluated at the momentum transfer of each microscopic interaction, and leads to a temperature dependence of the average coupling [34], and hence a qualitatively different temperature dependence of the electric conductivity is obtained. The interaction strength decreases with increasing temperature, and the effective cross section decreases accordingly.

The filled dark red circles are results for the most realistic scenario. Here we employ elastic $2 \leftrightarrow 2$ and inelastic $2 \leftrightarrow 3$ scatterings, and running coupling. The LPM effect is modeled as described in Ref. [34]. The inclusion of inelastic collisions accounts for an overall higher effective cross section than in the elastic scenarios. Therefore, the electric conductivity decreases by about 40%, whereas the temperature dependence seems to be dominated by the running of α_s.

In Fig. 1, we contrast the electric conductivity obtained using BAMPS with recent lQCD results, the transport model PHSD, a conformal, and a nonconformal holographic computation.

This study allows us in a unique way to study the overall effective scattering rates for a hot QCD plasma microscopically, including all leading-order elastic and inelastic processes. The electric conductivity reflects in a profound way the effect of inelastic pQCD scattering and the running of α_s. We believe that this is an important result of pQCD, and comparisons with other theories are reasonable.

4. Conclusion and Outlook

In this work we extracted the electric conductivity coefficient for a dilute gas of massless and classical particles numerically by solving the relativistic Boltzmann equation. We use two independent methods to extract the transport coefficient and compare with analytic results whenever possible. Running coupling is explicitly seen in a rise of $\sigma_{\rm el}/T$ with temperature. Inclusion of inelastic collisions decreases the conductivity. We see a similar functional dependence of the electric conductivity on temperature compared to recent lQCD calculations. Interaction properties of the QGP are reflected in the electric conductivity. More refined analytic and lQCD calculations, compared with microscopic transport simulations, will further restrain the value of the conductivity and shed light upon the microscopic interactions inside the QGP.

Acknowledgements

The authors are thankful to V. Greco and A. Puglisi for fruitful discussions. The authors thank G.S. Denicol, H. van Hees, J. Uphoff, and F.Senzel for constant interest in the subject. The authors are grateful to the Center for Scientific Computing Frankfurt for the computing resources. M.G. is grateful to the "Helmhotz Graduate School for Heavy Ion Research". I.B. acknowledges support by BMBF. Z.X. is supported by the NSFC and the MOST under Grants No. 11275103, No. 11335005, and No. 2014CB845400. This work was supported by the Helmholtz International Center for FAIR within the framework of the LOEWE program launched by the State of Hesse.

References

[1] Wesp C, El A, Reining F, Xu Z, Bouras I and Greiner C 2011 *Phys. Rev. C* **84** 054911
[2] Reining F, Bouras I, El A, Wesp C, Xu Z and Greiner C 2012 *Phys. Rev. E* **85** 026302
[3] Plumari S, Puglisi A, Scardina F and Greco V 2012 *Phys. Rev. C* **86** 054902
[4] Greif M, Reining F, Bouras I, Denicol G, Xu Z and Greiner C 2013 *Phys. Rev. E* **87** 033019
[5] Arnold P, Moore G D and Yaffe L G 2000 *J. High Energy Phys.* **11** 001
[6] Arnold P, Moore G D and Yaffe L G 2003 *J. High Energy Phys.* **05** 051
[7] Gupta S 2004 *Phys. Lett. B* **597** 57–62
[8] Aarts G, Allton C, Foley J, Hands S and Kim S 2007 *Phys. Rev. Lett.* **99** 022002
[9] Buividovich P V, Chernodub M N, Kharzeev D E, Kalaydzhyan T, Luschevskaya E V and Polikarpov M I 2010 *Phys. Rev. Lett.* **105** 132001
[10] Ding H T, Francis A, Kaczmarek O, Karsch F, Laermann E and Soeldner W 2011 *Phys. Rev. D* **83** 034504
[11] Burnier Y and Laine M 2012 *Eur. Phys. J. C* **72** 1902
[12] Brandt B B, Francis A, Meyer H B and Wittig H 2013 *J. High Energy Phys.* **2013** 100
[13] Amato A, Aarts G, Allton C, Giudice P, Hands S and Skullerud J I 2013 *Phys. Rev. Lett.* **111** 172001
[14] Cassing W, Linnyk O, Steinert T and Ozvenchuk V 2013 *Phys. Rev. Lett.* **110** 182301
[15] Steinert T and Cassing W 2014 *Phys. Rev. C* **89** 035203
[16] Puglisi A, Plumari S and Greco V (*Preprint* 1407.2559)
[17] Finazzo S I and Noronha J 2014 *Phys. Rev. D* **89** 106008
[18] Moore G D and Robert J M 2006 *hep-ph/0607172*
[19] Baym G and Heiselberg H 1997 *Phys. Rev. D* **56** 5254
[20] Tuchin K 2013 *Adv. High Energy Phys.* 490495
[21] Fernández-Fraile D and Gomez Nicola A 2006 *Phys. Rev. D* **73** 045025
[22] Greif M, Bouras I, Xu Z and Greiner C 2014 *Phys. Rev. D* **90** 094014
[23] Xu Z and Greiner C 2005 *Phys. Rev. C* **71** 064901
[24] Bouras I, Molnár E, Niemi H, Xu Z, El A, Fochler O, Greiner C and Rischke D H 2009 *Phys. Rev. Lett.* **103** 032301
[25] Fochler O, Xu Z and Greiner C 2010 *Phys. Rev. C* **82** 024907
[26] Bouras I, Molnár E, Niemi H, Xu Z, El A, Fochler O, Greiner C and Rischke D H 2010 *Phys. Rev. C* **82** 024910
[27] Fochler O, Xu Z and Greiner C 2011 *Nucl. Phys. A* **855** 420–423
[28] Uphoff J, Fochler O, Xu Z and Greiner C 2011 *Phys. Rev. C* **84** 024908
[29] Uphoff J, Fochler O, Xu Z and Greiner C 2012 *Phys. Lett. B* **717** 430–435
[30] Fochler O, Uphoff J, Xu Z and Greiner C 2013 *Phys. Rev. D* **88** 014018
[31] Senzel F, Fochler O, Uphoff J, Xu Z and Greiner C (*Preprint* 1309.1657)
[32] Uphoff J, Fochler O, Xu Z and Greiner C 2013 *Nucl. Phys. A* **910-911** 401–404
[33] Gunion J F and Bertsch G 1982 *Phys. Rev. D* **25** 746–753
[34] Uphoff J, Fochler O, Senzel F, Wesp C, Xu Z and Greiner C (*Preprint* 1401.1364)
[35] Green M S 1952 *J. Chem. Phys.* **20** 1281
[36] Kubo R 1957 *J. Phys. Soc. Jpn* **12** 570–586
[37] Huot S C, Kovtun P, Moore G D, Starinets A and Yaffe L G 2006 *J. High Energy Phys.* **12** 015–015

Electric Conductivity of the QGP

Armando Puglisi[1,2], **Salvatore Plumari**[1,2] **and Vincenzo Greco**[1,2]

[1]Department of Physics and Astronomy, University of Catania, Via Santa Sofia 64, I-95125 Catania, Italy

[2] Laboratorio Nazionale del Sud, INFN-LNS, Via Santa Sofia 63, I-95125 Catania, Italy

E-mail: [1]puglisia@lns.infn.it,[2]plumari@ct.infn.it,[3]greco@lns.infn.it

Abstract. The transport coefficients of strongly interacting matter have attracted a great interest in the field of Quark-Gluon Plasma (QGP). In this work we compute electric conductivity σ_{el} solving numerically the Relativistic Boltzmann Transport (RBT) equation in a uniform box with periodic boundary conditions considering 2−body scatterings. We compare numerical results obtained using two methods, Green-Kubo correlator and E-field method, with analytic formulas in Relaxation Time Approximation (RTA). We present results for the realistic case of the QGP system considering both a quasi-particle model tuned to lattice QCD thermodynamics as well as the case of a pQCD gas with a running coupling constant. Calculations based on RTA underestimate σ_{el} of about 60%.

1. Introduction

Relativistic Heavy Ion Collisions (HICs) experiments performed by Relativistic Heavy Ion Collider (RHIC) at BNL and by Large Hadron Collider (LHC) at CERN have reached the same conditions of temperature and energy density as they have existed in the early universe. A system of strongly interacting particles above the critical temperature $T_c \sim 160$ MeV [1, 2] is expected to undergo to a phase transition from hadron matter to Quark-Gluon Plasma (QGP) [3]. The phenomenological studies by viscous hydrodynamics [4, 5, 6] and parton transport [7, 8, 9] about the collective behaviour of such matter have shown that the QGP has a very small shear viscosity to entropy density ratio η/s, quite close to the conjectured lower bound for a strongly interacting system in the limit of infinite coupling $\eta/s = 1/4\pi$ [10].

As the hot QCD matter is a plasma, another key transport coefficient, yet much less studied, is the electric conductivity σ_{el}. This coefficient represents the linear response of the system to an applied external electric field. HICs are expected to generate very high electric and magnetic field ($eE \simeq eB \simeq m_\pi^2$) in the very early stage of the collisions. A large value of σ_{el} would lead to a relaxation time for electromagnetic field of the order of $1-2$ fm/c [11, 12], which would be of fundamental importance for the strength of the Chiral-Magnetic Effect [13]. Also in mass asymmetric collisions, like $Cu + Au$, the electric field directed from Au to Cu induces currents resulting in charge asymmetric collective flow directly related to σ_{el} [12]. Furthermore σ_{el} can be directly related to the emission rate of soft photons [14]. Despite its relevance there is yet only a poor theoretical and phenomenological knowledge of σ_{el} and its temperature dependece. In this work we compute the electric conductivity σ_{el} solving numerically the Relativistic Boltzmann Transport (RBT) equation considering two body elastic cross section.

2. Electric conductivity from the Relativistic Boltzmann Equation

The electric conductivity σ_{el} is the transport coefficient that characterizes the response of a system to an external electric field: Ohm's law $\vec{j} = \sigma_{el}\vec{E}$ defines the electric conductivity in linear response theory. The starting point of our computation is the Relativistic Boltzmann Transport (RBT) equation that in the presence of an external field can be written as [15]:

$$p^\mu \partial_\mu f(x,p) + qF^{\alpha\beta} p_\beta \frac{\partial}{\partial p^\alpha} f(x,p) = \mathcal{C}_{22}(x,p) = \int_{2,1',2'} (f_{1'}f_{2'} - f_1 f_2)\, |\mathcal{M}_{1'2'\to 12}| \delta^4(p_1+p_2-p_1'-p_2') \tag{1}$$

where $f(x,p)$ is the distribution function, $F^{\alpha\beta}$ is the electromagnetic field strength tensor, q is the charge. $\mathcal{C}_{22}[f]$ is the collision integral considering only $2 \to 2$ scatterings: $\mathcal{M}$ is the transition matrix for the elastic process linked to the differential cross section $|\mathcal{M}|^2 = 16\pi s(s-4m^2)d\sigma/dt$ with s and t the Mandelstam variables and m the mass of particles. As the Boltzmann equation is an integro-differential equation for the function $f(x,p)$, it is necessary to approximate the right hand side of Eq. (1) in order to obtain an analytical solution. The most simple scheme is the Relaxation Time Approximation (RTA) which simplifies the collision integral as $\mathcal{C}[f] \simeq -\frac{p^\mu u_\mu}{\tau}(f - f_{eq})$ where τ is the so-called relaxation time which represents the time scale for the system to relax toward the equilibrium state characterized by f_{eq}. Following simple mathematical steps, one obtains the following analytical formula [16, 17]:

$$\sigma_{el} = \frac{e^2}{3T} \sum_{j=q,\bar{q}} q_j^2 \int \frac{d^3p}{(2\pi)^3} \frac{\vec{p}^2}{E^2} \tau_j f_{eq} = \frac{e_\star^2}{3T} \left\langle \frac{\vec{p}^2}{E^2} \right\rangle \tau_q \rho_q \tag{2}$$

where q_j is the quarks charge $(\pm 1/3, \pm 2/3)$, ρ_q is the quark density, $e_\star^2 = e^2 \sum_{j=\bar{q},q} q_j^2 = 4e^2/3$. τ_q is the quark transport relaxation time which for a quark of species i can be written as $\tau_{tr,i}^{-1} = \sum_{j=q,\bar{q},g} \langle \rho_j v_{rel}^{ij} \sigma_{tr}^{ij} \rangle$ where v_{rel} is the relative velocity of the two scattering particles and σ_{tr}^{ij} is the transport cross section.

We have computed the electric conductivity σ_{el} solving numerically the RBT equation (see Ref. [18] for details) in a box with periodic boundary conditions using two methods: *Green-Kubo correlator* and *E-field* methods. The first one, in the framework of Linear Response Theory, relates transport coefficients to correlation functions of the corresponding flux or tensor in thermal equilibrium. In this approach one obtains $\sigma_{el} = \frac{V}{T}\int_0^\infty dt\, \langle j_z(t) j_z(0) \rangle = \frac{V}{T} \langle j_z(0) j_z(0) \rangle \tau_{\sigma_{el}}$ where j_z is the z component of the electric current, $\langle \ldots \rangle$ is the thermal average at equilibrium, i.e. without any external electric field. The second method used is suggested by the definition itself of $\sigma_{el} = J/E$: taking the ratio between the electric current measured and the electric field applied one obtains σ_{el}. To simulate a constant electric field $\vec{E}$ in the box, it is sufficient to solve the following equation of motions for particles $dp_z^i/dt = q_i e E_z$ (assuming $\vec{E}$ along the z direction).

In Fig. 1 we show the Green-Kubo correlator as a function of time for a system of massless particles interacting via an isotropic cross-section $\sigma_{tot} = 3$ mb for different temperatures $T = 0.1 - 0.6$ GeV: the behaviour of such correlators is clearly an exponential function $\exp(-t/\tau_{\sigma_{el}})$. In Fig. 2 we plot x,y and z components of electric current $\vec{j}$ as a function of time for a system of particles with $m = 0.4$ GeV, $T = 0.2$ GeV, $\sigma_{tot} = 10$ mb and an external electric field $eE = 0.05$ GeV/fm along z direction: j_x and j_y fluctuate around zero while j_z reaches a saturation value proportional to E.

3. Electric conductivity of the QGP

In this section we consider the more realistic case of quarks, antiquarks and gluons interacting via different anisotropic and energy dependent cross section according to the pQCD-like scheme with a screening mass m_D as arising from HTL approach: $m_D \sim g(T)T$. The total cross section

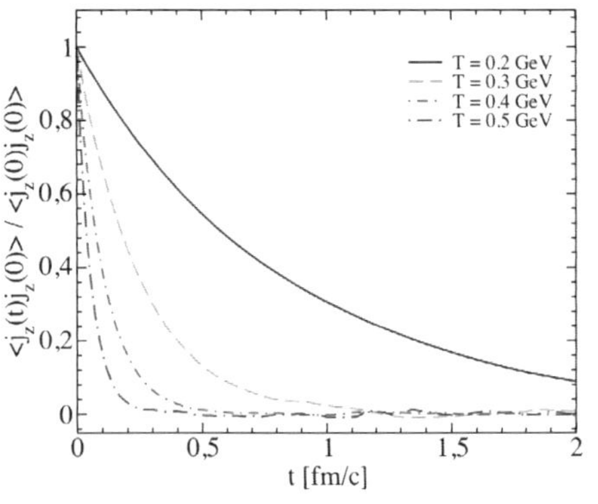

Figure 1. Green-Kubo correlators as a function of time for different temperatures for a system of massless particles interacting via isotropic scatterings with $\sigma_{tot} = 3$ mb.

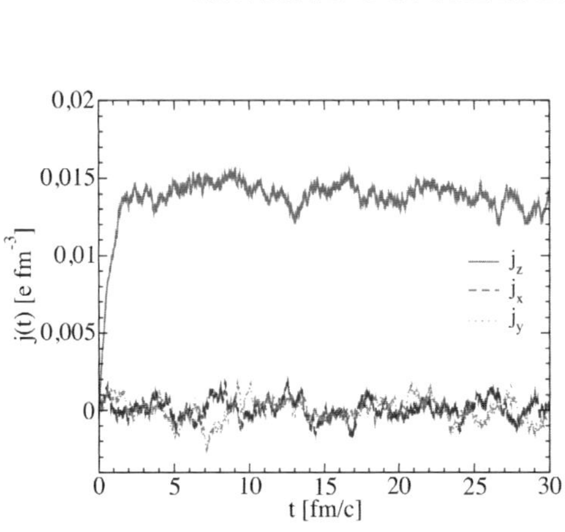

Figure 2. x, y and z components of electric current $\vec{j}$ for a system of massive particles ($m = 0.4$ GeV) at $T = 0.2$ GeV with an applied electric field $eE = 0.05$ GeV/fm.

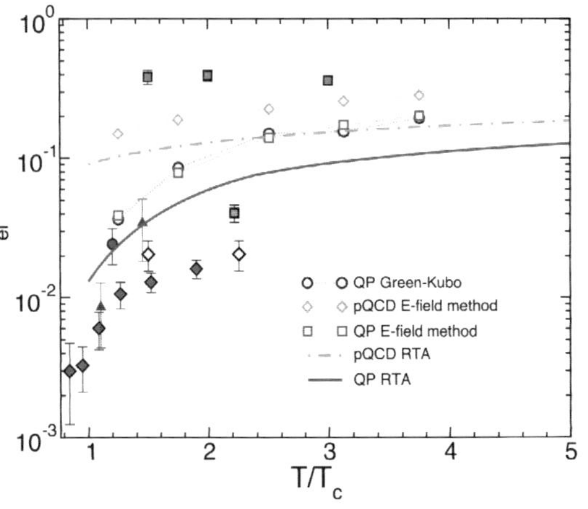

Figure 3. Electric conductivity σ_{el}/T as a function of T/T_c. Open circles are Green-Kubo results for QP model, green open squares are QP model results computed with E-field method, orange diamonds are pQCD results with E-filed method; green line and orange dot-dashed line are RTA for the QP model and pQCD case. Symbols are Lattice data (see text for details).

used has the following form: $\sigma_{tot}^{ij} = \beta^{ij}\sigma(s) = \beta^{ij}\frac{\pi\alpha_s^2}{m_D^2}\frac{s}{s+m_D^2}$ where $\alpha_s = g^2/(4\pi)$ and β^{ij} depends on the species of interacting particles: $\beta^{qq} = 16/9$, $\beta^{qq'} = 8/9$, $\beta^{qg} = 2$ and $\beta^{gg} = 9$.

In order to take into account the thermodynamics from lattice QCD computations, we employ the quasi-particle (QP) model. The aim of a QP model is to describe a strongly interacting system in terms of weakly interacting particles whose masses are generated by the non-perturbative effects. In this model the running coupling $g(T)$ can be parametrized by $g^2(T) = \frac{48\pi^2}{(11N_c-2N_f)\ln\left[\lambda\left(\frac{T}{T_c}-\frac{T_s}{T_c}\right)\right]^2}$ with $\lambda = 2.6$, $T_s/T_c = 0.57$, $T_c = 0.16$ GeV as in [19] similarly to [20]. Quarks' and gluons' masses are given by $m_q^2 = g^2T^2/3$ and $m_g^2 = 3g^2T^2/4$. We also study the behaviour of electric conductivity using the pQCD running coupling $g_{pQCD} = \frac{8\pi}{9}\ln^{-1}\left(\frac{2\pi T}{\Lambda_{QCD}}\right)$ considering massless particles: even if this case does not describe the phase transition, it is interesting to study the σ_{el} dependence on a different running coupling and also the asymptotic limit valid for $T \gg T_c$.

In Fig. 3 we show $\sigma_{el/T}$ as a function of T/T_c. Open circles are computed using Green-Kubo correlator, green squares with the E-field method (applying $eE = 0.02 \div 0.05$ GeV/fm to guarantee the saturation of electric current) for the quasi particle model, orange open diamonds represent result for the massless pQCD case computed only with the E-field method. Green line is RTA for QP model, orange dot-dashed line is RTA for the massless pQCD case. Symbols denotes Lattice data: grey squares [21], violet triangles [22], green circles [23], yellow diamonds [24], red diamonds [25] and orange square [26]. Green-Kubo results are in agreement with the E-field method. Numerical results predicted by the QP model are about a factor of 4 greater then recent Lattice QCD calculations [25]. Calculations based on RTA underestimate σ_{el}/T for both QP model and pQCD case of about a factor 1.7.

4. Conclusions

Transport coefficients characterize the response of a system to different kind of perturbations and determine the dynamics of the system toward the equilibrium state through dissipation. As the QGP created in Heavy Ions Collisions is a system far from equilibrium, the study of transport coefficients is mandatory. In this work we have computed σ_{el} solving numerically the RBT equation using two methods: the Electric field method, suggested by the definition itself ($J = \sigma_{el}E$), and Green-Kubo correlator. We find that the two methods are in very good agreement. Furthermore we find that RTA underestimates σ_{el} by about a factor of 1.7. Our results are quite general but they have been found considering only $2 \leftrightarrow 2$ collisions. It is very interesting to have similar information also when $2 \leftrightarrow 3$ collisions are included as in BAMPS [27].

Acknowledgments

V. G. acknowledges the support by the ERC-StG under QGPDyn grant.

References

[1] Jacak B V and Muller B 2012 *Science* **337** 310–314
[2] Shuryak E 2004 *Prog.Part.Nucl.Phys.* **53** 273–303
[3] Borsanyi S *et al.* (Wuppertal-Budapest Collaboration) 2010 *JHEP* **1009** 073
[4] Romatschke P and Romatschke U 2007 *Phys.Rev.Lett.* **99** 172301
[5] Schenke B, Jeon S and Gale C 2010 *Phys.Rev.* **C82** 014903
[6] Niemi H, Denicol G S, Huovinen P, Molnar E and Rischke D H 2011 *Phys.Rev.Lett.* **106** 212302
[7] Xu Z and Greiner C 2009 *Phys.Rev.* **C79** 014904
[8] Cassing W and Bratkovskaya E 2009 *Nucl.Phys.* **A831** 215–242
[9] Plumari S and Greco V 2012 *AIP Conf.Proc.* **1422** 56–61
[10] Kovtun P, Son D T and Starinets A O 2005 *Phys.Rev.Lett.* **94** 111601
[11] Tuchin K 2013 *Adv.High Energy Phys.* **2013** 490495
[12] Hirono Y, Hongo M and Hirano T 2012 (*Preprint* 1211.1114)
[13] Fukushima K, Kharzeev D E and Warringa H J 2008 *Phys.Rev.* **D78** 074033
[14] Turbide S, Rapp R and Gale C 2004 *Phys.Rev.* **C69** 014903
[15] Yagi K, Hatsuda T and Miake Y 2005 *Camb.Monogr.Part.Phys.Nucl.Phys.Cosmol.* **23** 1–446
[16] Fernandez-Fraile D and Gomez Nicola A 2006 *Phys.Rev.* **D73** 045025
[17] Puglisi A, Plumari S and Greco V 2014 (*Preprint* 1407.2559)
[18] Puglisi A, Plumari S and Greco V 2014 *Phys.Rev.* **D90** 114009
[19] Plumari S, Alberico W M, Greco V and Ratti C 2011 *Phys.Rev.* **D84** 094004
[20] Bluhm M, Kampfer B and Redlich K 2011 *Phys.Rev.* **C84** 025201
[21] Gupta S 2004 *Phys.Lett.* **B597** 57–62
[22] Kaczmarek O, Laermann E, Mller M, Karsch F, Ding H *et al.* 2012 *PoS* **ConfinementX** 185
[23] Brandt B B, Francis A, Meyer H B and Wittig H 2012 *PoS* **ConfinementX** 186
[24] Aarts G, Allton C, Foley J, Hands S and Kim S 2007 *Phys.Rev.Lett.* **99** 022002
[25] Amato A, Aarts G, Allton C, Giudice P, Hands S *et al.* 2013 *Phys.Rev.Lett.* **111** 172001
[26] Buividovich P, Chernodub M, Kharzeev D, Kalaydzhyan T, Luschevskaya E *et al.* 2010 *Phys.Rev.Lett.* **105** 132001
[27] Greif M, Bouras I, Greiner C and Xu Z 2014 *Phys.Rev.* **D90** 094014

Transport coefficients in second-order non-conformal viscous hydrodynamics

Radoslaw Ryblewski
The H. Niewodniczański Institute of Nuclear Physics, Polish Academy of Sciences, PL-31342 Kraków, Poland
E-mail: Radoslaw.Ryblewski@ifj.edu.pl

Abstract. Based on the exact solution of Boltzmann kinetic equation in the relaxation-time approximation, the precision of the two most recent formulations of relativistic second-order non-conformal viscous hydrodynamics (14-moment approximation and causal Chapman-Enskog method), standard Israel-Stewart theory, and anisotropic hydrodynamics framework, in the simple case of one-dimensional Bjorken expansion, is tested. It is demonstrated that the failure of Israel-Stewart theory in reproducing exact solutions of the Boltzmann kinetic equation occurs due to neglecting and/or choosing wrong forms of some of the second-order transport coefficients. In particular, the importance of shear–bulk couplings in the evolution equations for dissipative quantities is shown. One finds that, in the case of the bulk viscous pressure correction, such coupling terms are as important as the corresponding first-order Navier-Stokes term and must be included in order to obtain, at least qualitative, overall agreement with the kinetic theory.

1. Introduction

The theory of relativistic hydrodynamics turned out to be very successful in describing soft hadronic observables produced in ultra-relativistic heavy-ion collisions at RHIC and the LHC [1, 2]; for a review see also [3]. Due to existence of the lower bound of the shear viscosity to entropy density ratio resulting from quantum mechanical considerations, dissipative effects must be included in order to realistically model the quark-gluon plasma [4]. However, despite the successes of the first and most widely used, Israel-Stewart, formulation of causal relativistic second-order viscous hydrodynamics [5], the latter is known to suffer from certain approximations and assumptions. As a result, there are recently multiple developments focusing on constructing more complete and self-consistent methods for deriving the fluid-dynamical equations of motion. These methods include complete second-order treatments [6, 7], third-order treatments [8, 9] and the anisotropic hydrodynamics framework [10, 11, 12, 13, 14, 15, 16, 17, 18, 19, 20, 21, 22, 23, 24, 25, 26, 27, 28, 29]. The forms of the transport coefficients that are entering all of the dissipative hydrodynamics formulations have to be determined by matching the latter with the underlying microscopic kinetic theory. As a result, the minimal requirement for a reasonable non-conformal hydrodynamic theory is that it should be able to describe the evolution of a viscous medium at least in this regime.

In principle, all of the hydrodynamic studies based on the kinetic Boltzmann equation use the so-called relaxation-time approximation for the collisional kernel. Recently it was shown that in this case the exact solution of the Boltzmann kinetic equation can be constructed providing one restricts oneself to the case of Bjorken flow [30, 31, 32, 33]; for generalization to Gubser flow see [34, 35]. In the following we will show that this tool can be extremely useful for testing various dissipative hydrodynamics formulations available in literature against the underlying kinetic theory. We will show that among the cases considered herein, the Chapman-Enskog method leads to the best overall description of the medium in the general, non-conformal, case.

2. Relativistic dissipative hydrodynamics from kinetic theory

The evolution of the system within relativistic viscous hydrodynamics is governed by the energy and momentum continuity equations, $\partial_\mu T^{\mu\nu} = 0$, with the energy-momentum tensor given by

$$T^{\mu\nu} = \epsilon u^\mu u^\nu - (P + \Pi)\Delta^{\mu\nu} + \pi^{\mu\nu}, \tag{1}$$

where ϵ, P, Π and $\pi^{\mu\nu}$ are the energy density, pressure, bulk pressure correction and shear-stress tensor, respectively, and $\Delta^{\mu\nu} \equiv g^{\mu\nu} - u^\mu u^\nu$ is the projection operator orthogonal to the fluid four-velocity u^μ. One possible way to derive the remaining evolution equations for the dissipative quantities is to start with the underlying Boltzmann kinetic equation $p \cdot \partial f = \mathcal{C}[f]$. In the case where the collisional kernel is treated in the relaxation-time approximation, $\mathcal{C}[f] = -(p\cdot u)(f - f_{eq})/\tau_{eq}$, energy-momentum conservation equation (1) must be supplemented by the following relaxation-type equations [36, 37],

$$\tau_\Pi \dot{\Pi} + \Pi = -\zeta\theta - \delta_{\Pi\Pi}\Pi\theta + \lambda_{\Pi\pi}\pi^{\mu\nu}\sigma_{\mu\nu}, \tag{2}$$

$$\tau_\pi \dot{\pi}^{\langle\mu\nu\rangle} + \pi^{\mu\nu} = 2\eta\sigma^{\mu\nu} + 2\tau_\pi \pi_\gamma^{\langle\mu}\omega^{\nu\rangle\gamma} - \tau_{\pi\pi}\pi_\gamma^{\langle\mu}\sigma^{\nu\rangle\gamma} - \delta_{\pi\pi}\pi^{\mu\nu}\theta + \lambda_{\pi\Pi}\Pi\sigma^{\mu\nu}, \tag{3}$$

where τ_π and τ_Π are the shear and bulk relaxation time, η and ζ are the shear and bulk viscosities, $\omega^{\mu\nu} \equiv (\nabla^\mu u^\nu - \nabla^\nu u^\mu)/2$ is vorticity tensor, $\sigma^{\mu\nu} \equiv \nabla^{\langle\mu}u^{\nu\rangle}$ is the velocity stress tensor and $\theta \equiv \nabla_\mu u^\mu$ is the expansion scalar. We use the notation where $\nabla^\mu \equiv \Delta^\mu_\nu \partial^\nu$ is the projected spatial gradient and $A^{\langle\mu\nu\rangle} \equiv \Delta^{\mu\nu}_{\alpha\beta}A^{\alpha\beta}$, with $\Delta^{\mu\nu}_{\alpha\beta} \equiv (\Delta^\mu_\alpha\Delta^\nu_\beta + \Delta^\mu_\beta\Delta^\nu_\alpha - 2/3\Delta^{\mu\nu}\Delta_{\alpha\beta})/2$. In Eqs. (2) and (3) we also use shorthand notation for the proper-time derivative $\dot{(\ )} \equiv d/d\tau$. The quantities $\delta_{\Pi\Pi}$, $\lambda_{\Pi\pi}$, $\tau_{\pi\pi}$, $\delta_{\pi\pi}$ and $\lambda_{\pi\Pi}$ are second-order transport coefficients whose form depends on the method used to derive the evolution equations (2) and (3). Hereafter, we will consider three methods which are available in literature: Israel-Stewart theory [5], 14-moment approximation [7] and Chapman-Enskog method [9, 38]. Each of these formulations predict the same general form of equations (2) and (3), however, they lead to different forms of the second-order kinetic coefficients. In the following we will show which method provides the best agreement with the underlying kinetic theory equation in the simple case of purely-longitudinal boost-invariant (Bjorken) expansion. In this case, the results of the viscous hydrodynamics equations may be confronted with the exact solution of the Boltzmann kinetic equation [30, 31, 32]. Moreover, in this case, the six independent equations in (2) and (3) for dissipative fluxes reduce to two differential equations for bulk pressure correction Π and rapidity–rapidity coefficient of the shear-stress tensor $\pi^{\mu\nu}$.

3. Comparison with the exact solution of Boltzmann kinetic equation

In Fig. 1 we present (a) the pressure anisotropy $\mathcal{P}_L/\mathcal{P}_T \equiv (P + \Pi - \pi)/(P + \Pi + \pi/2)$ (directly related to shear stress tensor since $\pi \equiv -\tau^2\pi^{\eta\eta}$) and (b) bulk pressure correction scaled by proper time resulting from two second-order non-conformal viscous hydrodynamics formulations: Israel-Stewart [5] (purple dashed lines) and 14-moment approximation [7], with (brown dotted lines) and without (green dashed-dotted lines) shear–bulk couplings. One can observe that the results obtained within traditional Israel-Stewart theory provide the poorest description of the exact solution of the underlying Boltzmann kinetic equation [30, 31, 32] (red solid lines), for both shear and bulk viscous corrections. One observes also that, in order to obtain at least a qualitative

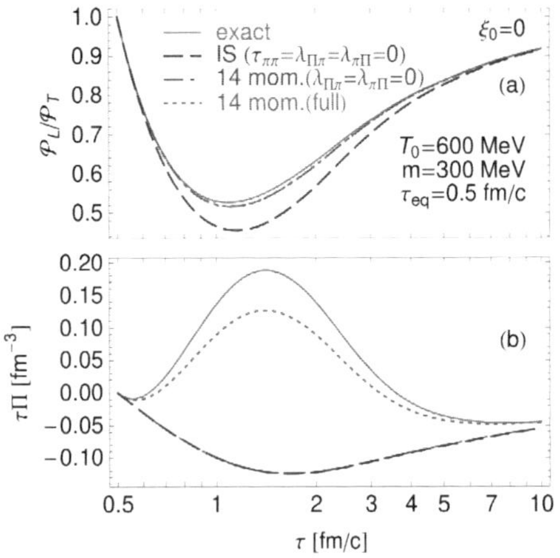

Figure 1. (Color online) Time evolution of (a) pressure anisotropy and (b) bulk pressure correction obtained using exact solution of the Boltzmann equation (red solid lines), Israel-Stewart viscous hydrodynamics (purple dashed lines), 14-moment approximation without (green dashed-dotted lines) and with (brown dotted lines) shear–bulk couplings.

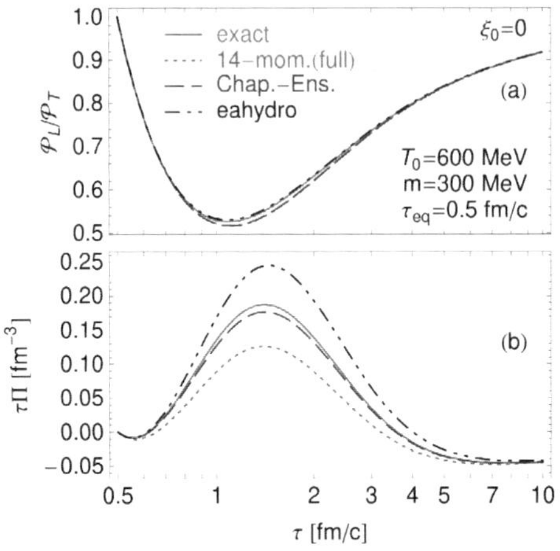

Figure 2. (Color online) Time evolution of (a) pressure anisotropy and (b) bulk pressure correction obtained using exact solution of the Boltzmann equation (red solid lines), 14-moment approximation (brown dotted lines), Chapman-Enskog method (blue dashed lines) and anisotropic hydrodynamics (black dashed-dotted-dotted lines).

description of the bulk pressure correction, one is forced to include the shear–bulk couplings ($\lambda_{\Pi\pi}$ and $\lambda_{\pi\Pi}$) in the two relaxation equations (2) and (3) for the dissipative quantities.

In Fig. 2, in addition to the previous results from 14-moment approximation (brown dotted lines) and the exact solution (red solid lines), we also show the results of the alternative formulation of second-order causal viscous hydrodynamics using Chapman-Enskog iterative method [38, 9] (blue dashed lines) and, for completeness, a novel formulation of anisotropic hydrodynamics framework [27] (black dashed-dotted-dotted lines). We can immediately see that, while the anisotropic hydrodynamics describes the bulk (isotropic) pressure correction at the same level of accuracy as the 14-moment approximation method, it provides a better description of the pressure anisotropy. Surprisingly, the Chapman-Enskog iterative procedure leads to equations providing the best overall description in the case of massive, thus non-conformal, fluid dynamics.

4. Conclusions

Herein we presented an analysis of the relative importance of the transport coefficients following from various second-order non-conformal viscous hydrodynamics formulations, *i.e.*, Israel-Stewart, 14-moment approximation and Chapman-Enskog method. As a benchmark, we used the new exact solution of the Boltzmann kinetic equation within relaxation-time approximation. We showed explicitly that the inclusion of shear–bulk couplings, which are ignored in Israel-Stewart theory, is crucial for capturing the characteristic evolution of the bulk pressure correction. We also showed that the Chapman-Enskog method leads to a set of equations that provide the best overall description of a non-conformal system. We note, however, that, in conformal case, the best description of the pressure anisotropy can be obtained within anisotropic hydrodynamics framework. We finish with the conclusion that the forms of the transport coefficients entering the relativistic second-order viscous hydrodynamics theory are crucial for the proper description of the quark-gluon plasma within this framework.

Acknowledgments

R.R. would like to thank collaborators: G. S. Denicol, W. Florkowski, A. Jaiswal and M. Strickland for useful discussions during preparation of this work and the Organizers of Hot Quarks 2014 conference for financial and technical support. R.R. was supported by Polish National Science Center Grant No. DEC-2012/07/D/ST2/02125 and DEC-2011/01/B/ST2/03915.

References

[1] Bozek P and Wyskiel I 2009 *Phys.Rev.* **C79** 044916
[2] Bozek P and Wyskiel-Piekarska I 2012 *Phys.Rev.* **C85** 064915
[3] Florkowski W 2010 *Phenomenology of Ultra-relativistic Heavy-ion Collisions* (World Scientific)
[4] Kovtun P, Son D and Starinets A 2005 *Phys.Rev.Lett.* **94** 111601
[5] Israel W and Stewart J M 1979 *Ann. Phys.* **118** 341–372
[6] Denicol G S, Niemi H, Molnár E and Rischke D H 2012 *Phys. Rev. D* **85**(11) 114047
[7] Denicol G, Jeon S and Gale C 2014 *Phys.Rev.* **C90** 024912
[8] El A, Xu Z and Greiner C 2010 *Phys. Rev.* **C81** 041901
[9] Jaiswal A 2013 *Phys.Rev.* **C88** 021903
[10] Martinez M and Strickland M 2010 *Nucl.Phys.* **A848** 183–197
[11] Florkowski W and Ryblewski R 2011 *Phys.Rev.* **C83** 034907
[12] Martinez M and Strickland M 2011 *Nucl.Phys.* **A856** 68–87
[13] Ryblewski R and Florkowski W 2011 *J.Phys.G* **G38** 015104
[14] Ryblewski R and Florkowski W 2011 *Eur.Phys.J.* **C71** 1761
[15] Florkowski W and Ryblewski R 2012 *Phys.Rev.* **C85** 044902
[16] Florkowski W, Martinez M, Ryblewski R and Strickland M 2012 *PoS* **ConfinementX** 221
[17] Martinez M, Ryblewski R and Strickland M 2012 *Phys.Rev.* **C85** 064913
[18] Ryblewski R and Florkowski W 2012 *Phys.Rev.* **C85** 064901
[19] Florkowski W, Maj R, Ryblewski R and Strickland M 2013 *Phys.Rev.* **C87** 034914
[20] Florkowski W, Martinez M, Ryblewski R and Strickland M 2013 *Nucl.Phys.* **A904-905** 803c–806c
[21] Rybczynski M and Florkowski W 2013 *J.Phys.* **G40** 025103
[22] Florkowski W and Ryblewski R 2013 *AIP Conf.Proc.* **1520** 160–165
[23] Tinti L and Florkowski W 2014 *Phys.Rev.* **C89** 034907
[24] Bazow D, Heinz U W and Strickland M 2014 *Phys.Rev.* **C90** 044908
[25] Strickland M 2014 *Nucl.Phys.* **A926** 92–101
[26] Florkowski W, Ryblewski R, Strickland M and Tinti L 2014 *Phys.Rev.* **C89** 054909
[27] Nopoush M, Ryblewski R and Strickland M 2014 *Phys.Rev.* **C90** 014908
[28] Nopoush M, Ryblewski R and Strickland M 2014 (*Preprint* 1410.6790)
[29] Ryblewski R 2013 *J.Phys.* **G40** 093101
[30] Florkowski W, Ryblewski R and Strickland M 2013 *Nucl.Phys.* **A916** 249–259
[31] Florkowski W, Ryblewski R and Strickland M 2013 *Phys.Rev.* **C88** 024903
[32] Florkowski W, Maksymiuk E, Ryblewski R and Strickland M 2014 *Phys.Rev.* **C89** 054908
[33] Florkowski W 2014 *EPJ Web Conf.* **71** 00045
[34] Denicol G S, Heinz U W, Martinez M, Noronha J and Strickland M 2014 (*Preprint* 1408.7048)
[35] Denicol G S, Heinz U W, Martinez M, Noronha J and Strickland M 2014 *Phys.Rev.Lett.* **113** 20, 202301
[36] Denicol G S, Florkowski W, Ryblewski R and Strickland M 2014 *Phys.Rev.* **C90** 044905
[37] Jaiswal A, Ryblewski R and Strickland M 2014 *Phys.Rev.* **C90** 044908
[38] Jaiswal A 2013 *Phys. Rev.* **C87** 051901

Hot Quarks 2014 IOP Publishing
Journal of Physics: Conference Series **612** (2015) 012059 doi:10.1088/1742-6596/612/1/012059

Initial state fluctuations and final state collectivity in high energy nuclear collisions: Status and Outlook

Björn Schenke
Physics Department, Brookhaven National Laboratory, Upton, NY 11973, USA
E-mail: bschenke@bnl.gov

Abstract. We review recent developments in the theoretical description of the initial state and collective dynamics in heavy ion and light-heavy ion collisions. After describing the current state of the art in event by event simulations of high energy nuclear collisions we will focus on open issues. These include early time non-equilibrium dynamics, small collision systems (p+p, p+A), hydrodynamic fluctuations, and lower beam energies ($\sqrt{s} < 200\,\text{GeV}$).

1. Introduction

A standard description of the bulk dynamics in heavy ion collisions has been established over the last several years. It includes the following components:

- An initial state description including fluctuations of nucleon positions and subnucleonic color charges
- A rapid transition to the applicability of viscous fluid dynamics
- Second order viscous relativistic fluid dynamics to describe the bulk of the evolution
- An equation of state provided by lattice Quantum Chromo Dynamics (QCD) calculations
- Description of the system at lower temperatures using microscopic transport simulations

We will discuss the various components in more detail below, focusing on the latest developments. We will then describe open issues and future directions.

2. Initial state

Over the last two years the modeling of the fluctuating initial state in high energy heavy ion collisions has advanced significantly. In the high energy limit, which is expected to be a good approximation at LHC energy (and possibly top RHIC energies), the initial energy momentum tensor can be computed from the gluon fields that are produced by two colliding sheets of color glass [1]. This description within an effective theory of quantum chromo dynamics (QCD) was established early on [2] and later implemented numerically [3, 4]. In [5, 6] it was significantly refined to cover finite extent and fluctuations in the transverse plane of the collision. Further, the color charge density distribution of the incoming nuclei was constrained using the IP-Sat model [7, 8] with parameters fit to HERA deeply inelastic scattering (DIS) data [9].

The big advantages of this IP-Glasma model are the reduction of free parameters compared to e.g. a Monte Carlo (MC) Glauber model [10] by using fits to DIS data, the inclusion of subnucleonic fluctuations that naturally lead to negative binomial distributions of multiplicities [5, 11], and the inclusion of pre-equilibrium evolution via classical Yang-Mills dynamics.

The NeXus [12] and EPOS [13] models have similar physics included, do not, however, describe the initial stage of the collision by classical gluon fields but strings.

Unfortunately, for lower energy collisions, the use of the high energy limit is problematic. Currently, low energy collisions are possibly best described by the non-equilibrium initial conditions generated by the initial collisions and string fragmentations in the microscopic UrQMD model [14]. Also AMPT [15, 16] may be used to determine initial states for lower energy collisions [17]. These models are less directly connected to a fundamental theory, and generally at the lower energies more modeling is required.

In addition to continuous work on deriving appropriate initial state models, some research is focused on determining constraints on the initial state from experimental data. An important part of this is to find good estimators of final flow observables that can be computed from properties of the initial state alone. The simplest estimators are the corresponding eccentricities, however, the non-linearity of hydrodynamics requires the inclusion of nonlinear contributions to the estimators, particularly for the higher harmonics [18]. In [19] constraints on eccentricity values that are compatible with experimental data were derived. This can help to determine the quality of an initial state model without having to run many event by event simulations. Viscous effects on this mapping between the initial and final state were discussed in [20].

Related studies focus on the probability distributions of initial eccentricities and how they characterize final state flow distributions and their moments. It is found in [21] that the initial rms anisotropy in harmonic n can be directly extracted from the measured ratio $v_n\{4\}/v_n\{2\}$, which gives direct access to a property of the initial density profile from experimental data. This holds for both heavy ion and p+A collisions.

3. Viscous fluid dynamics and transport

Following the success of ideal relativistic fluid dynamics [22] in describing the main characteristics of experimental data, second order[1] viscous relativistic fluid dynamics [23, 24, 25, 26, 27, 28, 29, 30, 31] allowed to refine simulations and improve the comparison with experimental data, in particular when also including realistic equations of state, determined from lattice QCD calculations [32, 33]. Most importantly, the combination of sophisticated initial state models and event by event viscous fluid dynamics allowed to determine the most precise constraints on the transport properties of the quark gluon plasma to date [34, 35, 36].

The mentioned "second order" refers to a gradient expansion of the kinetic equations. There exist different ways to perform this expansion in the literature. Israel and Stewart [37, 38, 39] used an expansion of the distribution function around its local equilibrium value in a series of reducible Lorentz tensors composed of particle four-momentum k^μ. They then truncated the expansion at second order in momentum, which leaves 14 unknown coefficients. Matching these coefficients to the 14 components of the particle four-current and energy momentum tensor (using the zeroth, first, and second moment of the Boltzmann equation) leads to the fluid dynamic equations. This approach however lacks a parameter that allows a systematic power-counting of corrections to the local equilibrium distribution function.

In a more recent derivation from kinetic theory [40, 41] an orthonormal basis is used in the expansion and exact relations between the expansion parameters and irreducible moments of the deviations of the distribution function from equilibrium are obtained. All terms in the expansion are kept, and the reduction of the degrees of freedom is done by identifying the microscopic time scales of the Boltzmann equation and keeping only the slowest ones. In addition, the equations

[1] The first order, or Navier Stokes formalism is conceptually simpler but introduces unphysical super-luminal signals that lead to numerical instabilities.

of motion for the dissipative quantities are truncated according to a systematic power-counting scheme in Knudsen and inverse Reynolds numbers. The resulting fluid dynamic equations are the current state of the art for simulations of heavy ion collisions.

Another recently developed approach is anisotropic hydrodynamics, where the gradient expansion is not done around an isotropic thermal distribution but one that has an arbitrarily large momentum anisotropy. This allows to systematically take into account the initially large difference between longitudinal and transverse pressure in heavy ion collisions [42, 43, 44].

Employing fluctuating IP-Glasma initial conditions $v_1 - v_5$ at both RHIC and LHC energies were computed using viscous relativistic fluid dynamics in Ref. [35]. Best agreement with experimental data is achieved for shear viscosity values $\eta/s = 0.12$ at RHIC and 0.2 at LHC. This hints at a temperature dependence of η/s, making the QGP a less perfect liquid at higher temperatures. A detailed study of the effect of a temperature dependent η/s was conducted in [45]. A specific parametrization of $(\eta/s)(T)$ in [35] lead to reasonable simultaneous agreement of computed v_n coefficients with both RHIC and LHC data.

Bulk viscosity is now more commonly being included in fluid-dynamic simulations. The simulation in [46] uses a constant ζ/s, while bulk viscosity has been studied in detail in Ref. [47, 48], where it was found that its inclusion increases all v_n coefficients, opposite to the effect of shear viscosity. For ultra central events, where hydrodynamic calculations typically over predict the ratio of v_2/v_3, bulk viscous effects [49] and the usually neglected nucleon-nucleon correlations in the incoming nuclei [50] help to improve agreement with experimental data.

Finally, it is worth noting that new exact solutions including radial expansion [51] have now been used for detailed tests of numerical implementations in certain limits [52], giving additional validation to the numerical codes that are being actively used.

4. Open issues

In the following we discuss some pressing outstanding issues in the description of high energy nuclear collisions whose solution will lead to significant advances in our understanding of nuclear matter under extreme conditions.

4.1. Pre-equilibrium dynamics

In the discussion above, we have assumed that we can directly go from the initial state to a fluid dynamic description of the medium, which involves the assumption of local thermal equilibrium and momentum isotropy (up to viscous corrections). However, the initial state is far from equilibrated and the initial dynamics will be out of equilibrium. The question remains if and how quickly the system can become thermalized and whether one can describe the process in the weak coupling limit.

In this limit, instabilities [53], triggered by quantum fluctuations, and subsequent strong scattering of over-occupied fields, may lead to rapid isotropization and quenching of $\Pi^{\mu\nu}$ to reasonable values justifying the use of viscous hydrodynamics already at early times. These unstable dynamics require a full 3+1 dimensional simulation including a realistic description of quantum fluctuations. Significant progress is being made in this direction [54, 55, 56, 57, 58], however, some disagreement remains about the correct procedure to obtain results for realistic couplings within classical statistical real time lattice simulations [59].

In the strong coupling limit, which is accessible for a conformal field theory via the conjectured AdS/CFT correspondence [60], it was found that the system can be described by viscous hydrodynamics for times as early as $\tau \sim 0.25\,\mathrm{fm}$ [61]. Unfortunately, the detailed QCD process that leads to such "hydrodynamization" is completely inaccessible in this approach.

4.2. Small collision systems

High-multiplicity events of proton-lead, deuteron-gold, and ^{3}He-gold collisions produce v_n values strikingly similar to those in heavy-ion collisions of comparable final-state particle multiplicity [62, 63, 64, 65, 66]. This observation has triggered strong interest in determining whether the same mechanism as in heavy ion collisions, namely collective flow, is responsible for generating the measured v_n, or whether multi-particle correlations from the initial state can explain the phenomenon.

Several calculations that use fluctuating initial states and final state collective effects can successfully explain the observed v_n in p+A collisions [67, 68, 69, 70]. In particular the observed mass hierarchy of identified particle elliptic flow is naturally reproduced by models based on fluid dynamics [71, 69]. Viscous hydrodynamic calculations using the IP-Glasma model under-predict v_2 and v_3 significantly [72]. The main reason for this is the assumption of a mostly spherical proton which leads to insufficient initial shape fluctuations. This becomes even clearer when studying d+Au and ^{3}He+Au collisions in the same model [73]. Here the obtained v_2 and v_3 are significantly larger. Introducing a more granular substructure of the proton as suggested in [74] could help improve the agreement with experimental data in p+Pb collisions.

One should question whether the system size and viscous corrections in such small systems allow for the application of hydrodynamics. In [75] it was found that in p+Pb collisions Knudsen numbers are beyond the applicability limit almost during the whole evolution even when using a small constant $\eta/s = 0.08$. This does not mean that the main physics driving the measured anisotropies is not that of final state effects. It means, however, that results obtained using viscous fluid dynamic simulations are not necessarily to be trusted.

There are alternative theoretical explanations for the measured multi-particle correlations and momentum anisotropies. Saturation models have reproduced correlation data over a wide kinematic range, [76, 77] showing that additional hydrodynamic flow is in fact required to describe the data from heavy-ion collisions, but has no room in smaller systems like p+p or p+A [76]. Various other calculations [78, 79, 80] also indicate that momentum anisotropies can be generated without the need of final state collective effects.

It is still not settled what effect mainly produces the observed long range correlations and momentum anisotropies in small systems like p+p and p+A. It is however clear that the study of light-heavy ion collisions is a rich subject that will lead to deep insight into the mechanism of particle production in QCD.

4.3. Hydrodynamic fluctuations

Due to the finite number of particles, additional fluctuations should occur during the hydrodynamic evolution. Owing to the fluctuation-dissipation theorem, the amplitude of these hydrodynamic fluctuations is governed by the viscosities [81, 82]. This offers the possibility to constrain the viscosity of the QGP independently from the discussed anisotropic flow analysis.

Recently, hydrodynamic fluctuations have been implemented in a fully 3+1 dimensional viscous fluid dynamic simulation [83]. The results obtained in this work show that the contribution of hydrodynamic fluctuations to event by event fluctuations of observables in most situations is sub-leading compared to the contribution from initial state fluctuations. However, in ultra central collisions the average values and variances of computed v_n including only hydrodynamic fluctuations are on the order of the values obtained experimentally. Hence, a calculation combining initial state fluctuations and thermal fluctuations for ultra central events, could potentially explain the data and provide an alternative measurement of η/s [83].

4.4. Low energy collisions

Describing lower energy collisions using fluid dynamics comes with several challenges. First, the assumption of zero baryon chemical potential μ_B, which is good at high energies, does no longer hold. It is necessary to include baryon number conservation and an equation of state with finite μ_B. The latter can not be directly computed on the lattice because of the sign problem. However, at small values of the baryon chemical potential, the construction of a such an equation of state using a Taylor expansion of the pressure [84] is possible [85, 86, 87]. Another challenge is the determination of a realistic initial state. Most likely the high energy limit and hence the color glass condensate and glasma framework will not provide a good description for relatively low energy collisions. Thus, one will have to resort to string models as mentioned above. Further, baryon diffusion can play an important role in addition to shear and bulk viscosity [88] and microscopic hadronic descriptions [89, 90, 91, 92, 93] potentially gain in importance relative to the hydrodynamic stage. All of these aspects make the description of low energy collisions a complex task that needs significant theoretical development. Comprehensive simulations including all relevant physics and sources of fluctuations will be necessary to determine the detailed phase structure of QCD by comparison of observables with experimental data.

Acknowledgments
BPS is supported under DOE Contract No. DE-AC02-98CH10886 through a DOE Office of Science Early Career Award.

References

[1] Gelis F, Iancu E, Jalilian-Marian J and Venugopalan R 2010 *Ann.Rev.Nucl.Part.Sci.* **60** 463–489
[2] Kovner A, McLerran L D and Weigert H 1995 *Phys. Rev.* **D52** 6231–6237
[3] Krasnitz A and Venugopalan R 2000 *Phys. Rev. Lett.* **84** 4309–4312
[4] Lappi T 2003 *Phys. Rev.* **C67** 054903
[5] Schenke B, Tribedy P and Venugopalan R 2012 *Phys. Rev. Lett.* **108** 252301
[6] Schenke B, Tribedy P and Venugopalan R 2012 *Phys. Rev.* **C86** 034908
[7] Bartels J, Golec-Biernat K J and Kowalski H 2002 *Phys. Rev.* **D66** 014001
[8] Kowalski H and Teaney D 2003 *Phys. Rev.* **D68** 114005
[9] Rezaeian A H, Siddikov M, Van de Klundert M and Venugopalan R 2013 *Phys.Rev.* **D87** 034002
[10] Miller M L, Reygers K, Sanders S J and Steinberg P 2007 *Ann. Rev. Nucl. Part. Sci.* **57** 205–243
[11] Schenke B, Tribedy P and Venugopalan R 2014 *Phys.Rev.* **C89** 024901 (*Preprint* `1311.3636`)
[12] Drescher H, Ostapchenko S, Pierog T and Werner K 2002 *Phys.Rev.* **C65** 054902 (*Preprint* `hep-ph/0011219`)
[13] Werner K, Karpenko I, Pierog T, Bleicher M and Mikhailov K 2010 *Phys.Rev.* **C82** 044904
[14] Petersen H, Steinheimer J, Burau G, Bleicher M and Stocker H 2008 *Phys.Rev.* **C78** 044901
[15] Zhang B, Ko C M, Li B A and Lin Z w 2000 *Phys. Rev.* **C61** 067901
[16] Pang L, Wang Q and Wang X N 2014 *Phys.Rev.* **C89** 064910 (*Preprint* `1309.6735`)
[17] Solanki D, Sorensen P, Basu S, Raniwala R and Nayak T K 2013 *Phys.Lett.* **B720** 352–357
[18] Gardim F G, Grassi F, Luzum M and Ollitrault J Y 2012 *Phys.Rev.* **C85** 024908 (*Preprint* `1111.6538`)
[19] Retinskaya E, Luzum M and Ollitrault J Y 2014 *Nucl.Phys.* **A926** 152–158 (*Preprint* `1401.3241`)
[20] Gardim F G, Noronha-Hostler J, Luzum M and Grassi F 2014 (*Preprint* `1411.2574`)
[21] Yan L and Ollitrault J Y 2014 *Phys.Rev.Lett.* **112** 082301 (*Preprint* `1312.6555`)
[22] Kolb P F, Sollfrank J and Heinz U W 2000 *Phys. Rev.* **C62** 054909
[23] Baier R, Romatschke P, Son D T, Starinets A O and Stephanov M A 2008 *JHEP* **04** 100
[24] Romatschke P and Romatschke U 2007 *Phys. Rev. Lett.* **99** 172301
[25] Dusling K and Teaney D 2008 *Phys. Rev.* **C77** 034905
[26] Luzum M and Romatschke P 2008 *Phys. Rev.* **C78** 034915
[27] Song H and Heinz U W 2009 *J.Phys.* **G36** 064033 (*Preprint* `0812.4274`)
[28] Song H, Bass S A, Heinz U, Hirano T and Shen C 2011 *Phys. Rev. Lett.* **106** 192301
[29] Schenke B, Jeon S and Gale C 2011 *Phys. Rev. Lett.* **106** 042301
[30] Schenke B, Jeon S and Gale C 2011 *Phys. Lett.* **B702** 59–63
[31] Bozek P 2012 *Phys.Rev.* **C85** 034901
[32] Huovinen P and Petreczky P 2010 *Nucl. Phys.* **A837** 26–53
[33] Bazavov A, Bhattacharya T, Cheng M, DeTar C, Ding H *et al.* 2012 *Phys.Rev.* **D85** 054503
[34] Luzum M and Ollitrault J Y 2013 *Nucl.Phys.* **A904-905** 377c–380c (*Preprint* `1210.6010`)
[35] Gale C, Jeon S, Schenke B, Tribedy P and Venugopalan R 2013 *Phys.Rev.Lett.* **110** 012302
[36] Gale C, Jeon S and Schenke B 2013 *Int.J.Mod.Phys.* **A28** 1340011 (*Preprint* `1301.5893`)
[37] Israel W 1976 *Ann. Phys.* **100** 310–331

[38] Stewart J 1977 *Proc.Roy.Soc.Lond.* **A357** 59
[39] Muronga A 2002 *Phys. Rev. Lett.* **88** 062302
[40] Denicol G, Niemi H, Molnar E and Rischke D 2012 *Phys.Rev.* **D85** 114047 (*Preprint* 1202.4551)
[41] Denicol G, Niemi H, Bouras I, Molnar E, Xu Z *et al.* 2012 (*Preprint* 1207.6811)
[42] Martinez M and Strickland M 2010 *Nucl.Phys.* **A848** 183–197
[43] Ryblewski R and Florkowski W 2012 *Phys.Rev.* **C85** 064901 (*Preprint* 1204.2624)
[44] Florkowski W, Ryblewski R and Strickland M 2013 *Nucl.Phys.* **A916** 249–259 (*Preprint* 1304.0665)
[45] Niemi H, Denicol G S, Huovinen P, Molnar E and Rischke D H 2011 *Phys.Rev.Lett.* **106** 212302
[46] Bozek P and Wyskiel-Piekarska I 2012 *Phys.Rev.* **C85** 064915 (*Preprint* 1203.6513)
[47] Noronha-Hostler J, Denicol G S, Noronha J, Andrade R P G and Grassi F 2013 *Phys.Rev.* **C88** 044916
[48] Noronha-Hostler J, Noronha J and Grassi F 2014 *Phys.Rev.* **C90** 034907 (*Preprint* 1406.3333)
[49] Rose J B, Paquet J F, Denicol G S, Luzum M, Schenke B *et al.* 2014 (*Preprint* 1408.0024)
[50] Denicol G, Gale C, Jeon S, Paquet J F and Schenke B 2014 (*Preprint* 1406.7792)
[51] Gubser S S 2010 *Phys.Rev.* **D82** 085027 (*Preprint* 1006.0006)
[52] Marrochio H, Noronha J, Denicol G S, Luzum M, Jeon S *et al.* 2013 (*Preprint* 1307.6130)
[53] Mrowczynski S 1994 *Phys. Rev.* **C49** 2191–2197
[54] Dusling K, Epelbaum T, Gelis F and Venugopalan R 2011 *Nucl.Phys.* **A850** 69–109
[55] Dusling K, Gelis F and Venugopalan R 2011 *Nucl.Phys.* **A872** 161–195
[56] Dusling K, Epelbaum T, Gelis F and Venugopalan R 2012 *Phys.Rev.* **D86** 085040 (*Preprint* 1206.3336)
[57] Berges J and Schlichting S 2013 *Phys.Rev.* **D87** 014026 (*Preprint* 1209.0817)
[58] Epelbaum T and Gelis F 2013 *Phys.Rev.Lett.* **111** 232301 (*Preprint* 1307.2214)
[59] Berges J, Schenke B, Schlichting S and Venugopalan R 2014 (*Preprint* 1409.1638)
[60] Maldacena J M 1999 *Int.J.Theor.Phys.* **38** 1113–1133 (*Preprint* hep-th/9711200)
[61] Heller M P, Janik R A and Witaszczyk P 2012 *Phys.Rev.Lett.* **108** 201602 (*Preprint* 1103.3452)
[62] Aad G *et al.* (ATLAS Collaboration) 2013 *Phys.Rev.Lett.* **110** 182302 (*Preprint* 1212.5198)
[63] Aad G *et al.* (ATLAS Collaboration) 2013 *Phys.Lett.* **B725** 60–78 (*Preprint* 1303.2084)
[64] Chatrchyan S *et al.* (CMS Collaboration) 2013 *Phys.Lett.* **B724** 213–240 (*Preprint* 1305.0609)
[65] Abelev B B *et al.* (ALICE Collaboration) 2013 *Phys.Lett.* **B726** 164–177 (*Preprint* 1307.3237)
[66] Adare A *et al.* (PHENIX Collaboration) 2014 (*Preprint* 1404.7461)
[67] Bozek P 2012 *Phys.Rev.* **C85** 014911 (*Preprint* 1112.0915)
[68] Bozek P and Broniowski W 2013 *Phys.Lett.* **B718** 1557–1561 (*Preprint* 1211.0845)
[69] Werner K, Bleicher M, Guiot B, Karpenko I and Pierog T 2014 *Phys.Rev.Lett.* **112** 232301
[70] Kozlov I, Luzum M, Denicol G, Jeon S and Gale C 2014 (*Preprint* 1405.3976)
[71] Bozek P, Broniowski W and Torrieri G 2013 *Phys.Rev.Lett.* **111** 172303 (*Preprint* 1307.5060)
[72] Schenke B and Venugopalan R 2014 *Phys.Rev.Lett.* **113** 102301 (*Preprint* 1405.3605)
[73] Schenke B and Venugopalan R 2014 (*Preprint* 1407.7557)
[74] Schlichting S and Schenke B 2014 *Phys.Lett.* **B739** 313–319 (*Preprint* 1407.8458)
[75] Niemi H and Denicol G 2014 (*Preprint* 1404.7327)
[76] Dusling K and Venugopalan R 2012 *Phys.Rev.Lett.* **108** 262001 (*Preprint* 1201.2658)
[77] Dusling K and Venugopalan R 2013 *Phys.Rev.* **D87** 054014 (*Preprint* 1211.3701)
[78] Gyulassy M, Levai P, Vitev I and Biro T 2014 *Phys.Rev.* **D90** 054025 (*Preprint* 1405.7825)
[79] Dumitru A and Giannini A V 2014 , arXiv:1406.5781 (*Preprint* 1406.5781)
[80] Dumitru A and Skokov V 2014 (*Preprint* 1411.6630)
[81] Kapusta J, Muller B and Stephanov M 2012 *Phys.Rev.* **C85** 054906 (*Preprint* 1112.6405)
[82] Springer T and Stephanov M 2013 *Nucl.Phys.* **A904-905** 1027c–1030c (*Preprint* 1210.5179)
[83] Young C, Kapusta J, Gale C, Jeon S and Schenke B 2014 (*Preprint* 1407.1077)
[84] Allton C, Doring M, Ejiri S, Hands S, Kaczmarek O *et al.* 2005 *Phys.Rev.* **D71** 054508
[85] Huovinen P, Petreczky P and Schmidt C 2012 *Central Eur.J.Phys.* **10** 1385–1387 (*Preprint* 1202.3104)
[86] Cheng M, Hegde P, Jung C, Karsch F, Kaczmarek O *et al.* 2009 *Phys.Rev.* **D79** 074505
[87] Borsanyi S, Endrodi G, Fodor Z, Katz S, Krieg S *et al.* 2012 *JHEP* **1208** 053 (*Preprint* 1204.6710)
[88] Monnai A 2012 *Phys.Rev.* **C86** 014908 (*Preprint* 1204.4713)
[89] Bass S and Dumitru A 2000 *Phys. Rev.* **C61** 064909
[90] Teaney D, Lauret J and Shuryak E V 2001 *Phys. Rev. Lett.* **86** 4783–4786
[91] Hirano T, Heinz U W, Kharzeev D, Lacey R and Nara Y 2006 *Phys. Lett.* **B636** 299–304
[92] Nonaka C and Bass S A 2007 *Phys. Rev.* **C75** 014902
[93] Karpenko I, Bleicher M, Huovinen P and Petersen H 2014 *J.Phys.Conf.Ser.* **503** 012040 (*Preprint* 1311.0133)

Hot Quarks 2014 IOP Publishing
Journal of Physics: Conference Series **612** (2015) 012060 doi:10.1088/1742-6596/612/1/012060

Soft probes of p+Pb and Pb+Pb collisions in the ATLAS experiment at the LHC

S. Radhakrishnan for the ATLAS collaboration.
Department of Physics, Stony Brook University, NY, USA

Abstract. Collective flow and azimuthal correlation measurements of particles in the p_{T} region $\lesssim$ 10 GeV provide valuable information in constraining the initial conditions and expansion dynamics of the system produced in heavy-ion collisions. ATLAS collaboration has recently measured correlations between flow harmonics in Pb+Pb collisions at $\sqrt{s_{\mathrm{NN}}} = 2.76$ TeV, using an event-shape selection procedure, and has separated the components arising from linear and non-linear hydrodynamic evolution in higher order harmonics (v_4 and v_5). A brief overview of these results are presented in these proceedings. Recently, azimuthal correlations extending to large pseudorapidity differences ("ridge"), similar to that seen in heavy-ion collisions, was observed in p+Pb collisions at LHC. These proceedings also present a measurement of the Fourier harmonics associated with the ridge correlations (v_1 to v_5) in p+Pb collisions at $\sqrt{s_{\mathrm{NN}}}$ = 5.02 TeV by ATLAS collaboration. The results are compared with v_n in Pb+Pb collisions at $\sqrt{s_{\mathrm{NN}}} = 2.76$ TeV with similar event multiplicities. Reasonable agreement is observed after accounting for the difference in the average p_{T} of particles produced in the two collision systems, consistent with a recent hydrodynamic calculation.

1. Introduction

The particles produced in heavy-ion collisions at RHIC and LHC show azimuthal anisotropy that extend over large pseudorapidity (η) [1, 2]. In AA collisions these long-range correlations are well described by a hydrodynamic expansion resulting from the anisotropic initial density distribution [3, 4]. The azimuthal anisotropy in particle production is often expanded in a Fourier series, $\frac{dN}{d\phi} \sim 1 + \sum_{n=1}^{\infty} 2v_n \cos(n(\phi - \Psi_n))$. The second and third harmonics in the Fourier expansion arise from a linear response to the corresponding initial eccentricities, ϵ_n, i.e $v_n \propto \epsilon_n$ [5]. However higher-order harmonics contain non-linear contributions from lower-order harmonics [5, 6]. Using an event-shape selection procedure, ATLAS collaboration has measured the correlations between flow harmonics in Pb+Pb collisions at $\sqrt{s_{\mathrm{NN}}} = 2.76$ TeV. The correlation between higher and lower-order harmonics show non-linear correlations, consistent with expectations from hydrodynamics. The linear and non-linear components are separated from the measured correlations for different centrality intervals. A brief discussion of these results are presented in these proceedings.

Recently, azimuthal correlations extending to large pseudorapidity differences (referred to as "ridge"), similar to that observed in heavy-ion collisions, was observed in p+Pb collisions at LHC [7, 8, 9] and d+Au collisions at RHIC [10]. The theoretical understanding of the origin of these correlations is still a matter of debate [11, 12]. ATLAS collaboration has done a detailed measurement of the first five Fourier harmonics ($v_1 - v_5$) associated with the ridge correlations in p+Pb collisions at $\sqrt{s_{\mathrm{NN}}} = 5.02$ TeV. These proceedings discuss the ATLAS v_n results in p+Pb collisions. A comparison between v_n from p+Pb collisions and Pb+Pb collisions at $\sqrt{s_{\mathrm{NN}}}$

= 2.76 TeV with similar event multiplicity is also included. Reasonable agreements are observed after accounting for the difference in the average p_T of particles produced in the two collision systems, consistent with a recent hydrodynamic calculation, thus suggesting a similar or perhaps common origin of these correlations in both systems.

2. Flow correlations in Pb+Pb collisions

Correlation between flow harmonics in heavy-ion collisions can be studied using an event-shape analysis. The event-shape selection is done using the transverse energy (E_T) distribution in the ATLAS [13] forward calorimeter (FCal) in $3.3 < |\eta| < 4.9$. For each event, the energy distribution is expanded in a Fourier series, $2\pi\frac{dE_T}{d\phi} = (\sum E_T)(1 + 2\sum_{n=1}^{\infty} q_n cos(n(\phi - \psi_n)))$. The reduced flow vector, q_n, represents the E_T weighted observed flow coefficients. In each centrality class, events are classed into different ellipticity bins according to the value of the observed elliptic flow, q_2. The v_n values are then calculated for each of these q_2 classes using tracks in the Inner Detector (ID, $|\eta| < 2.5$), using a two-particle correlation method.

The left panel in Figure 1 shows the v_4 values measured in the ID, as a function of the v_2 values measured in the ID for different q_2 classes. The results for different centrality classes are shown. The observed non-linear correlation between v_4 and v_2 is found to be well described by a two component fit, motivated by hydrodynamics [5], of the form $v_4 = \sqrt{c_0^2 + (c_1 v_2^2)^2}$. The middle panel shows the extracted linear ($v_n^L = c_0$) and non-linear components ($v_n^{NL} = \sqrt{v_n^2 - c_0^2}$), for $n = 4$, as a function of the number of participants. The values from a similar decomposition using the results from event-plane correlation measurements [14] are also shown. Good agreement can be seen between both, suggesting that the correlations between flow magnitudes arise from the correlations between the flow angles. Right panel shows a similar decomposition for v_5, from a fit of the form $v_5 = \sqrt{c_0^2 + (c_1 v_2 v_3)^2}$ to the measured correlation between v_5 and v_2. More results related to the analysis along with a detailed description of the procedure and systematics can be found elsewhere [15].

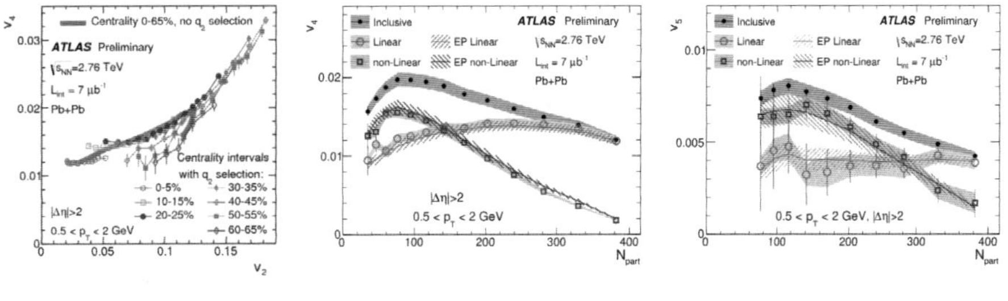

Figure 1. (Left) Correlation between v_2 and v_4, both measured in $0.5 < p_T < 2$ GeV, for different centrality intervals. The data points in each centrality interval correspond to different q_2 bins. (Middle) N_{part} dependence of v_4 and the extracted linear and non-linear components associated with it. Also shown are the linear and non-linear components calculated from event-plane correlations. (Right) Similar plot for v_5. Error bars and shaded bands represent statistical and systematic uncertainties, respectively. [15]

3. Ridge in p+Pb collisions

The long-range correlations in p+Pb are studied using two-particle correlation constructed using the charged particle tracks reconstructed in the ID. A recoil subtraction procedure, similar to that described in [9], is applied at the per-trigger yield level by subtracting the per-trigger yield above uncorrelated pairs in a "peripheral" event class – defined as having $E_T < 10$ GeV in the FCal on the Pb-going side. The recoil subtracted distribution is then expanded in a Fourier series, and the two particle harmonic coefficients $v_{n,n}$ in the expansion are used to compute the single particle coefficients v_n, assuming factorization: $v_n(p_\mathrm{T}^\mathrm{a}) = v_{n,n}(p_\mathrm{T}^\mathrm{a}, p_\mathrm{T}^\mathrm{b})/\sqrt{v_{n,n}(p_\mathrm{T}^\mathrm{b}, p_\mathrm{T}^\mathrm{b})}$. A detailed description of the analysis procedure and systematic uncertainties along with a complete set of results can be found in [16].

The v_n values extracted from the subtracted distribution, for n = 2, 3, 4 and 5, as a function of the transverse momentum are shown in the left panel of Figure 2, for events with $220 \leq N_\mathrm{ch}^\mathrm{rec} < 260$. The v_n values increase with p_T in the low p_T region, reach a maximum around 3 − 5 GeV and then decrease with further increase in p_T, but remain positive in the measured range. This is similar to the p_T dependence of v_n harmonics measured in heavy-ion collisions [17]. The magnitude of v_n is found to decrease with increasing harmonic number n. The second panel of Figure 2 shows the first order harmonic, v_1, as a function of p_T^a. The values are obtained assuming factorization, but accounting for the sign change of v_1 at low p_T. Good agreements are found between the v_1 values extracted using three different p_T^b ranges. The v_1 is negative at low p_T, crosses zero ~ 1.5 GeV and then increase to 0.1 in the 4 − 6 GeV range. A similar p_T dependence for v_1 (with $v_1(p_\mathrm{T})$ crossing zero ~ 1.1 GeV) was observed in Pb+Pb collisions at $\sqrt{s_\mathrm{NN}} = 2.76$ TeV [17].

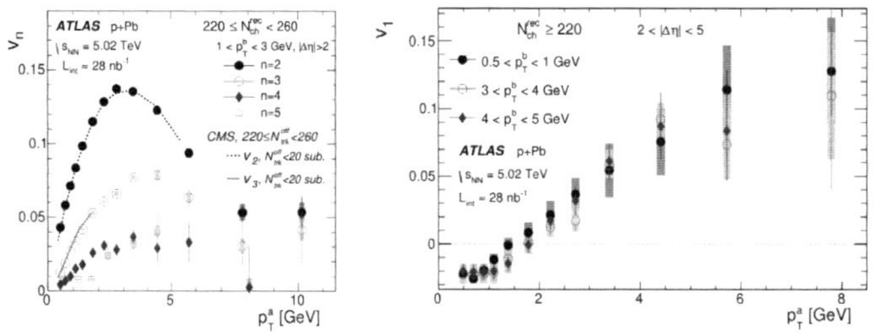

Figure 2. (Left) v_n for n = 2−5 as a function of p_T^a for associated particles in the range 1 < p_T^b < 3 GeV and 2 < $|\Delta\eta|$ < 5 for events with $220 \leq N_\mathrm{ch}^\mathrm{rec} < 260$. (Right) v_1 as a function of p_T^a for different associated p_T selections for events with $220 \leq N_\mathrm{ch}^\mathrm{rec} < 260$. The error bars and shaded boxes represent the statistical and systematic uncertainties, respectively [16].

The left-panels of Figure 3 compares the v_n values from p+Pb collision in the event-activity class $220 \leq N_\mathrm{ch}^\mathrm{rec} < 260$ with those from Pb+Pb collisions in 55−60% centrality from [17]. The two event classes are selected such that they have similar efficiency corrected number of tracks with $p_\mathrm{T} > 0.5$ GeV and $|\eta| < 2.5$. The larger v_2 values in Pb+Pb collisions can be attributed to the elliptic collision geometry of the Pb+Pb system, while the larger v_4 values could be due to the non-linear coupling between v_2 and v_4 in the collective expansion [15]. The v_3 data for Pb+Pb collisions are similar in magnitude to those in p+Pb collisions which could be expected if v_3 arises from density fluctuations in both systems. A scaling relation between the v_n of p+Pb and Pb+Pb systems was proposed recently in [18]. It is argued that the $v_n(p_\mathrm{T})$ in p+Pb and $v_n(p_\mathrm{T}/K)$ in Pb+Pb systems should be proportional to each other at a given p_T, with $K = 1.25$ being the ratio of mean p_T in the two collision systems. The right panels of Figure 3 show the comparison between $v_n(p_\mathrm{T})$ in p+Pb and $v_n(p_\mathrm{T}/K)$ in Pb+Pb. The $v_2(p_\mathrm{T}/K)$ and $v_4(p_\mathrm{T}/K)$ in Pb+Pb are also scaled vertically by an empirical factor of 0.66. The two sets of values are found to agree well with each other after the scaling of p_T axis.

4. Conclusions

Event-shape selection are used to measure correlations between flow harmonics and to separate linear and non-linear components in higher harmonics (v_4 and v_5). The linear and non-linear

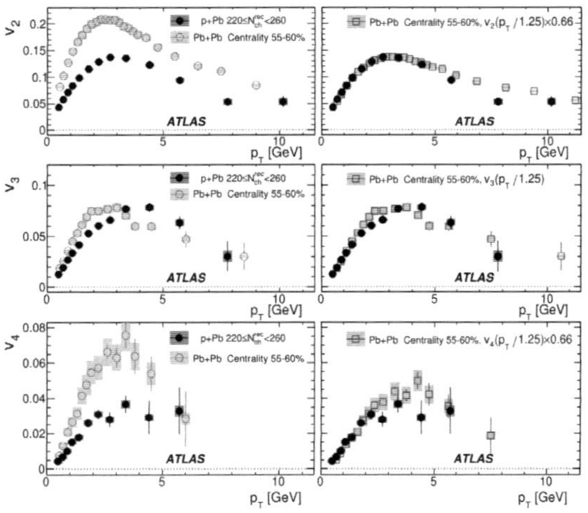

Figure 3. v_2 (top row), v_3 (middle row) and v_4 (bottom row) data compared between p+Pb collisions with $220 \leq N_{\mathrm{ch}}^{\mathrm{rec}} < 260$ in this analysis and Pb+Pb collisions in 55-60% centrality from [17]. Left column shows the original data with their statistical (error bars) and systematic uncertainties (shaded boxes). In right column, the same Pb+Pb data are rescaled horizontally by a constant factor of 1.25, and the v_2 and v_4 are also down-scaled by an empirical factor of 0.66 to match the p+Pb data [16].

components are extracted as function of centrality and show good agreement with the linear and non-linear components evaluated from event-plane correlation measurements. The first five Fourier harmonics associated with the ridge correlations in p+Pb collisions are measured. The v_n values show similar p_{T} dependence as the v_n from heavy-ion collisions. The $v_n(p_{\mathrm{T}})$ values from p+Pb collisions (for n =2, 3 and 4) are compared to those from Pb+Pb collisions for event class with similar average multiplicity. After applying a scale factor of $K = 1.25$, that accounts for the difference of mean p_{T} in the two collision systems, the $v_n(p_{\mathrm{T}}/K)$ from Pb+Pb is found to have similar shape as the $v_n(p_{\mathrm{T}})$ from p+Pb collisions. This could suggest that the long-range correlations in p+Pb and Pb+Pb systems are driven by similar dynamics.

5. References

[1] Ackermann K *et al.* [STAR Collaboration], *Phys.Rev.Lett.* **86**, 402 (2001)
[2] Aamodt K *et al.* [ALICE Collaboration], *Phys.Rev.Lett.* **107** (2011) 032301
[3] Heinz U W and Snellings R, *Ann.Rev.Nucl.Part.Sci.* **63** (2013) 123
[4] Alver B and Roland G, *Phys.Rev.C* **81** (2010) 054905
[5] Gardim F G, Grassi F, Luzum M, and Ollitrault J-Y, *Phys.Rev.C* **85** (2012) 024908
[6] Teaney D and Yan L, *Phys.Rev.C* **90** 024902
[7] Abelev B *et al.* [ALICE Collaboration], *Phys.Lett.B* **719**, (2013) 29-41
[8] ATLAS Collaboration, *Phys.Rev.Lett.* **110**, 182302 (2013)
[9] CMS Collaboration, *Phys.Lett.B* **724** (2013) 213
[10] Adare A *et al.* [PHENIX Collaboration], *Phys.Rev.C* **78** (2008) 014901
[11] Bozek P, *Phys.Rev.C* **85** (2012) 014911
[12] Dusling K and Venugopalan R, *Phys.Rev.D* **87** no. 5, (2013) 051502
[13] ATLAS Collaboration, 2008 *JINST* **3** S08003
[14] ATLAS Collaboration, *Phys.Rev.C*, **90** (2014) 024905.
[15] ATLAS Collaboration, ATLAS-CONF-2014-022, http://cds.cern.ch/record/1702980
[16] ATLAS Collaboration, *Phys.Rev.C* **90** 044906 (2014)
[17] ATLAS Collaboration, *Phys.Rev.C* **86** (2012) 014907
[18] Basar G and Teaney D, *Phys.Rev.C* **90** 054903

New formulation of leading order anisotropic hydrodynamics

LeonardoTinti
Institute of Physics, Jan Kochanowski University, PL-25406 Kielce, Poland
E-mail: tinti@fi.infn.it

Abstract.
Anisotropic hydrodynamics is a reorganization of the relativistic hydrodynamics expansion, with the leading order already containing substantial momentum-space anisotropies. The latter are a cause of concern in the traditional viscous hydrodynamics, since large momentum anisotropies generated in ultrarelativistic heavy-ion collisions are not consistent with the hypothesis of small deviations from an isotropic background, i.e., from the local equilibrium distribution.

We discuss the leading order of the expansion, presenting a new formulation for the (1+1)–dimensional case, namely, for the longitudinally boost invariant and cylindrically symmetric flow. This new approach is consistent with the well established framework of Israel and Stewart in the close to equilibrium limit (where we expect viscous hydrodynamics to work well). If we consider the (0+1)–dimensional case, that is, transversally homogeneous and longitudinally boost invariant flow, the new form of anisotropic hydrodynamics leads to better agreement with known solutions of the Boltzmann equation than the previous formulations, especially when we consider massive particles.

1. Introduction

Relativistic hydrodynamics plays a fundamental role in modeling of relativistic heavy-ion collisions, see for instance Refs. [1, 2, 3, 4, 5, 6]. Early calculations were based on perfect fluid hydrodynamics, however, nowadays viscous codes are preferred. Both because they provide a better description of the data and because of general arguments that the fluid viscosity cannot be zero, which follows from quantum mechanical considerations [7] as well as from the AdS/CFT correspondence [8]. Despite its obvious success, there are still fundamental issues with the ordinary viscous hydrodynamics expansion. A new approach to treat these problems is *anisotropic hydrodynamics* (aHydro) [9, 10, 11, 12, 13, 14], where the large momentum anisotropy, providing large pressure corrections, is treated in a non perturbative way starting from the leading order of the hydrodynamics expasion.

2. The hydrodynamics expansion

The most common assumption for deriving hydrodynamics from relativistic kinetic theory is that the particle distribution function $f(x,p)$ is very close to local equilibrium. Ignoring conserved charges and in the Boltzmann limit we have

$$f(x,p) = f_{\rm eq.}(x,p) + \delta f(x,p), \qquad f_{\rm eq.}(x,p) = k \exp\left[-\frac{p\cdot U(x)}{T(x)}\right], \tag{1}$$

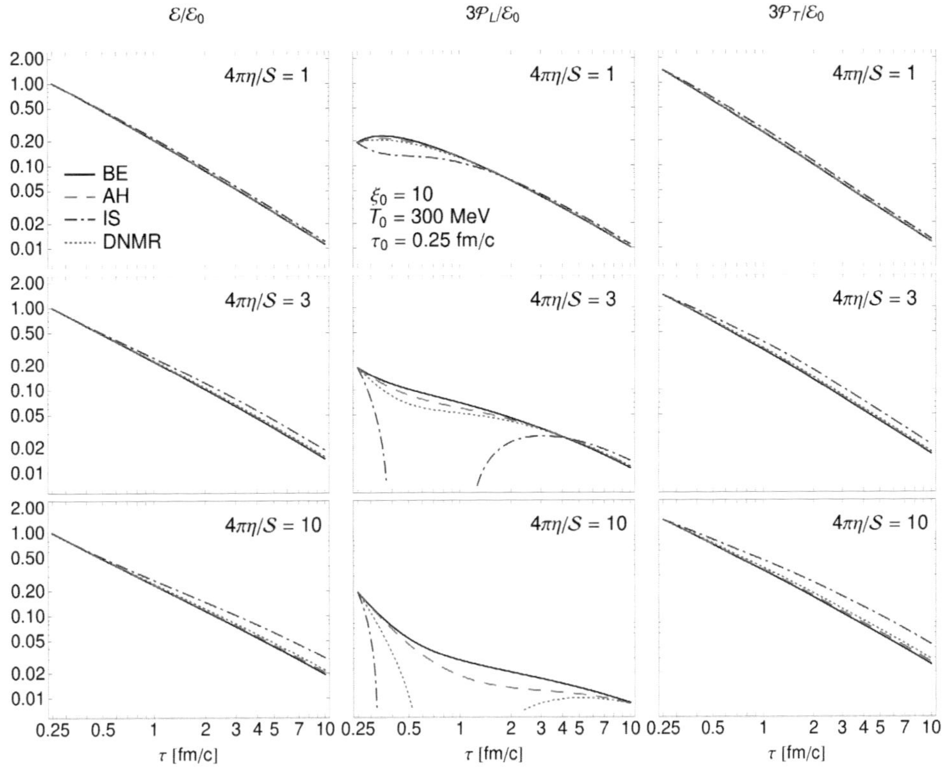

Figure 1. (Color online) Comparison of anisotropic hydrodynamics and second-order viscous hydrodynamics with the exact solution of the Boltzmann equation, for energy density, longitudinal pressure, and transverse pressure evolution(figure taken from [16]).

with T and U^μ being the effective temperature and the fluid four velocity, respectively. The leading order in (1), $f_{\text{eq.}}$, describes the perfect fluid. The viscous correction depends only on δf which is treated as a small perturbation. However, when we consider an (almost) boost invariant flow like the one we expect in the early stages of heavy ions collisions, we encounter fundamental problems. The four velocity gradients are inversely proportional to the proper time, therefore, the pressure corrections become close to the equilibrium pressure, questioning the validity of the perturbative treatment.

The main feature of anisotropic hydrodynamics is to treat the large momentum anisotropy in a non perturbative way starting from the leading order, namely, we write

$$f(x,p) = f_{\text{aniso.}}(x,p) + \delta\tilde{f}(x,p). \tag{2}$$

In this way, the deviation $\delta\tilde{f}$ from the (non isotropic and dissipative) background $f_{\text{aniso.}}$ can be small enough to justify a perturbative treatment. The first formulation of aHydro used the point dependent version of the Romatschke-Strickland form (presented in [15]) for the leading order of the anisotropic expansion, which in the local rest frame (LRF) reads

$$f_{\text{aniso.}}(x,p) = k \exp\left[-\frac{1}{\Lambda(x)}\sqrt{p_T^2 + \zeta(x)p_L^2}\right]. \tag{3}$$

Here Λ is the momentum scale (the effective temperature T is defined using the Landau matching and is different from Λ in general), p_T and p_L are the transverse and longitudinal momenta, and ζ is the anisotropy parameter. In order to close the system of equations for the leading order of the anisotropic expansion, one has used the four momentum conservation (the first moment of the Boltzmann equation and the Landau matching) and the particle creation equation (the zeroth moment of the Boltzmann equation). In addition, the collisional kernel was treated in the relaxation time approximation.

For a longitudinally boost invariant and transversely homogeneous system there is an exact solution of the relativistic Boltzmann equation [16]. We show in Fig. 1 one of the plots in [16]. The comparison is done between the exact solution (BE), Israel-Stewart theory, the new formulation of second-order viscous hydrodynamics presented in [6] (DNMR), and anisotropic hydrodynamics (AH). Anisotropic hydrodynamics is always very close to the exact solution, while IS is providing unphysical vanishing longitudinal pressure $\mathcal{P}_L$, and significant deviations from the exact evolution of the temperature T and the transverse pressure $\mathcal{P}_T$. In the most extreme case (large shear viscosity over entropy ratio $\eta/\mathcal{S}$), even the DNMR approach is not reliable.

3. New formulation of the leading order of anisotropic expansion

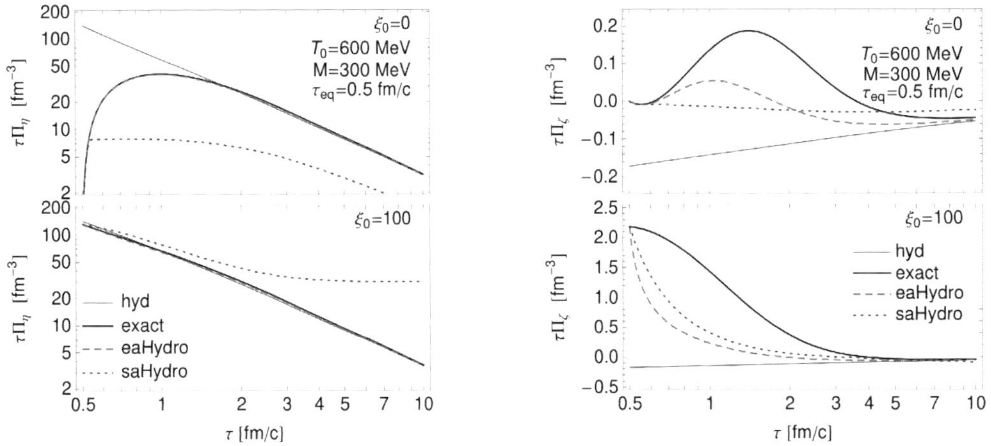

Figure 2. (Color online) Time dependence of shear and bulk viscous pressure multiplied by τ (figure taken from [19]).

The anisotropic background (3) takes into account differences between the longitudinal pressure $\mathcal{P}_L$ and the transverse pressure $\mathcal{P}_T$, only. However if there is a non-vanishing radial flow we expect anisotropies even in the transverse plane. As the system evolves toward equilibrium, these corrections become more important. One way to handle non trivial transverse dynamics is to treat $\delta\tilde{f}$ in the anisotropic expansion (2) in a perturbative way [17]. Alternatively, we propose here to include more dynamic effects connected with anisotropy in the leading order itself.

In Ref. [18] we extended the formalism of anisotropic hydrodynamics to the (1+1)-dimensional case. We started from a generalization of the Romatschke-Strickland form, which in the local rest frame reads

$$f_{\rm aniso.}(x,p) = k \exp\left[-\frac{1}{\lambda(x)}\sqrt{(1+\xi_X)p_X^2 + (1+\xi_y)p_Y^2 + (1+\xi_Z)p_Z^2}\right], \qquad (4)$$

where Z is the longitudinal direction, and X is the direction of the transverse flow. We used the second moment of the Boltzmann equation, in addition to the energy and momentum conservation, in order to obtain a closed set of equations. We proved that these equations reduce to the Israel-Stewart equations in the close to equilibrium limit, where we know that second-order viscous hydrodynamics is justified.

We later compared this new set of equations with the solution of the Boltzmann equation and the original prescription for anisotropic hydrodynamics [19]. There is a large improvement of the agreement with the exact solution, especially for massive particles. In Fig. 2 we show the comparison between the new formulation (eaHydro) and the original one (saHydro). The shear evolution $\tau\Pi_\eta$ is very well reproduced, while the bulk evolution $\tau\Pi_\zeta$ still shows some deviations from the exact solution. Note that τ is the (longitudinal) proper time, $\mathcal{P}_{\rm eq.}$ is the equilibrium pressure, $\Pi_\eta = \frac{2}{3}\left(\mathcal{P}_T - \frac{1}{3}\mathcal{P}_L\right)$, and $\Pi_\zeta = \frac{1}{3}\left(2\mathcal{P}_T + \mathcal{P}_L - 3\mathcal{P}_{\rm eq.}\right)$.

4. Conclusions

Anisotropic hydrodynamics is a reorganization of the hydrodynamic expansion around a non-isotropic background. The leading order already provides large longitudinal pressure corrections, justifying the perturbative treatment of the next to leading order in heavy ion collisions. The original prescription for the leading order of anisotropic hydrodynamics does not take into account pressure anisotropies in the transverse plane, therefore requiring a next to leading order treatment in presence of transverse expansion. We extended the original treatment allowing for cylindrically symmetric expansion already in the leading order. The agreement with the exact solution in the case of vanishing transverse flow has been largely improved. Bulk dynamics is not well reproduced, however, an interesting proposal is to introduce an extra degree of freedom taking into account the isotropic pressure corrections, see Ref. [20].

Acknowledgements

This work has been supported by Polish National Science Center grant No. DEC-2012/06/A/ST2/00390.

References

[1] A. Muronga, Phys. Rev. C **69**, 034903 (2004).
[2] M. Luzum and P. Romatschke, Phys. Rev. C **78**, 034915 (2008) [Erratum-ibid. C **79**, 039903 (2009)].
[3] B. Schenke, S. Jeon and C. Gale, Phys. Lett. B **702**, 59 (2011).
[4] C. Shen, U. Heinz, P. Huovinen and H. Song, Phys. Rev. C **84**, 044903 (2011).
[5] P. Bozek and I. Wyskiel-Piekarska, Phys. Rev. C **85**, 064915 (2012).
[6] G. S. Denicol, H. Niemi, E. Molnar and D. H. Rischke, Phys. Rev. D **85**, 114047 (2012).
[7] P . Danielewicz and M. Gyulassy, Phys. Rev. **D** 31, 53 (1985).
[8] P. Kovtun, D. T. Son and A. O. Starinets, Phys. Rev. Lett. **94**, 111601 (2005).
[9] M. Martinez, R. Ryblewski and M. Strickland, Phys. Rev. C **85**, 064913 (2012).
[10] R. Ryblewski and W. Florkowski, Phys. Rev. C **85**, 064901 (2012).
[11] R. Ryblewski, J. Phys. G **40**, 093101 (2013).
[12] W. Florkowski, R. Ryblewski and M. Strickland, Phys. Rev. D **86**, 085023 (2012).
[13] W. Florkowski and O. Madetko, Acta Phys. Polon. B**45**, 1103 (2014).
[14] W. Florkowski, E. Maksymiuk, R. Ryblewski, and M. Strickland, Phys. Rev. C **89**, 054908 (2014).
[15] P. Romatschke and M. Strickland, Phys. Rev. D **68**, 036004 (2003).
[16] W. Florkowski, R. Ryblewski and M. Strickland, Phys. Rev. C **88**, 024903 (2013)
[17] D. Bazow, U. W. Heinz and M. Strickland, arXiv:1311.6720 [nucl-th].
[18] L. Tinti and W. Florkowski, Phys. Rev. C **89**, 034907 (2014)
[19] W. Florkowski, R. Ryblewski, M. Strickland, and L. Tinti, Phys. Rev. C **89**, 054909 (2014)
[20] M. Nopoush, R. Ryblewski and M. Strickland, Phys. Rev. C **90**, 014908 (2014).

Multi-strange baryon production in pp, p–Pb and Pb–Pb collisions measured with ALICE

Domenico Colella for the ALICE Collaboration
Dipartimento Interateneo di Fisica "M. Merlin" and Sezione INFN, Via Orabona 4, 70126 Bari, Italy
E-mail: domenico.colella@cern.ch

Abstract. Transverse momentum spectra and yields of charged Ξ and Ω at mid-rapidity in pp, p–Pb and Pb–Pb collisions at the LHC have been measured by the ALICE Collaboration. These baryons are identified by reconstruction of their weak decay topology, in modes with only charged decay products, using the excellent tracking and particle identification capabilities of the detector. The recent measurements of the multi-strange baryon production relative to non-strange particles in p–Pb collisions are presented: this would help to understand the change in relative strangeness production from pp collisions to Pb–Pb collisions. Results on the nuclear modification factors for the charged Ξ and Ω particles, compared with those for other light particles, are also reported.

1. Introduction

Measuring strange and multi-strange particle production in relativistic heavy-ion interactions give a unique insight into the properties of the hot and dense matter created in the collision, as there is no net strangeness content in the initially colliding nuclei. Moreover, the high energy and luminosity available at the LHC grant a relative abundant production of such particles.

An enhanced production of strange particles in A–A compared to pp interactions was one of the earliest proposed signatures of the formation of a deconfined Quark-Gluon Plasma (QGP) [1]. In pp collisions, the production of strange particles is limited by the local canonical strangeness conservation. This is not the case in central heavy-ion collisions, where strange quarks are expected to be produced more abundantly, resulting in an enhanced production of strange baryons at the end of the hadronisation stage. Strangeness enhancement in heavy-ion collisions with respect to pp (p–Be) has been observed in Pb–Pb collisions at the NA57 experiment [2], at STAR [3] in Au–Au collisions, and indeed confirmed by ALICE in central Pb–Pb collisions at $\sqrt{s_{\rm NN}} = 2.76$ TeV [4].

One of the most striking differences between the A–A and pp collisions first observed at RHIC was the suppression of the hadron production at high transverse momentum ($p_{\rm T}$) in central Au–Au collisions at $\sqrt{s_{\rm NN}} = 200$ GeV when compared to expectations from an incoherent superposition of nucleon-nucleon interactions [5]. The effect, confirmed by the charged-particle nuclear modification factor ($R_{\rm AA}$) as a function of $p_{\rm T}$ measured in Pb–Pb collisions at the LHC [6], is interpreted as the result of the energy loss suffered by partons when traversing the hot and dense matter created in an ultra-relativistic heavy-ion collision. The larger suppression observed at LHC compared to that at RHIC can be explained by the higher density of the medium created in collisions at higher centre-of-mass energy.

2. Multi-strange decay reconstruction with the ALICE detector

The ALICE detector was designed to study heavy-ion physics at the LHC. It consists of a central barrel with a large solenoid providing a 0.5 T field for tracking and particle identification and forward detectors for triggering and centrality selection. Tracking and vertexing are performed using the Inner Tracking System (ITS), consisting of six layers of silicon detectors, and the Time Projection Chamber (TPC). The two innermost layers of the ITS and the V0 detector (scintillation hodoscopes placed on either side of the interaction region) are used for triggering. The V0 also provides the centrality determination in Pb–Pb collisions as well as the multiplicity class selection in p–Pb collisions. A complete description of the ALICE sub-detectors can be found in [7].

Multi-strange baryons are reconstructed through their weak decay topologies, namely $\Xi^- \to \pi^- + \Lambda$, $\Omega^- \to K^- + \Lambda$ (with $\Lambda \to \pi^- + p$), and the corresponding charge conjugate decays for the anti-particles. The branching ratios are 63.9% and 43.3%, for the Ξ and the Ω, respectively. The Ξ and the Ω candidates are found by combining reconstructed charged tracks: cuts on geometry and kinematics are applied to first select the Λ candidate and then to match it with all the remaining secondary tracks (bachelor candidates). In addition, particle identification is performed by a selection on the specific energy loss in the TPC for the three daughter tracks.

3. Results

The strangeness enhancement is defined as the ratio between the yields in Pb–Pb collisions and those in pp interactions at the same energy, both normalized to the mean number of participants ($\langle N_{part} \rangle$). The measured enhancement for multi-strange baryons is shown in Figure 1a and 1b as a function of $\langle N_{part} \rangle$: it increases with centrality and with the strangeness content of the particle as already observed at lower energies, and decreases as the centre-of-mass energy increases, continuing the trend established between SPS and RHIC energies.

An alternative way to look at the strangeness enhancement, in order to factor out the general increase in multiplicity with N_{part} from the actual increase in strange particle production is to study the hyperon-to-pion ratios, namely $(\Xi^- + \overline{\Xi}^+)/(\pi^- + \pi^+)$ and $(\Omega^- + \overline{\Omega}^+)/(\pi^- + \pi^+)$, for A–A and pp collisions. These ratios are shown in Figure 1c as a function of $\langle N_{part} \rangle$, both at LHC and RHIC energies. The relative production of strangeness in pp collisions at the LHC is larger than at RHIC energy while it is almost the same at the two energies for A–A collisions. The decrease of enhancement with increasing energy is then caused by the change in pp collisions, in agreement with the hypothesis of progressive relaxing of canonical suppression in pp collisions with increasing energy. The hyperon-to-pion ratio increases when going from pp to A–A, showing a relative enhancement of strangeness production in A–A collisions (normalized to the pion yield) which is about half of that seen when normalizing to $\langle N_{part} \rangle$. Figure 2 shows the recently measured p–Pb hyperon-to-pion ratios as a function of the charged-particle multiplicity density, compared with results from pp and Pb–Pb collisions. An enhancement of the multi-strange baryons with multiplicity in p–Pb data is observed; it still increases with the strangeness content of the baryon, considering also the Λ/π ratios published in [8].

In Figure 1c the predictions for the hyperon-to-pion ratios from thermal models, based on a grand canonical approach are also shown: these predictions are provided by the GSI-Heidelberg model described in [9] (full line, with a chemical freeze-out temperature parameter T = 164 MeV) and THERMUS described in [10] (dashed line, with T = 170 MeV). The corresponding temperatures were obtained from extrapolations of RHIC results to LHC energies. Recently, a global fit of all the hadron yields, measured by the ALICE Collaboration, has also been performed [11] with the thermal grand canonical model; the chemical equilibrium temperature parameter obtained from the simultaneous fit is lower than the one extrapolated from low energy results and corresponds to about T = 156 MeV. The solid (GSI-Heidelberg

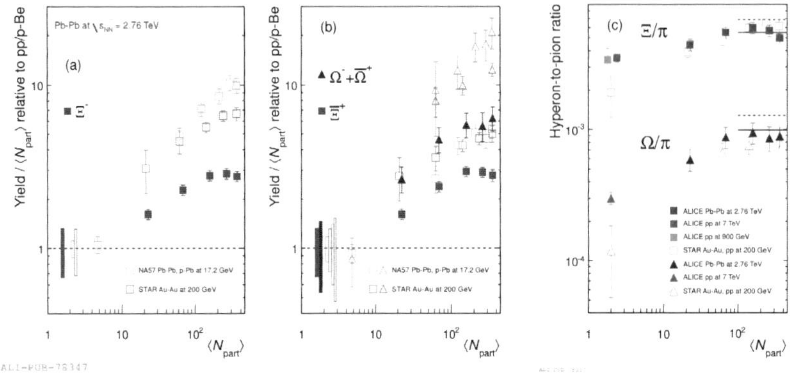

Figure 1. (a), (b) Enhancements in $|y| < 0.5$ as a function of $\langle N_{\text{part}}\rangle$ measured by ALICE and compared to SPS and RHIC data. The bars on the dotted line indicate the systematic uncertainties on the pp reference. (c) Hyperon-to-pion ratios as a function of $\langle N_{\text{part}}\rangle$, for A–A and pp collisions at LHC and RHIC energies.

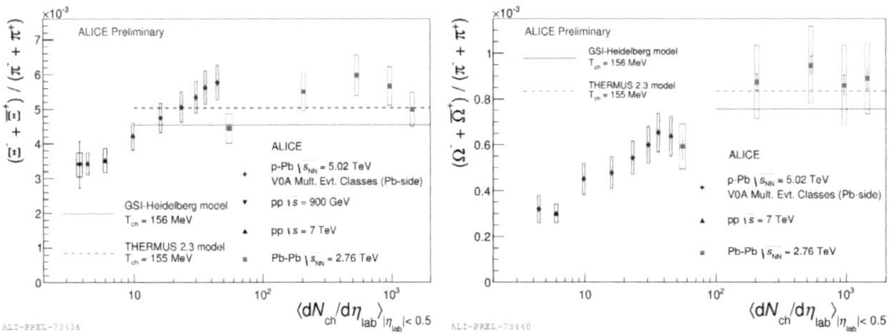

Figure 2. $(\Xi^- + \overline{\Xi}^+)/(\pi^- + \pi^+)$ and $(\Omega^- + \overline{\Omega}^+)/(\pi^- + \pi^+)$ ratios as a function of charged-particle multiplicity density in pp, p–Pb and Pb–Pb collisions.

model) and dashed (THERMUS) green lines in Figure 2 show ratios corresponding to chemical equilibrium temperatures of 156 MeV and 155 MeV, respectively.

The nuclear modification factor (R_{AA}) is defined as the ratio of the p_{T} spectra in Pb–Pb and in pp collisions scaled by the number of nucleon-nucleon collisions [12]. In Figure 3 the R_{AA} as a function of the p_{T} for multi-strange baryons, in the most central (0-10%) and most peripheral (60-80%) collisions, are shown and compared with those for lighter hadrons (π, K and p). In the most central class the reference spectra in pp collisions have been extended to the region $6 < p_{\text{T}} < 8$ GeV/c for the Ξ and $5 < p_{\text{T}} < 7$ GeV/c for the Ω, using a Lévy-Tsallis fit function, to match the p_{T} reach of the spectra in Pb–Pb collisions. Looking at the most central events, the R_{AA} for Ξ follows the same trend as the proton at high p_{T} (> 6 GeV/c), indicating that the suppression does not depend on the particle mass. At intermediate p_{T} there are indications of mass-ordering among the baryons, that can be interpreted as consequence of the radial flow [13]. The R_{AA} for the Ω is above unity, which might be the result of the larger contribution of strangeness enhancement compared to Ξ. When going from the most central to the most peripheral events the R_{AA} values and the relative differences between the particles are reduced. As expected, the strangeness enhancement effect for the Ω also becomes weaker in the most peripheral events.

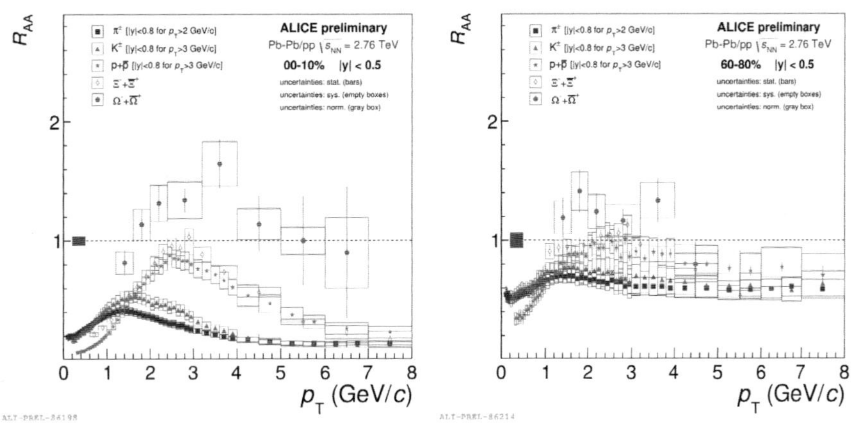

Figure 3. Ξ and Ω nuclear modification factors compared to those for π, K and p in 0-10% (left) and 60-80% (right) centrality $\sqrt{s_{NN}}$ = 2.76 TeV Pb–Pb collisions.

4. Conclusions

The production of the multi-strange baryons Ξ^- and Ω^- and their anti-particles has been measured by the ALICE Collaboration in pp, p–Pb and Pb–Pb collisions. A strangeness enhancement has been observed in Pb–Pb collisions with similar features as in lower energy measurements. The decreasing trend with increasing centre-of-mass energy has been confirmed. The hyperon-to-pion ratios measured in p–Pb collisions bridge the values in pp to those in Pb–Pb and reveal a dependency of the strangeness enhancement on the charged-particle multiplicity. The comparison to the grand canonical thermal model expectations for T = 155 MeV in Pb–Pb collisions yields, within a broad agreement with the measurements in Pb–Pb collisions, a tension between ratios for Ξ and Ω in p–Pb collisions: while Ω/π only approaches the saturation level at the highest measured multiplicity, the Ξ/π values at high multiplicity are comparable to the Pb–Pb values and lie above the equilibrium limits. Finally, the R_{AA} shows similar suppression for all the species at $p_T > 6$ GeV/c. There is evidence of mass ordering at intermediate p_T and indication of a R_{AA} increase due to the larger strangeness enhancement for the Ω.

References

[1] Rafelski J and Müller B 1982 *Phys. Rev. Lett.* **48** 1066
Koch P, Müller B and Rafelski J 1986 *Phys. Rep.* **142** 167
[2] Antinori F *et al.* 2006 *J. Phys.* G **32** 427
Antinori F *et al.* 2010 *J. Phys.* G **37** 045105
[3] Abelev B *et al.* 2008 *Phys. Rev.* C **77** 044908
[4] Abelev B *et al.* 2014 *Phys. Lett. B* **728** 216
Abelev B *et al.* 2014 *Phys. Lett. B* **734** 409
[5] Adcox K *et al.* 2002 *Phys. Rev. Lett.* **88** 022301
Adler C *et al.* 2002 *Phys. Rev. Lett.* **89** 202301
[6] Aamodt K *et al.* 2011 *Phys. Lett.* B **696** 30
[7] Aamodt K *et al.* 2008 *JINST* **3** S08002
[8] Abelev B *et al.* 2014 *Phys. Lett.* B **728** 25-38
[9] Andronic A, Braun-Munzinger P and Stachel J 2009 *Phys. Lett.* B **673** 142
Andronic A, Braun-Munzinger P and Stachel J 2009 *Phys. Lett.* B **678** 516 (Erratum).
[10] Cleymans J, Kraus I, Oeschler H, Redlich K and Wheaton S 2006 *Phys. Rev.* C **74** 034903
[11] Stachel J, Andronic A, Braun-Munzinger P and Redlich K 2014 *J. Phys. Conf. Ser.* **509** 012019
[12] Abelev B *et al.* 2013 *Phys. Rev.* C **88** 044909
[13] Abelev B *et al.* 2014 *Phys. Lett.* B **736** 196-207

Hot Quarks 2014 IOP Publishing
Journal of Physics: Conference Series **612** (2015) 012063 doi:10.1088/1742-6596/612/1/012063

Study of strange mesons in p+p, d+Au and Cu+Cu collisions at $\sqrt{s_{NN}}$ = 200 GeV

Priyanka Sett for the PHENIX Collaboration
Nuclear Physics Division, Bhabha Atomic Research Center, Mumbai-400085, India
E-mail: sett.priyanka@gmail.com

Abstract. The PHENIX experiment at RHIC has measured invariant transverse momentum (p_T) spectra and nuclear modification factor (R_{AB}) of strange mesons K^{*0} and K^0_S, in p+p, d+Au and Cu+Cu collisions at center of mass energy 200 GeV. The p_T range of these measurements spans from 1.1 to 8.5 GeV/c for K^{*0} and from 2 to 13 GeV/c for K^0_S. Similar to the other light-quark mesons measurements, both these strange mesons show no cold-nuclear-matter effects in the measured p_T range in d+Au collisions. The nuclear modification factor for d+Au (R_{dAu}) $\approx$ 1 and is almost constant as a function of p_T. In case of Cu+Cu peripheral collisions, no suppression is registered with respect to the p+p yields scaled with number of binary collisions. In central Cu+Cu collisions, both mesons suffer substantial amount of suppression at high p_T (> 5 GeV/c), which is similar to the suppression suffered by light-quark mesons. In the intermediate p_T range ($2 < p_T < 5$ (GeV/c)), the strange mesons are less suppressed than light-quark mesons (π^0) and more suppressed than baryons (p, $\bar{p}$).

1. Introduction

Lattice QCD calculations predict phase transition from hadronic phase to quark gluon plasma (QGP) above an energy density of 1 GeV/fm^3 [1]. The high energy heavy ion colliders, RHIC and LHC, provide the means to create a quark gluon plasma in the laboratory and study its properties. Particles traversing this hot/dense medium suffer significant energy loss which results in the modification of fragmentation functions and softening of the particle spectra. The modification of the spectra due to the medium can be quantified by the "nuclear modification factor", defined as the ratio of the yield in heavy ion collisions (A+B) to the yield in p+p collisions scaled by the number of binary collisions ($N_{\rm coll}$). Analytically, it is expressed as, $R_{AB} = \frac{d^2N_{AB}/dy\,dp_T}{N_{coll}\cdot d^2N_{pp}/dy\,dp_T}$. The deviation of the value of R_{AB} from 1, is a manifestation of medium effects. In heavy ion collisions, the energy loss by high p_T partons is reflected by the suppression of the yield (i.e. $R_{AB} < 1$). Figure 1 shows the R_{AA} for different particles for the most central Au+Au collisions at $\sqrt{s_{NN}} = 200$ GeV. It is seen that for $p_T > 5$ GeV/c, the R_{AA} value is around 0.2 which is similar for all particles irrespective of quark content. At intermediate p_T (2 < p_T GeV/c < 5), the medium-effects for particles with different quark content are different. Different bulk phenomenons (e.g. flow, recombination, Cronin effect) may play important roles towards hadron production in this p_T range. The measurement of strange hadrons is important to study collective effects in the intermediate p_T and the parton energy-loss effects in the high p_T range. These proceedings present the overview of K^0_S and K^{*0} measurements at midrapidity by PHENIX detector in p+p, d+Au and Cu+Cu collisions at $\sqrt{s_{NN}}$ = 200 GeV. The measurements

Journal of Physics: Conference Series **612** (2015) 012063 doi:10.1088/1742-6596/612/1/012063

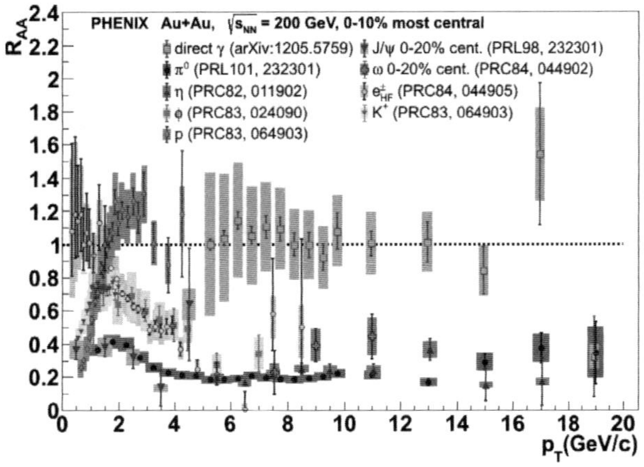

Figure 1. (color online) Nuclear modification factor as a function of p_T for π^0, p, γ, η, ϕ, J/ψ, ω, $e^{\pm}$, $K^{\pm}$ for the most central Au+Au collisions at $\sqrt{s_{NN}} = 200$ GeV. Plot taken from Ref. [3].

include the invariant yield transverse momentum spectra and nuclear modification factor for different centrality bins [2]. This article mainly focuses on the results related to the nuclear modification factor obtained in d+Au and Cu+Cu collisions.

2. PHENIX detector

The results presented here are obtained with the help of global, tracking and PID detectors of PHENIX [4]. The event information is obtained from the Beam Beam Counters (BBC), located at $|\eta| < 0.35$ and covering 2π in azimuth. The track reconstruction and momentum determination are done with the help of the Drift Chambers (DC) and first layer of the Pad Chambers (PC). The second and third layers of PC help to suppress contribution of the secondary tracks originating from the decay of long-lived particles or from the interaction of tracks with the detector material. The Time of Flight (TOF) detector identifies charged hadrons. The Electro-Magnetic Calorimeter (EMCal) measures the position and the energy of photons and electrons.

3. Analysis

The K^0_S and K^{*0} mesons are reconstructed from $K^0_S \to \pi^0(\to \gamma\gamma)\pi^0(\to \gamma\gamma)$ and $K^{*0} \to K^{\pm}\pi^{\mp}$ channels respectively. K^0_S reconstruction is done in two steps. Initially, π^0s are reconstructed from photon pairs after applying kinematic and analysis cuts. Then K^0_S meson is reconstructed from π^0 pairs within the same event. For K^{*0} reconstruction, a pair of oppositely charged tracks with $p_T > 0.3$ GeV/c is required. To have measurement in a wider p_T range, three different exclusive and statistically independent techniques are used for K^{*0} analysis. In one of the techniques both tracks are identified in TOF as pions and kaons; it is used to measure K^{*0} production at low p_T. In other approach, tracks are not identified but matched in PC3 and given the masses of kaons and pions. This technique extends the measurement to highest possible p_T. The intermediate p_T range is obtained by identifying kaon in TOF and matching the other track in PC3 before assigning the mass of pion. For analysis details one may refer [2].

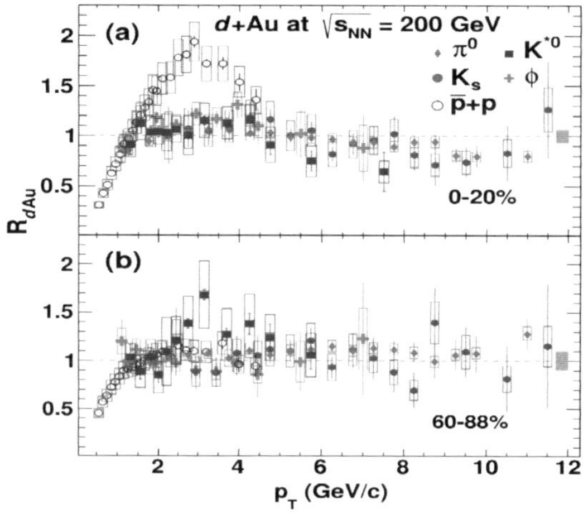

Figure 2. (color online) Nuclear modification factor as a function of p_T for K^0_S and K^{*0} for (a) the most central (0-20%) and (b) the most peripheral (60-88%) d+Au collisions at $\sqrt{s_{NN}}$ = 200 GeV. Results from π^0 [5], ϕ [6] and protons [7] are also shown.

4. Results

The nuclear modification factor for K^0_S and K^{*0} in d+Au collisions at $\sqrt{s_{NN}}$ = 200 GeV for the most central and the most peripheral bins are shown in Fig. 2. A comparison with π^0 [5], ϕ [6] and protons [7] are also shown. The statistical and systematic errors are shown with error bars and boxes respectively. It is observed that R_{dAu} is consistent with unity within uncertainties for all measured centrality bins for $p_T > 1$ GeV/c. In the most central bin, there is a slight hint of Cronin enhancement in the intermediate p_T (2-5 GeV/c) range and suppression at $p_T >$ 6-8 GeV/c. The comparisons with π^0 and ϕ mesons show that cold nuclear matter effects do not play a significant role in these mesons productions. However, the proton shows an enhancement in the central collisions at intermediate p_T, which can be explained by recombination models [8].

The nuclear modification factor for K^0_S and K^{*0} along with π^0 [9] and ϕ [6] in Cu+Cu collisions at $\sqrt{s_{NN}}$ = 200 GeV for the most central (0-20%), minimum bias (0-94%) and the most peripheral (60-94%) collisions are shown in the left panel of Fig. 3. The statistical and systematic errors are shown with error bars and boxes respectively. Particles are suppressed in minimum-bias and most central collisions for $p_T > 5$ GeV/c, whereas, in the most central collisions, R_{CuCu} reaches a value around 0.5. In the intermediate p_T (2-5 GeV/c) range, K^{*0} and ϕ are less suppressed than π^0 and below 2 GeV/c, neither of K^{*0} and ϕ are suppressed. Peripheral collisions follow binary scaling within uncertainties for all these mesons in the measured p_T range. Figure 3 (right panel), compares the suppression of mesons and baryons for the central and peripheral collisions. The data for K^{*0}, ϕ [6] and π^0 [9] are shown for Cu+Cu collisions and the data for $K^{\pm}$ [7] and protons [7] are shown for Au+Au collisions for similar $\langle N_{part}\rangle$ values. The statistical and systematic errors are shown with error bars and boxes respectively. For the central collisions, π^0 is the most suppressed meson and has a flat behavior in the measured p_T range. In the intermediate p_T, the mesons, K^{*0} and ϕ are less suppressed than π^0 but more suppressed than protons. At high p_T, all particles are suppressed and seem to have similar R_{AA} values within uncertainties, irrespective of their valence quark content. In peripheral collisions, R_{AA} for all particles, except protons are consistent with unity within error bars for $p_T > 2$

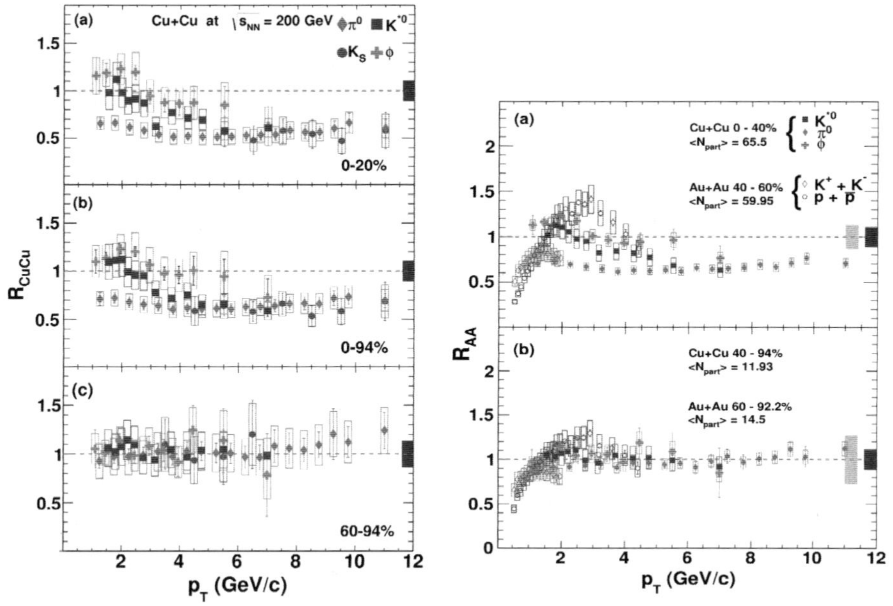

Figure 3. (color online) Left panel : Nuclear modification factor as a function of p_T for K_S^0, K^{*0} for different centralities in Cu+Cu collisions at $\sqrt{s_{NN}} = 200$ GeV. Results from π^0 [9] and ϕ [6] are also shown. Right panel : Comparison of the nuclear modification factor of π^0 [9], ϕ [6], and K^{*0} in Cu+Cu collisions and proton [7] and kaon [7] in Au+Au collisions at $\sqrt{s_{NN}} = 200$ GeV. For details see text.

GeV/c.

5. Conclusions

Measurements of K_S^0 and K^{*0} mesons via $\pi^0\pi^0$ and $K^{\pm}\pi^{\mp}$ decays respectively, in p+p, d+Au and Cu+Cu collisions at $\sqrt{s_{NN}} = 200$ GeV, with PHENIX are presented. In d+Au collisions, in the measured p_T range, no significant cold nuclear matter effect is seen for both mesons. In Cu+Cu collisions, no suppression is observed in peripheral collisions. In the most central collisions, for $p_T > 5$ GeV/c, K_S^0 and K^{*0} show a suppression similar to π^0 and ϕ within uncertainties. However, for $2 < p_T$ (GeV/c) < 5, baryons show no suppression, π^0s are the most suppressed species and particles with strange quark content show an intermediate behavior. These results provide additional constraints to the modelisation of hadron energy loss in medium.

References

[1] Shuryak E 1980 *Phys. Repts.* **61** 71
[2] Adare A *et al* [PHENIX Collaboration] 2014 *Phys. Rev. C* **90** 054905 (*Preprint* `1405.3628[hep-ex]`)
[3] Tserruya I 2014 *EPJ Web Conf* **70** 00022 (*Preprint* `1212.6722[nucl-ex]`)
[4] Adox K *et al* [PHENIX Collaboration] 2003 *Nucl. Instrum. Methods.* **499** 469
[5] Adler S S *et al* [PHENIX Collaboration] 2007 *Phys. Rev. Lett.* **98** 172302
[6] Adare A *et al* [PHENIX Collaboration] 2011 *Phys. Rev. C* **83** 024909
[7] Adare A *et al* [PHENIX Collaboration] 2013 *Phys. Rev. C* **88** 024906
[8] Fries R, Greco V and Sorensen P 2008 *Ann. Rev. Nucl. Part. Sci* **58** 177
[9] Adare A *et al* [PHENIX Collaboration] 2008 *Phys. Rev. Lett.* **101** 162301

Hot Quarks 2014 IOP Publishing
Journal of Physics: Conference Series **612** (2015) 012064 doi:10.1088/1742-6596/612/1/012064

Resonances as Probes of Heavy-Ion Collisions at ALICE

A. G. Knospe (for the ALICE Collaboration)
The University of Texas at Austin, Department of Physics, Austin, TX, USA
E-mail: anders.knospe@cern.ch

Abstract. Hadronic resonances serve as unique probes in the study of the hot and dense nuclear matter produced in heavy-ion collisions. Properties of the hadronic phase of the collision can be extracted from measurements of the suppression of resonance yields. A comparison of the transverse-momentum spectra of the ϕ(1020) meson and the proton (which have similar masses) can be used to study particle production mechanisms. Resonance measurements in pp collisions provide input for tuning QCD-inspired particle production models and serve as reference measurements for other collision systems. Measurements of resonances in p–Pb collisions allow nuclear effects in the absence of a hot and dense final state to be studied. The ALICE Collaboration has measured resonances in pp, p–Pb, and Pb–Pb collisions. These measurements will be discussed and compared to results from other experiments and to theoretical models.

Resonances serve as useful probes that allow the characteristics of heavy-ion collisions to be studied at different stages of their evolution. While the yields of stable hadrons are fixed at chemical freeze-out, the yields of resonances can be modified by hadronic scattering processes after chemical freeze-out [1–3]. Regeneration, in which resonance decay products scatter pseudo-elastically through a resonance state (*e.g.*, $\pi K \rightarrow K^{*0} \rightarrow \pi K$), can increase the measured yield of the intermediate resonance state without changing the yields of the stable hadrons. Elastic re-scattering of a resonance decay product can smear the invariant mass resolution and may impede reconstruction of the original resonance. Pseudo-elastic scattering of a resonance decay product through a different resonance state (*e.g.*, a pion from a K^{*0} decay scattering through a ρ state) will prevent reconstruction of the first resonance [4]. Regeneration and re-scattering are expected to be most important for $p_T \lesssim 2$ GeV/c [1,4]. The final resonance yields at kinetic freeze-out will be determined by the chemical freeze-out temperature, the time between chemical and kinetic freeze-out, the resonance lifetime, and the scattering cross sections of its decay products with other hadrons. Theoretical models that take these effects into account can be used to estimate the properties of the hadronic phase using measured resonance yields (or their ratios to stable particles) as input [2,5,6].

In addition, partial restoration of chiral symmetry around the phase transition between partonic and hadronic matter may lead to changes in the masses or the widths of resonances [7–10]. Mechanisms that determine the shapes of particle p_T spectra, including the relative strengths of quark recombination [11,12] and hydrodynamical effects [13–15], are studied experimentally using many different particle species; the ϕ(1020), a meson with a mass similar to the proton, provides valuable information regarding the effects of mass and baryon number on the shapes of particle p_T spectra. Measurements of the nuclear modification factor R_{AA} of hadrons allows the in-medium energy loss of partons to be studied. Resonance R_{AA} measurements will provide constraints against which the results of theoretical models can be tested. The effects of ordinary nuclear matter on particle p_T spectra can be studied through measurement of the nuclear modification factor in p–Pb collisions (R_{pPb}).

These proceedings describe recent results from the ALICE experiment related to these topics, with the focus on the K^{*0} (denoting the average of $K^*(892)^0$ and $\overline{K}^*(892)^0$) and the ϕ (abbreviating $\phi(1020)$). Further details of these results can be found in [16–19]. The main components of the ALICE detector [20] used in the analyses described here are the Inner Tracking System (for tracking and vertex finding), Time Projection Chamber [21] (for tracking and particle identification), the Time-of-Flight detector (for particle identification), and the VZERO detector (used to determine event multiplicity and centrality classes).

The K^{*0} and ϕ are measured through invariant-mass reconstruction of their charged hadronic decay channels ($K^{*0} \to \pi^{\pm}K^{\mp}$ and $\phi \to K^-K^+$). The combinatorial background is estimated using an event-mixing technique. After subtraction of this background, the resonance signals are fitted using a peak fit function (a relativistic Breit-Wigner function for K^{*0}, the convolution of a Breit-Wigner function and a Gaussian for ϕ) added to a first- or second-order polynomial, which describes the residual background. The p_T-dependent yields, masses, and widths of K^{*0} and ϕ are extracted from these fits. The p_T spectra are fitted using Lévy-Tsallis functions for pp and p–Pb collisions and Boltzmann-Gibbs blast-wave functions for Pb–Pb collisions. These fit functions are used to extract the mean transverse momentum $\langle p_T \rangle$ and to extrapolate the resonance yields below the lowest measured p_T bin so that the total p_T-integrated yield can be calculated. These quantities have been measured for Pb–Pb collisions at $\sqrt{s_{NN}} = 2.76$ TeV [18], p–Pb collisions at $\sqrt{s_{NN}} = 5.02$ TeV [19], and pp collisions at $\sqrt{s} = 0.9$, 2.76, and 7 TeV [16, 17, 19].

The K^{*0} and ϕ masses and widths have been measured as functions of p_T for central and peripheral Pb–Pb collisions at $\sqrt{s_{NN}} = 2.76$ TeV [18]. These quantities are consistent with the values extracted from Monte-Carlo simulations (in which particles are produced using HIJING and their interactions with the ALICE detector are simulated using GEANT3) and no centrality dependence is observed.

The ratios of p_T-integrated particle yields K^{*0}/K and ϕ/K are shown in Fig. 1(a) as a function of $(dN_{ch}/d\eta)^{1/3}$ for pp, p–Pb, and Pb–Pb collisions. The abscissa $(dN_{ch}/d\eta)^{1/3}$ is used as a proxy for the system radius, following a practice used in femtoscopy studies [22]. The K^{*0}/K^- ratio in central collisions is suppressed with respect to the ratio in pp, p–Pb, and peripheral Pb–Pb collisions. The K^{*0}/K^- ratio is also suppressed with respect to the value extracted from a thermal model fit of ALICE data [23] with a chemical freeze-out temperature of 156 MeV and zero baryo-chemical potential (the K^{*0} is excluded from this fit). The centrality-dependent suppression of the K^{*0}/K^- ratio in Pb–Pb collisions is consistent with a scenario in which the reconstructible K^{*0} yield is reduced due to re-scattering (with re-scattering being stronger than regeneration). As described in [18], the measured K^{*0}/K^- ratio in central Pb–Pb collisions can be used along with a theoretical model [2, 5, 6] (a thermal model with resonance yields modified by re-scattering) to obtain an estimate of 2 fm/c for the lower limit of the time between chemical and kinetic freeze-out. In contrast, the ϕ/K ratio is not suppressed in central collisions and only a weak centrality dependence is observed. This suggests that re-scattering effects are not important for the ϕ, which has a lifetime an order of magnitude longer than the K^{*0} and longer than many estimates of the lifetime of the hadronic phase. The K^{*0}/K and ϕ/K ratios measured in p–Pb collisions are consistent with the trends observed in pp and peripheral Pb–Pb collisions. The K^{*0}/K^- ratios measured in pp and A–A collisions at $\sqrt{s_{NN}} = 62.4$ and 200 GeV [24] follow the same trend as the values observed at LHC energies. The ϕ/K^- ratios observed in pp, d–Au, and A–A collisions at $\sqrt{s_{NN}} = 200$ GeV [25] are consistent within uncertainties with the values observed at the LHC for similar values of $(dN_{ch}/d\eta)^{1/3}$, although the RHIC measurements tend to be larger than the values observed at LHC.

The p/π ratio shows a pronounced increase as a function of p_T for $p_T < 3$ GeV/c in Pb–Pb collisions [26]. Different explanations have been proposed as the cause of this increase. An increasing baryon-to-meson ratio in this p_{T} range suggests quark recombination as the particle production mechanism. However, hydrodynamical effects could also lead to an increasing p/π ratio because of the different masses of the two particles. The ϕ, a meson that has a mass similar to the proton (within 9%), allows the baryon-to-meson ratio to be studied without the added complication of a mass difference. Fig. 1(b) shows the p/ϕ ratio as a function of p_{T} for different centrality and multiplicity intervals in pp, p–Pb, and Pb–Pb collisions [19]. For central Pb–Pb collisions, the ratio is flat for $p_{\mathrm{T}} < 3-4$ GeV/c. This is consistent with hydrodynamical effects: the shapes of the p_{T} spectra are determined by the particle masses and not the quark content. In peripheral Pb–Pb collisions, the p/ϕ ratio takes on an increasingly steep slope, suggesting that the p and ϕ may have different production mechanisms in these collisions. The values and slopes of the p/ϕ ratio observed in pp and low-multiplicity p–Pb collisions are similar to those observed in peripheral Pb–Pb collisions. The p/ϕ ratio in p–Pb collisions does not depend on multiplicity for $p_{\mathrm{T}} > 1$ GeV/c. A possible flattening in this ratio is observed for $p_{\mathrm{T}} < 1$ for high-multiplicity p–Pb collisions, which could be interpreted as a hint for the onset of collective behavior in these collisions.

The mean transverse momentum $\langle p_{\mathrm{T}}\rangle$ has been measured for $\pi^{\pm}$, $\mathrm{K}^{\pm}$, (anti)protons, K^{*0}, and ϕ in pp, p–Pb, and Pb–Pb collisions and for Λ in p–Pb collisions [17–19, 26, 27]. Mass ordering in $\langle p_{\mathrm{T}}\rangle$ is observed in central Pb–Pb collisions. The K^{*0}, p, and ϕ, which have similar masses, are observed to have similar $\langle p_{\mathrm{T}}\rangle$ values. This is expected from hydrodynamical models. However, a splitting is observed between the p and ϕ for peripheral Pb–Pb collisions, with the ϕ having large $\langle p_{\mathrm{T}}\rangle$ values. The same effect is observed in the p/ϕ ratio. In p–Pb and pp collisions, the mass ordering is only approximate: $\langle p_{\mathrm{T}}\rangle$ for the resonances is larger than $\langle p_{\mathrm{T}}\rangle$ for p and Λ. A plot of $\langle p_{\mathrm{T}}\rangle$ as a function of particle mass suggests the possibility of two different trends: one for the mesons (including the resonances) and another for the baryons. The values of $\langle p_{\mathrm{T}}\rangle$ for p–Pb follow different trends from the values for Pb–Pb collisions as functions of $(\mathrm{d}N_{\mathrm{ch}}/\mathrm{d}\eta)^{1/3}$ [19]. The values of $\langle p_{\mathrm{T}}\rangle$ for the highest-multiplicity p–Pb collisions reach (or even exceed) the values observed for central Pb–Pb collisions, despite originating from a smaller system. This suggests the possibility of different particle production mechanisms in Pb–Pb and p–Pb collisions.

The nuclear modification factors of these resonances have been computed for Pb–Pb collisions at $\sqrt{s_{\mathrm{NN}}} = 2.76$ TeV (R_{AA}) and p–Pb collisions at $\sqrt{s_{\mathrm{NN}}} = 5.02$ TeV (R_{pPb}, for ϕ only) [19]. For central Pb–Pb collisions at high p_{T} ($\gtrsim 5$ GeV/c), both K^{*0} and ϕ are strongly suppressed ($R_{\mathrm{AA}} \approx 0.15-0.2$), with similar suppression observed for charged hadrons and identified $\pi^{\pm}$, $\mathrm{K}^{\pm}$, and p. At low p_{T} (< 2 GeV/c), the K^{*0} is more suppressed than charged hadrons in central Pb–Pb collisions, which may be an indication that re-scattering effects are of greater importance in this p_{T} range. For $2 < p_{\mathrm{T}} < 6$ GeV/c in central collisions, the ϕ R_{AA} is greater than R_{AA} for the lighter mesons ($\pi^{\pm}$ and $\mathrm{K}^{\pm}$), but less than R_{AA} for protons, which have a similar mass. Since no pp reference p_{T} spectra have yet been measured at $\sqrt{s} = 5.02$ TeV, the reference for R_{pPb} is calculated by interpolating between measured ϕ spectra for $\sqrt{s_{\mathrm{NN}}} = 2.76$ and 7 TeV. The values of R_{pPb} for ϕ indicate a moderate Cronin peak (reaching a value of ≈ 1.2) for $3 < p_{\mathrm{T}} < 6$ GeV/c. In contrast, $\pi^{\pm}$ have no significant Cronin peak, while protons appear to have a stronger Cronin peak in the same p_{T} range. For $p_{\mathrm{T}} \gtrsim 8$ GeV/c, R_{pPb} does not deviate from unity for $\pi^{\pm}$, p, or ϕ.

In summary, the K^{*0} and ϕ resonance have been measured in different collision systems at LHC energies. The K^{*0}/K ratio is suppressed in central collisions, consistent with re-scattering of its decay daughters, while the ϕ/K ratio is not suppressed due to its longer lifetime. The ϕ/p ratio is flat for $p_{\mathrm{T}} < 3-4$ GeV/c in central Pb–Pb collisions, which is expected from hydrodynamical models since the particles have similar masses. The values of $\langle p_{\mathrm{T}}\rangle$ in Pb–Pb and p–Pb collisions follow different trends as a function of $(\mathrm{d}N_{\mathrm{ch}}/\mathrm{d}\eta)^{1/3}$ and strict mass ordering of $\langle p_{\mathrm{T}}\rangle$ values is violated for K^{*0}, p, ϕ, and Λ in p–Pb collisions. The nuclear modification factor

R_{AA} of resonances at high p_T shows large suppression, consistent with the suppression observed for stable hadrons. At high p_T, the ϕ meson R_{pPb} is consistent with unity.

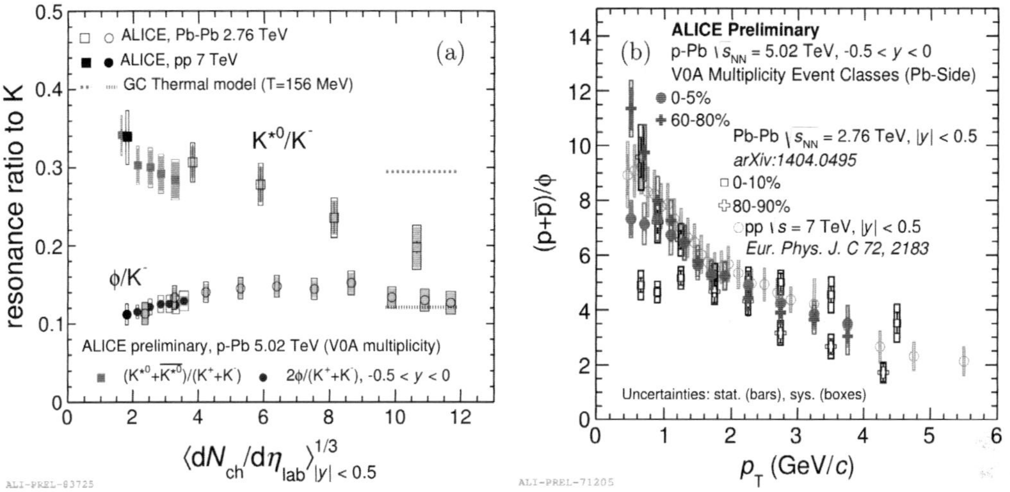

Figure 1. For pp, p–Pb, and Pb–Pb collisions: (a) ratios K^{*0}/K and ϕ/K as functions of $(dN_{ch}/d\eta)^{1/3}$, (b): Ratio p/ϕ as a function of p_T in different centrality and multiplicity intervals.

References

[1] Bleicher M and Stöcker H 2004 *J. Phys.* G **30** S111–8
[2] Markert C *et al* 2002 *Proc. of PASI 2002* ed Elze H T *et al* pp 533–52 (*Preprint* `arXiv:hep-ph/0206260`)
[3] Vogel S and Bleicher M 2005 *Proc. of the 43rd International Winter Meeting on Nuclear Physics, Bormio, Italy, 2005* ed Iori I and Moroni A (INSPIRE) (*Preprint* `arXiv:nucl-th/0505027`)
[4] Bleicher M and Aichelin J 2002 *Phys. Lett.* B **530** 81–87
[5] Torrieri G and Rafelski J 2001 *Phys. Lett.* B **509** 239–45
[6] Rafelski J, Letessier J and Torrieri G 2001 *Phys. Rev.* C **64** 054907
[7] Brown G E and Rho M 2002 *Rhys. Rep.* **363** 85–171
[8] Rapp R *et al* 2010 *Relativistic Heavy Ion Physics* ed Stock R (Springer Berlin Heidelberg) pp 134–75
[9] Brodsky S J and de Teramond G F 1988 *Phys. Rev. Lett.* **60** 1924–7
[10] Eletsky V L, Belkacem M, Ellis P J and Kapusta J I 2001 *Phys. Rev.* C **64** 035202
[11] Fries R, Muller B, Nonaka C and Bass S 2003 *Phys. Rev. Lett.* **90** 202303
[12] Fries R, Greco V and Sorensen P 2008 *Annu. Rev. Nucl. Part. Sci* **58** 177–205
[13] Qui Z, Shen C and Heinz U 2012 *Phys. Lett.* B **707** 151–5
[14] Shen C, Heinz U, Huovinen P and Song H 2011 *Phys. Rev.* C **84** 044903
[15] Bożek P and Wyskiel-Piekarska I 2012 *Phys. Rev.* C **85** 064915
[16] Aamodt K *et al* (ALICE Collaboration) 2011 *Eur. Phys. J.* C **71** 1594
[17] Abelev B *et al* (ALICE Collaboration) 2012 *Eur. Phys. J.* C **72** 2183
[18] Abelev B *et al* (ALICE Collaboration) 2014 *CERN-PH-EP-2014-060* (*Preprint* `arXiv:1404.0495`)
[19] Bellini F (for the ALICE Collaboration) 2014 *Nucl. Phys.* A **931** 846–50
[20] Aamodt K *et al* (ALICE Collaboration) 2008 *J. Inst.* **3** No. S08002 i–245
[21] Alme J *et al* 2010 *Nucl. Inst. Meth. in Phys. Res.* A **622** 316–67
[22] Lisa M A *et al* 2005 *Annu. Rev. Nucl. Part. Sci.* **55** 357
[23] Stachel J, Andronic A, Braun-Munzinger P and Redlich K 2014 *J. Phys.: Conf. Ser.* **509** 012019
[24] Aggarwal M M *et al* (STAR Collaboration) 2011 *Phys. Rev. C* **84**(3) 034909
[25] Abelev B I *et al* (STAR Collaboration) 2009 *Phys. Rev.* C **79** 064903
[26] Abelev B *et al* (ALICE Collaboration) 2013 *Phys. Rev.* C **88** 044910
[27] Abelev B *et al* (ALICE Collaboration) 2014 *Phys. Lett.* B **728** 25–38

(Hyper)nuclei and anti-(hyper)nuclei production in Pb–Pb collisions in ALICE at LHC

Manuel Colocci (for the ALICE Collaboration)
Dipartimento di Fisica dell'Università di Bologna and INFN-sezione di Bologna. Via Irnerio 46, 40126 Bologna, Italy
E-mail: manuel.colocci@cern.ch

Abstract. The ALICE detector at the LHC is used to study Pb–Pb collisions at $\sqrt{s_{\rm NN}} = 2.76$ TeV. At this energy almost equal amounts of matter and anti-matter are produced in the central rapidity region. This in turn enables the production of (hyper)nuclei and anti-(hyper)nuclei, which are measured with nearly identical abundances. Thanks to its high quality tracking and particle identification capabilities, the ALICE detector allows the investigation of these rarely produced (anti-)matter states. Preliminary results on the production of light (anti-)nuclei and (anti-)hypernuclei, and a search for exotic bound states are discussed in this article.

1. Introduction
In relativistic heavy-ion collisions a large amount of (kinetic) energy may be deposited in a small volume, allowing for creation of the Quark-Gluon Plasma[1] (QGP), a phase of matter in which the quark and gluon degrees of freedom normally confined within hadrons are mostly liberated. The created system expands and cools down after its production, and the transition into an hadron gas occurs. In the final state of this process a large amount of light (anti-)nuclei and (anti-)hypernuclei is also formed[2, 3]. Thanks to its precise tracking and particle identification capabilities, the ALICE experiment[4] successfully records these produced (anti-)matter states giving the opportunity to investigate their production mechanisms: actually, an (hyper)nucleus is supposedly formed by a coalescence process of two or more (anti-)baryons at the last stage of the collision[5] and/or during the hadronization process at the chemical freeze-out of the QGP[6]. Furthermore, the large amount and similar abundance of matter and anti-matter nuclei observed by the ALICE experiment, allows the investigation of other key features of light nuclei as mass and lifetime. The measurement of the production and properties of light nuclei could also provide useful information for the search of the anti-matter in the universe and, in particular, for the potential detection of the dark matter with hypothetical decays into anti-deuteron[7].

2. (Anti-)nuclei
The identification of (anti-)nuclei is possible via their specific energy loss measurement ($\mathrm{d}E/\mathrm{d}x$) provided by the Time Projection Chamber[8] (TPC) of the ALICE experiment. ^{3}He nucleus is identified in a wide range of the transverse momenta, up to $p_{\rm T} \simeq 8$ GeV/c, the deuteron up to $p_{\rm T} \simeq 1.5$ GeV/c, requiring that the energy deposit of each track is compatible, within 3σ (where σ is the standard deviation), with the expected value for a given species. To extend the deuteron

identification up to about $p_{\rm T} \simeq 4.5$ GeV/c, the measurement of its velocity provided by the Time Of Flight[9] (TOF) detector is used as well. The ALICE experiment is also equipped with the High Momentum Particle Identification (HMPID) system which is a ring-imaging Cherenkov detector covering a limited acceptance region. Employing it, it is possible to further enlarge the transverse momentum region where the deuteron is identified, up to $p_{\rm T} \simeq 8$ GeV/c in the most central events (in the $0-10\%$ centrality class), where significant statistics are collected. A large amount of the nuclei detected in this way at low $p_{\rm T}$ (up to about 1.5 GeV/c for deuteron and about 2 GeV/c for ^{3}He) is created in secondary collisions (material "knock out") and not in the primary interactions. In order to separate this component, the measurement of the Distance of Closest Approach (DCA) to the primary vertex provided by the Inner Tracking System[10] (ITS) is used. A cut of 1 cm on the projection of the DCA along the beam-pipe direction is applied and its second component lying in the transverse plane is fitted with two different Monte-Carlo templates to remove the remaining contamination. Fig. 1 and Fig. 2 show the preliminary spectra of deuteron and ^{3}He nucleus as a function of $p_{\rm T}$ and the centrality class of events, already corrected for tracking efficiency and acceptance. The corresponding anti-particle spectra are not shown due to large uncertainties on the description of their interaction with the detector material. The spectra are fitted with a blast-wave function[11], that assumes a thermal emission of particles from an expanding source. The mean $\langle p_{\rm T} \rangle$ of deuterons and ^{3}He increases with centrality as already observed for lighter particles such as pions, kaons and protons. This trend can generally be explained as a consequence of the expanding source[12].

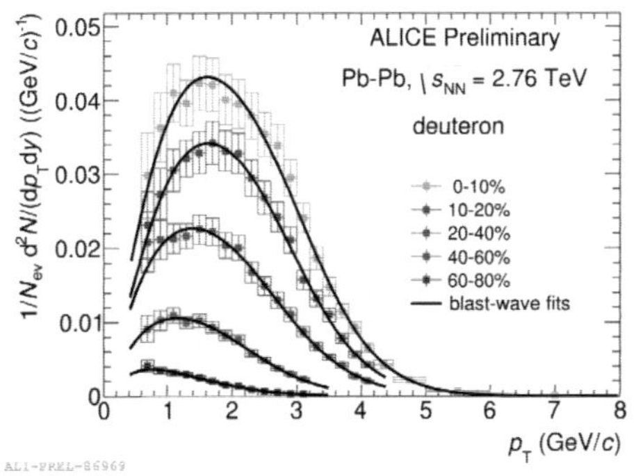

Figure 1. Deuteron spectra. The statistical and systematic uncertainties are reported with a vertical line and an open box, respectively.

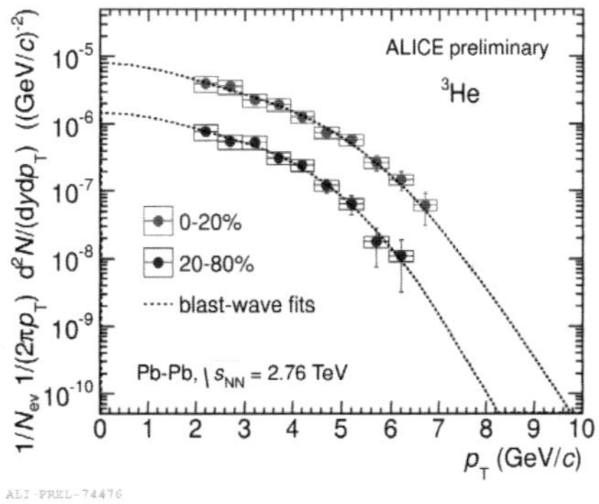

Figure 2. ^{3}He spectra. As done for the deuteron, also the fit with an hydrodynamical model (blast-wave) is performed.

An analysis of the 2011 Pb–Pb data, which is the bulk of the statistics, also reveals 10 candidate $^4\overline{\rm He}$ nuclei. The identification strategy is based on an offline trigger which accepts only tracks with a TPC dE/dx corresponding to a $^3\overline{\rm He}$ (within 3σ) or to a heavier anti-nucleus, and on the measurement of the mass-over-charge of the anti-particles, obtained using the TOF detector. The $p_{\rm T}-$integrated yields of p, d, ^{3}He and $^4\overline{\rm He}$ as a function of their mass are reported in Fig. 3. A decreasing exponential behavior is observed and a preliminary reduction factor for each additional (anti-)nucleon of about 300 is obtained. This implies that the generation of the next stable anti-matter nucleus ($^6\overline{\rm Li}$) is not reachable with the current available statistics.

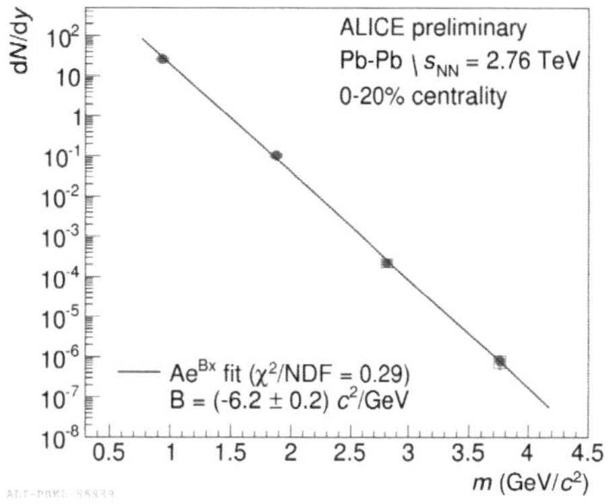

Figure 3. $p_{\rm T}-$integrated yields of p, d, ^{3}He and $^4\overline{\rm He}$ in the $0-10\%$ centrality class. The blue line is the fit with the function Ae^{Bm}.

3. (Anti-)hypermatter

In high energy Pb–Pb collisions strange quarks are produced abundantly compared to elementary pp collisions. Consequently, an enhanced production of hypernuclei is expected. The lightest known hypermatter nucleus is the hypertriton ($^3_\Lambda$H), formed by a proton, a neutron and a Λ particle. This nucleus and the corresponding anti-particle, the anti-hypertriton ($^3_{\overline{\Lambda}}\overline{\rm H}$), are reconstructed via their decay channel $^3_\Lambda{\rm H} \rightarrow {}^3{\rm He} + \pi^-$ ($^3_{\overline{\Lambda}}\overline{\rm H} \rightarrow {}^3\overline{\rm He} + \pi^+$). Both daughter particles are identified via the TPC dE/dx measurement and with a series of topological cuts: a maximum DCA between the ^{3}He and π candidates ($< 0.7\sigma$), a minimum DCA of pions from the primary vertex (> 0.4 cm), and a maximum pointing angle between the line joining the primary and secondary vertex and the total momentum (< 0.05 rad). In Fig. 4 the invariant mass distribution of (^{3}He$+\pi^-$)+($^3\overline{\rm He}$$+\pi^+$) is shown. The signal (filled black points) is extracted from the data by subtracting the corresponding background (red histogram). This background is estimated from the like-sign distribution ($^3{\rm He} + \pi^+$) + ($^3\overline{\rm He} + \pi^-$). With this technique the $^3_\Lambda$H and $^3_{\overline{\Lambda}}\overline{\rm H}$ are clearly identified in central and semi-central collisions in a wide momentum range ($2 < p_{\rm T} < 10$ GeV/c). The comparison of the hypertriton yield with predictions from statistical thermal models[13, 14] with various assumptions of freeze-out conditions is shown in Fig. 5. The preliminary comparison favours a model that requires thermal equilibrium during the hadronization process with a chemical freeze-out temperature $T_{ch} = 156$ MeV. The lifetime of hypertriton is preliminary measured: $\tau = 181 \pm 44$(stat.) $\pm\, 29$(syst.) ps, compatible with the existing experimental results. This is obtained via the decay law $N(t) = N(0)e^{-t/\tau}$ where $t = l/(\beta\gamma c)$ and $\beta\gamma c = p/m$, l being the decay length, p the momentum and m the $^3_\Lambda$H mass.

The investigation of (anti-)hypermatter is also extended to two exotic hypothetical states: the $\overline{\Lambda{\rm n}}$ and the H-dibaryon. The first is searched for the expected decay $\overline{\Lambda{\rm n}} \rightarrow \overline{\rm d} + \pi^-$ with a similar experimental technique employed as for the hypertriton measurement. The investigation is made only for the $\overline{\Lambda{\rm n}}$ state excluding its matter counterpart (Λn) because of the lower background corresponding to the particle "knock-out" from the detector material. The second, the H-dibaryon, is a hypothetical state of uuddss ($\Lambda\Lambda$) first predicted in [15] via bag model calculations. Its search is made in the channel H($\Lambda\Lambda$) $\rightarrow \Lambda + {\rm p} + \pi$. The analysis strategy is analogous to the $\overline{\Lambda{\rm n}}$ state, with the exception that a second V^0 decay has to be identified. No significant signal is observed in the invariant mass distributions of these exotic states. It is therefore only possible to set upper limits on the production yields of these hypothetical states. They are more than a factor 10 below the statistical thermal model predictions which, on the other hand, are in agreement with the measurement of the production yield of d, ^{3}He and (anti-)$^3_\Lambda$H[16].

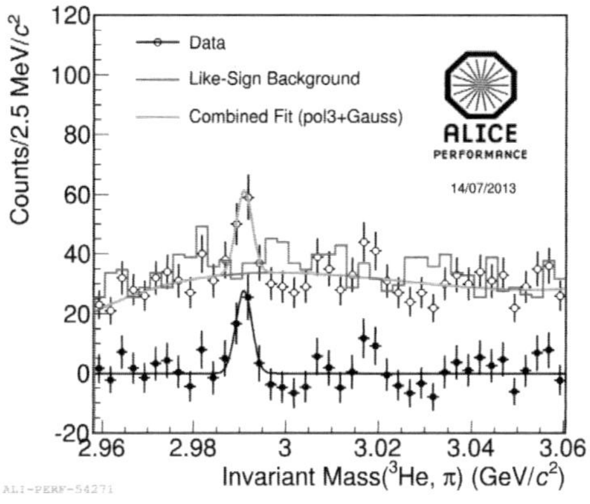

Figure 4. Invariant mass distribution of (^{3}He + π^-) + ($^3\overline{\text{He}}$ + π^+). See text for more details.

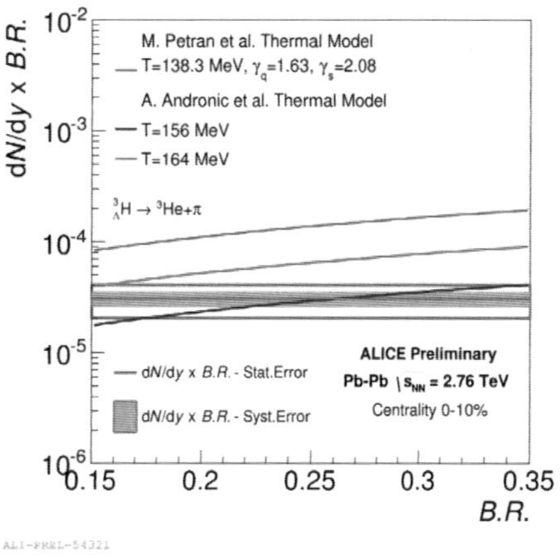

Figure 5. Production yield dN/dy of $^3_\Lambda$H (not corrected for the B.R.) compared to expectations from statistical thermal models.

4. Summary and perspectives

Preliminary ALICE results for light (hyper)nuclei in Pb–Pb collisions at $\sqrt{s_{\rm NN}}$ = 2.76 TeV have been presented. During the forthcoming LHC Run-II data taking the precision of the current measurements could significantly improve due to the expected increase of event statistics. The measurement of the (anti-)hypertriton will be performed in other decay channels and multi-strange anti-hypernuclei could be observed. The search for (anti-)hypermatter will be extended to other hypothetical exotic states.

[1] Braun-Munzinger P, Stachel J 2007 *Nature* **448** 302.
[2] Agakishiev H *et al.* (STAR Collaboration) 2011 *Nature* **473** 353.
[3] Abelev B I *et al.* (STAR Collaboration) 2010 *Science* **328** 58.
[4] Abelev B I *et al.* (ALICE Collaboration) 2014 *Int. J. Mod. Phys. A* **29** 1430044.
[5] Xue L, Ma Y G, Chen J H, Zhang S 2012 *Phys. Rev. C* **85** 064912.
[6] Andronic A, Braun-Munzinger P, Stachel J, Stocker H 2011 *Phys. Lett. B* **697** 203.
[7] Donato F, Fornengo N, Salati P 2000 *Phys. Rev. D* **62** 043003.
[8] Alme J *et al.* 2010 *Nucl. Instrum. Meth. A* **622** 316.
[9] Akindinov A *et al.* 2013 *Eur. Phys. J. Plus* **128** 44.
[10] Contin G 2012 *JINST* **7** C06007.
[11] Schnedermann E, Sollfrank J, Heinz U W 1993 *Phys. Rev. C* **48** 2462.
[12] Abelev B I *et al.* (ALICE Collaboration) 2012 *Phys. Rev. Lett* **109** 252301.
[13] Petran M, Latessier J, Petracek V, Rafelski J 2013 *Phys. Rev. C* **88** 034907.
[14] Andronic A, Braun-Munzinger P, Stachel J 2009 *Phys. Lett. B* **673** 142.
[15] Jaffe R L 1977 *Phys. Rev. Lett.* **38** 195, 1977 *erratum ibid.* **38** 617.
[16] Floris M 2014 *Nucl. Phys. A* **931** 103.